Millikan
QC454.P48 V47 2007
Very high resolution
photoelectron spectroscopy

Lecture Notes in Physics

Editorial Board

R. Beig, Wien, Austria
W. Beiglböck, Heidelberg, Germany
W. Domcke, Garching, Germany
B.-G. Englert, Singapore
U. Frisch, Nice, France
P. Hänggi, Augsburg, Germany
G. Hasinger, Garching, Germany
K. Hepp, Zürich, Switzerland
W. Hillebrandt, Garching, Germany
D. Imboden, Zürich, Switzerland
R. L. Jaffe, Cambridge, MA, USA
R. Lipowsky, Golm, Germany
H. v. Löhneysen, Karlsruhe, Germany
I. Ojima, Kyoto, Japan
D. Sornette, Zürich, Switzerland
S. Theisen, Golm, Germany
W. Weise, Garching, Germany
J. Wess, München, Germany
J. Zittartz, Köln, Germany

The Lecture Notes in Physics

The series Lecture Notes in Physics (LNP), founded in 1969, reports new developments in physics research and teaching – quickly and informally, but with a high quality and the explicit aim to summarize and communicate current knowledge in an accessible way. Books published in this series are conceived as bridging material between advanced graduate textbooks and the forefront of research to serve the following purposes:

- to be a compact and modern up-to-date source of reference on a well-defined topic;
- to serve as an accessible introduction to the field to postgraduate students and nonspecialist researchers from related areas;
- to be a source of advanced teaching material for specialized seminars, courses and schools.

Both monographs and multi-author volumes will be considered for publication. Edited volumes should, however, consist of a very limited number of contributions only. Proceedings will not be considered for LNP.

Volumes published in LNP are disseminated both in print and in electronic formats, the electronic archive is available at springerlink.com. The series content is indexed, abstracted and referenced by many abstracting and information services, bibliographic networks, subscription agencies, library networks, and consortia.

Proposals should be sent to a member of the Editorial Board, or directly to the managing editor at Springer:

Dr. Christian Caron
Springer Heidelberg
Physics Editorial Department I
Tiergartenstrasse 17
69121 Heidelberg/Germany
christian.caron@springer.com

Stefan Hüfner (Ed.)

Very High Resolution Photoelectron Spectroscopy

 Springer

Editor

Professor Stefan Hüfner
Universität des Saarlandes
FR 7.2 Experimentalphysik
Am Stadtwald, Geb. 22
66123 Saarbrücken
E-mail: huefner@mx.uni-saarland.de

Stefan Hüfner, *Very High Resolution Photoelectron Spectroscopy*, Lect. Notes Phys. 715
(Springer, Berlin Heidelberg 2007), DOI 10.1007/b11942573

Library of Congress Control Number: 2006938038

ISSN 0075-8450
ISBN-10 3-540-68130-2 Springer Berlin Heidelberg New York
ISBN-13 978-3-540-68130-4 Springer Berlin Heidelberg New York

This work is subject to copyright. All rights are reserved, whether the whole or part of the material is concerned, specifically the rights of translation, reprinting, reuse of illustrations, recitation, broadcasting, reproduction on microfilm or in any other way, and storage in data banks. Duplication of this publication or parts thereof is permitted only under the provisions of the German Copyright Law of September 9, 1965, in its current version, and permission for use must always be obtained from Springer. Violations are liable for prosecution under the German Copyright Law.

Springer is a part of Springer Science+Business Media
springer.com
© Springer-Verlag Berlin Heidelberg 2007

The use of general descriptive names, registered names, trademarks, etc. in this publication does not imply, even in the absence of a specific statement, that such names are exempt from the relevant protective laws and regulations and therefore free for general use.

Typesetting: by the author and techbooks using a Springer LaTeX macro package
Cover design: WMXDesign GmbH, Heidelberg

Printed on acid-free paper   SPIN: 11942573   54/techbooks   5 4 3 2 1 0

Preface

The present volume attempts to give a concise overview over the field of high-resolution PES. The most important message is that now (finally) this technique yields not only a good deal of information that can be compared with results obtained by other experimental techniques but also insight that cannot be obtained by other techniques.

At this stage it is a pleasure to acknowledge the editors of the Springer Lecture Notes series, to give me the possibility to edit this volume and contribute to it. A special word of thanks in this respect goes to Hilbert v. Loehneysen for his help and also for his friendship over the years. Much from my knowledge on photoelectron spectroscopy comes from collaborations over a long period of time. The people I had the privilege to collaborate with are too numerous to mention them all. I would however like to mention those that have worked with me in the last years on very high resolution PES, namely D. Ehm, B. Eltner, G. Nicolay, F. Reinert, S. Schmidt and P. Steiner.

A special word of thanks goes to S. Schmidt who worked intensively and competently on the technical details of this volume. Without his expertise this book would not exist.

Finally I have to thank the Deutsche Forschungsgemeinschaft for very generous financial support by contract HU149-19-1 and within the Sonderforschungsbereich SFB 277 in project B5.

Saarbrücken, Germany *Stefan Hüfner*

Contents

1 Introduction
S. Hüfner 1
References 9

Part I Many-Body Effects

2 Photoemission Spectroscopy with Very High Energy Resolution: Studying the Influence of Electronic Correlations on the Millielectronvolt Scale
F. Reinert, S. Hüfner 13
2.1 Introduction 13
2.2 Experimental Considerations 15
2.3 Theory of the Photoemission Spectrum 22
2.4 Scattering at Phonons, Electrons, and Impurities 25
2.5 Surface Modification and the Influence on Shockley States 41
2.6 Another High-resolution Paradigm: The Kondo Resonance 44
2.7 Summary and Conclusions 48
References 49

3 Photoemission as a Probe of the Collective Excitations in Condensed Matter Systems
P. D. Johnson, T. Valla 55
3.1 Introduction 55
3.2 The Photoemission Process 56
3.3 Electron–Phonon Coupling in Metallic Systems 59
3.4 Studies of the Dichalcogenides 63
3.5 Magnetic Systems 71
3.6 Studies of the High-T_C Superconductors 75
3.7 Summary and Outlook 81
References 82

4 High-resolution Photoemission Spectroscopy of Solids Using Synchrotron Radiation
K. Shimada 85
4.1 Introduction 85
4.2 Inelastic Mean Free Path, Energy
 and Angular Resolution 86
4.3 High-Resolution Photoemission Spectroscopy
 in the VUV and SX Regions 88
4.4 High Energy Resolution Photoemission Spectroscopy
 with HX Combined with VUV and SX 104
4.5 Summary 107
References 109

Part II Low-Dimensional Systems

5 Photoemission on Quasi-One-Dimensional Solids: Peierls, Luttinger & Co.
R. Claessen, J. Schäfer, and M. Sing 115
5.1 Introduction 115
5.2 Electronic Instabilities in One Dimension 116
5.3 Photoemission of Quasi-1D CDW Systems 121
5.4 Electronic Correlation Effects in 1D 130
5.5 Conclusions and Open Questions 142
References 143

6 Atomic Chains at Surfaces
J. E. Ortega, F. J. Himpsel 147
6.1 Introduction to One-Dimensional Systems 147
6.2 One-Dimensional Quantum Wells at Metal Surfaces 151
6.3 Atomic Chains on Semiconductor Surfaces:
 The Ultimate Nanowires 163
6.4 Summary and Future Avenues 179
References 182

Part III Ultimate Resolution

7 High-Resolution Photoemission Spectroscopy of Low-T_c Superconductors
T. Yokoya, A. Chainani, and S. Shin 187
7.1 Introduction 187
7.2 High-Resolution and Low-Temperature Photoemission
 Spectroscopy 190
7.3 Superconducting DOS 191

7.4	Photoemission Results of Superconducting Gap and Strong-coupling Line Shape	193
7.5	Anomalous SC Gap Form	201
7.6	Fermi Surface Sheet Dependence	205
7.7	Summary and Future Prospects	208
References		211

Part IV Molecules

8 Very-High-Resolution Laser Photoelectron Spectroscopy of Molecules
K. Kimura 215

8.1	Introduction	215
8.2	REMPI Photoelectron Spectroscopy	217
8.3	Compact cm^{-1}-Resolution ZEKE Photoelectron Analyzers	222
8.4	Application	230
8.5	Concluding Remarks	236
References		237

Part V High-Temperature Superconductors and Transition-Metal Oxides

9 Doping Evolution of the Cuprate Superconductors from High-Resolution ARPES
K. M. Shen, Z.-X. Shen 243

9.1	Introduction	243
9.2	High-Temperature Superconductivity	244
9.3	Photoemission Studies of the Lightly Doped Cuprates	247
9.4	Conclusions	267
References		268

10 Many-Body Interaction in Hole- and Electron-Doped High-T_c Cuprate Superconductors
T. Takahashi, T. Sato, and H. Matsui 271

10.1	Introduction	271
10.2	Experiments	273
10.3	Results and Discussion	273
References		292

11 Dressing of the Charge Carriers in High-T_c Superconductors
*J. Fink, S. Borisenko, A. Kordyuk, A. Koitzsch, J. Geck,
V. Zabolotnyy, M. Knupfer, B. Büchner, and H. Berger* 295

11.1	Introduction	295

11.2	High-T_c Superconductors	297
11.3	Angle-resolved Photoemission Spectroscopy	300
11.4	The Bare-particle Dispersion	308
11.5	The Dressing of the Charge Carriers at the Nodal Point	311
11.6	The Dressing of the Charge Carriers at the Antinodal Point	315
11.7	Conclusions	322
References		323

12 High-Resolution Photoemission Spectroscopy of Perovskite-Type Transition-Metal Oxides
H. Wadati, T. Yoshida, and A. Fujimori 327

12.1	Introduction	327
12.2	Electronic Structure	328
12.3	Samples	329
12.4	Case Studies	331
References		346

Part VI High Energy and High Resolution

13 High-Resolution High-Energy Photoemission Study of Rare-Earth Heavy Fermion Systems
A. Sekiyama, S. Imada, A. Yamasaki, and S. Suga 351

13.1	Introduction	351
13.2	Experimental	352
13.3	High-Resolution Soft X-ray Photoemission Study of Ce Compounds	352
13.4	High-Energy Photoemission Study of Pr Compounds	361
References		371

14 Hard X-Ray Photoemission Spectroscopy
Y. Takata 373

14.1	Introduction	373
14.2	Experimental Aspects	374
14.3	Performance and Characteristics	376
14.4	Applications	380
14.5	Summary	395
References		396

Index 399

List of Contributors

Helmut Berger
École Politechnique Fédérale
de Lausanne
Institut de Physique de
la Matière Complex
CH-1015 Lausanne, Switzerland
helmuth.berger@epfl.ch

Sergey Borisenko
IFW Dresden
P.O. Box 270116
D-01171 Dresden, Germany
S.Borisenko@ifw-dresden.de

Bernd Büchner
IFW Dresden
P.O. Box 270016
D-01171 Dresden, Germany

Ashish Chainani
RIKEN SPring-8 Center
Sayo-gun
Hyogo 679-5148, Japan
chainani@spring8.or.jp

Ralph Claessen
Universität Würzburg
Experimentelle Physik IV
D-97074 Würzburg, Germany
claessen@physik.uni-wuerzburg.de

Jörg Fink
IFW Dresden
P.O. Box 270116
D-01171 Dresden, Germany
and
Iowa State University
Ames Laboratory
Ames, Iowa 50011, USA
j.fink@ifw-dresden.de

Atsushi Fujimori
Department of Physics and
Department of Complexity
Science and Engineering
University of Tokyo
Kashiwa, Chiba 277-8561, Japan
fujimori@k.u-tokyo.ac.jp

Jochen Geck
IFW Dresden
P.O. Box 270016
D-01171 Dresden, Germany
J.Geck@ifw-dresden.de

Franz J. Himpsel
University of Wisconsin Madison
Department of Physics
1150 University Ave.
Madison, Wisconsin 53706-1390
USA
fhimpsel@wisc.edu

List of Contributors

Stefan Hüfner
Universität des Saarlandes
FR 7.2 Experimentalphysik
Postfach 151150
D-66041 Saarbrücken, Germany
huefner@mx.uni-saarland.de

Shin Imada
Osaka University
Department of Material Physics
Graduate School of Engineering
Science
Toyonaka, Osaka 560-8531, Japan

Peter D. Johnson
Condensed Matter and
Materials Science Department
Brookhaven National Laboratory
Upton, NY, 11973, USA
pdj@bnl.gov

Katsumi Kimura
Institute for Molecular Science
Okazaki 444-8585, Japan
and
Japan Advanced Institute
of Science and Technology
Nomi 923-01292, Japan
k-kimura@ims.ac.jp

Martin Knupfer
IFW Dresden
P.O. Box 270016
D-01171 Dresden, Germany
M.Knupfer@ifw-dresden.de

Andreas Koitzsch
IFW Dresden
P.O. Box 270016
D-01171 Dresden, Germany
A.Koitzsch@ifw-dresden.de

Alexander Kordyuk
IFW Dresden
P.O. Box 270016
D-01171 Dresden, Germany
and
Institute of Metal Physics
of the National Academy of
Sciences of Ukraine
03142 Kyiv, Ukraine
kordyuk@gmail.com

Hiroaki Matsui
Tohoku University
Department of Physics
Sendai 980-8578, Japan
h.matsui@arpes.phys.tohoku.ac.jp

Jose Enrique Ortega
Departamento de Física Aplicada I
Universidad del País Vasco
Plaza de Oñate 2
E-20018 San Sebastian, Spain
and
DIPC and Centro Mixto CSIC/UPV
Paseo Manuel Lardizabal 4
E-20018 San Sebastian, Spain
enrique.ortega@ehu.es

Friedrich Reinert
Universität Würzburg
Experimentelle Physik II
Am Hubland
D-97074 Würzburg, Germany
reinert@physik.uni-wuerzburg.de

Takafumi Sato
Tohoku University
Department of Physics
Sendai 980-8578, Japan
t-sato@arpes.phys.tohoku.ac.jp

Jörg Schäfer
Universität Würzburg
Experimentelle Physik IV
D-97074 Würzburg, Germany
joerg.schaefer@physik.uni-wuerzburg.de

Akira Sekiyama
Osaka University
Department of Material Physics
Graduate School of
Engineering Science
Toyonaka Osaka 560-8531, Japan
sekiyama@mp.es.osaka-u.ac.jp

Kyle M. Shen
Laboratory of Atomic and
Solid State Physics
Cornell University
Ithaca NY 14853, USA
kmshen@ccmr.cornell.edu

Zhi-Xun Shen
Stanford University
Department of Applied Physics and
Stanford Synchrotron Radiation
Laboratory
Stanford, California 94305, USA
zxshen@stanford.edu

Kenya Shimada
Hiroshima University
Hiroshima Synchrotron Radiation
Research Center
2-313, Kagamiyama
Higashi-Hiroshima 739-8526, Japan
kshimada@hiroshima-u.ac.jp

Shik Shin
The University of Tokyo
The Institute for Solid State Physics
Kashiwa
Chiba 277-8581, Japan
and
Riken/SPring-8
Sayo-gun Hyogo 679-5148, Japan
shin@issp.u-tokyo.ac.jp

Michael Sing
Universität Würzburg
Experimentelle Physik IV
D-97074 Würzburg, Germany
sing@physik.uni-wuerzburg.de

Shigemasa Suga
Osaka University
Department of Material Physics
Graduate School of Engineering
Science
Toyonaka, Osaka 560-8531, Japan

Takashi Takahashi
Tohoku University
Department of Physics
Sendai 980-8578, Japan
t.takahashi@arpes.phys.tohoku.ac.jp

Yasutaka Takata
Riken/SPring-8
Sayo-gun
Hyogo 679-5148, Japan
takatay@spring8.or.jp

Tonica Valla
Condensed Matter and
Materials Science Department
Brookhaven National Laboratory
Upton, NY, 11973, USA

Hiroki Wadati
Department of Physics and
Department of Complexity
Science and Engineering
University of Tokyo
Kashiwa, Chiba 277-8561, Japan

Atsushi Yamasaki
Osaka University
Department of Material Physics
Graduate School of Engineering
Science
Toyonaka, Osaka 560-8531, Japan

Takayoshi Yokoya
Okayama University
The Graduate School of
Natural Science and Technology
3-1-1 Tsushima-naka
Okayama 700-8530 Japan
yokoya@cc.okayama-u.ac.jp

Teppei Yoshida
Department of Physics and
Department of Complexity
Science and Engineering
University of Tokyo
Kashiwa Chiba 277-8561, Japan

Volodymyr Zabolotnyy
IFW Dresden
P.O. Box 270016
D-01171 Dresden, Germany
V.Zabolotnyy@ifw-dresden.de

1

Introduction

S. Hüfner

Universität des Saarlandes, FR 7.2 Experimentalphysik, Postfach 151150, D-66041 Saarbrücken, Germany
huefner@mx.uni-saarland.de

Photoemission Spectroscopy (PES) is one of the most extensively used methods to study the electronic structure of atoms, molecules, solids and adsorbates [1]. Its effectiveness stems from the fact that the method is relatively straightforward and that it allows the simultaneous determination of energy and momentum of electrons. In recent years the energy resolution of this method has been improved considerably, namely for experiments using UV radiation to about 1 meV (corresponding to about 10 K), and in addition the energy resolution for experiments employing soft x-rays (1–5 keV) has reached values of about 50 meV. Both these improvements have opened new spectroscopic possibilities. In order to demonstrate the new domain in this field the editors of Springer Lecture Notes in Physics have suggested to document this new state of photoemission spectroscopy in a volume to allow the interested community a comprehensive view on this field.

In a one-electron description of a system, photoemission spectroscopy measures the k-resolved electronic structure.

Such a one-electron description is however hardly ever adequate because of the presence of electron–electron, electron–phonon or electron–magnon interactions where the latter two are usually smaller than the former. In order to estimate the possible magnitude of the electron–electron interaction a simple calculation is given. The binding energy of an electron in hydrogen is 13.6 eV; the Bohr radius being 0.53 Å, the electron–electron interaction for two electrons five times the Bohr radius apart – namely 2.65 Å (which equals almost the nearest neighbor distance in Cu metal) – is 2.72 eV, a non negligible number if compared to valence state energies of the order of 5 to 10 eV. Of course in many materials (also in Cu-metal) the electron–electron interaction is screened out to a large degree. However in many interesting new materials treated in this volume these interactions are large, and influence the electronic structure considerably, and must be taken into account in analyzing PES data, leading to large deviations from a one-electron interpretation of the PES data.

In the language of theoretical physics, the property measured in a photoemission experiment is the spectral function $A(E, k)$. This is related to the

Green's function of the system by

$$A(E, \boldsymbol{k}) = \frac{1}{\pi} |\mathrm{Im} G(E, \boldsymbol{k})| \;, \tag{1.1}$$

where

$$G(E, \boldsymbol{k}) = \frac{1}{E - E(\boldsymbol{k})} \;. \tag{1.2}$$

In the non-interacting electron case this yields

$$A^0(E, \boldsymbol{k}) = \frac{1}{\pi} \delta \left(E - E^0(\boldsymbol{k}) \right) \;, \tag{1.3}$$

namely a δ-function at $E^0(\boldsymbol{k})$ (Koopmans' Theorem). In the interacting electron case one has (Σ is the so-called self-energy)

$$E(\boldsymbol{k}) = E^0(\boldsymbol{k}) + \Sigma \;. \tag{1.4}$$

With $\Sigma = \mathrm{Re}\Sigma + i\mathrm{Im}\Sigma$, this is leading to

$$A(E, \boldsymbol{k}) = \frac{1}{\pi} \frac{\mathrm{Im}\Sigma}{(E - E^0(\boldsymbol{k}) - \mathrm{Re}\Sigma)^2 + (\mathrm{Im}\Sigma)^2} \;. \tag{1.5}$$

This means that PES measures the many-body properties of the sample under investigation. While usually it is assumed that Σ results from the electron–electron interaction there can also be contributions from the electron–phonon interaction (and electron–magnon interaction in magnetic systems) and impurities (electron–ion interaction).

Now a few examples for many-body interactions observed in PES will be presented in order to demonstrate their different manifestations. The examples will not be taken solely from very-high-resolution spectra in order to show also the wide variation in their magnitude.

It was pointed out by Sawatzky that already the PE spectrum of the simplest possible system, namely H_2 [2], must be viewed in a many-body picture (see Fig. 1.1).

This spectrum consists not of a single line ($1s^2 \xrightarrow{\hbar\omega} [1s, e]$ photoionization) but of many lines, which reflect the vibrational structure of the H_2^+ molecule and are excited by the PE process via the electron–phonon interaction.

Another, totally different manifestation of the electron–phonon interaction can bee seen in the PE spectra of a superconductor, namely V_3Si. In this case the the electron–phonon interaction leads to a lowering of the total energy of the system manifested by the opening of a gap, which can be directly viewed by PES (Fig. 1.2).

The electron–electron interaction can be observed in many aspects of the valence band PE spectra of Ni metal. The two-peak structure at the Fermi energy is produced by the exchange splitting (Fig. 1.3) while the extra structure at 6 eV is a so-called correlation satellite (Fig. 1.4), which is visible in the valence band but also in all core levels. This latter feature is produced by two screening channels for the photohole.

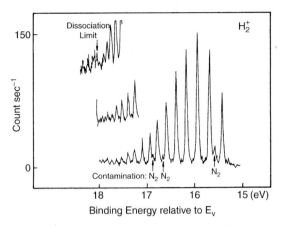

Fig. 1.1. UPS spectrum ($\hbar\omega = 21.2\,\text{eV}$) of molecular hydrogen (with a small amount of N_2 impurity) showing the vibrational structure up to the dissociation limit [1–3], E_V is the vacuum level

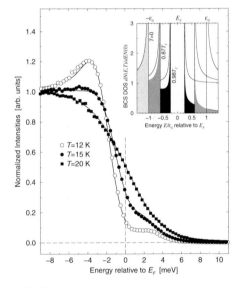

Fig. 1.2. Valence band photoemission in the energy regime around the Fermi energy for V_3Si, at temperatures around the superconducting transition temperature ($T_c = 17\,\text{K}$). The inset shows the theoretical BCS quasi-particle density of states at different temperatures [1,4]. *Shaded regions* represent the fraction of occupied states according to the Fermi function

The electron–phonon and the electron–electron interactions apart from giving distinct new features in the PE spectra, which require an interpretation beyond the one-electron picture, also are responsible for a renormalization of the electron dispersion curve and a finite lifetime for energies beyond the

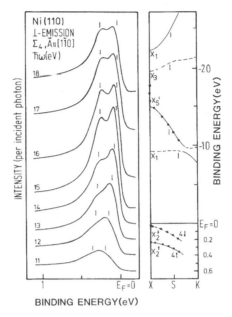

Fig. 1.3. Exchange splitting of Ni near the X point of the Brillouin zone measured by unpolarized PES. The *left panel* shows the spectra and the *right one* the interpretation in terms of the spin-split band structure. For $\hbar\omega \geq 15\,\mathrm{eV}$ the observed splitting stays constant, indicating a transition into a gap [1,5]

Fermi energy. This has been observed now in many substances and Fig. 1.5 shows an example for the (111) surface state in the L-gap of of Cu [7].

In summary: there are many and substantial deviations in the photoelectron spectra from a one-electron picture which are summarized under the label many-body interactions. While some, like the satellites in Ni-metal, can be observed by normal low-resolution experiments many more interesting cases like electron–phonon interaction or the Kondo resonance can only be investigated by very-high-resolution experiments, such as described in this volume. High resolution in this context means a resolution better than 10 meV.

The resolution of a PE spectrometer can be obtained by measuring the width of a narrow core line (e.g. the $4f$ line of metallic Au or the valence p line in a rare gas), or the width of the Fermi edge (of a noble metal) at low temperatures. Figure 1.6 shows the so far narrowest width of the Fermi edge of Au measured at low temperatures by laser excited PE [8]. The instrumental width obtained from this experimental Fermi edge is 360 µeV. This is by itself a remarkable experimental achievement. However it has to be contrasted with the narrowest rare-gas line obtained to date. This is the Ar $3p_{3/2}$ line measured by laser excitation and a time-of-flight spectrometer for the energy analysis (Fig. 1.7, [9]). This line has a width of 7 µeV yielding a resolution which is considerably better than that obtained in the experiment presented in Fig. 1.6.

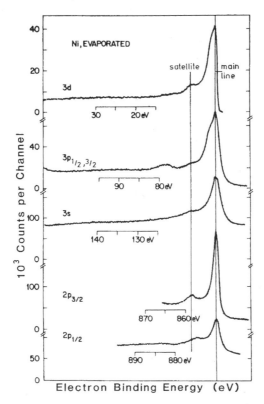

Fig. 1.4. XPS spectra of the $3d$-valence, $3p$, $3s$, $2p_{3/2}$ and $2p_{1/2}$ levels of Ni metal [6]. The main lines have been lined up to demonstrate the constant distance of the satellite position (even for the $3d$ valence band)

Finally, Fig. 1.8 shows how a particular spectral feature, namely that of the (111) surface state of Ag, changes with increasing resolution. Since the data from Nicolay in 2000 have been obtained with a resolution of 3 meV, while the total measured width in this experiment is 6 meV, one is obviously able to measure the intrinsic widths of this surface valence band state with present-day technology. This is a considerable achievement and has led to new insight particularly in the spectroscopy of solids (noteworthy the high-temperature superconductors) and merits a review which is presented in this volume.

One can look at the question of resolution also from another direction. The most important temperature scale on earth is the equivalent of room temperature, meaning $300\,\mathrm{K} \approx 25\,\mathrm{meV}$. If one wants to measure such an energy reasonably accurate, say with an instrumental width of a tenth of this energy, one needs a resolution of 3 meV. One can thus argue that a resolution of the order of 3 meV (or better) brings PES into the situation to investigate many important solid state phenomena in reasonable detail. This simple reasoning

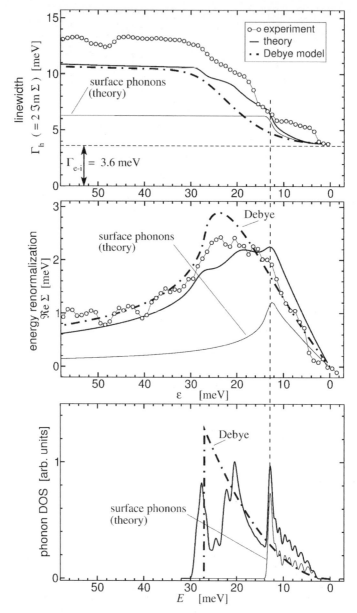

Fig. 1.5. *Top* and *middle*: comparison of the self-energy of the Cu(111) surface state as measured, with an 'exact' theory (*full line*). Also shown (*dash dot line*) are the results from a Debye model calculation ($\lambda = 0.115, \hbar\omega = 27\,\text{meV}$). While the results from the Debye model follow a smooth curve the results from experiment and theory show some distinct structures. The structure at $\varepsilon = 13\,\text{meV}$ is attributed to the influence of surface phonons as can bee seen by a comparison to the phonon density of states (DOS) in the lower panel [1,7]

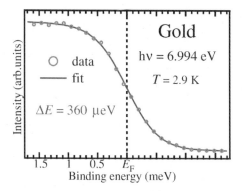

Fig. 1.6. Narrowest Fermi edge so far, measured with a hemispherical analyzer on Au at $T = 2.9\,\mathrm{K}$ using a laser radiation source. The instrumental resolution after subtraction of the temperature broadening is $\Delta E = 360\,\mu\mathrm{eV}$ [8]

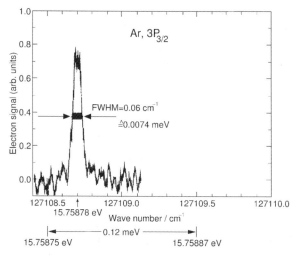

Fig. 1.7. High resolution ZEKE photoemission spectrum from gaseous Argon. Only the $3p_{3/2}$ line is shown with an inherent width of $0.0074\,\mathrm{meV}$ [9]

has proven surprisingly successful, because with the new generation of instruments that indeed achieve the mentioned resolution, PES has been able to connect its results to the information obtained by many other techniques and has broadened its appeal considerably.

The present volume which intends to survey the field of high resolution PES (high resolution meaning an energy resolution of about $10\,\mathrm{meV}$ or better) starts with an introductory chapter (Reinert and Hüfner) that gives a brief account of photoemission spectroscopy and also explains the effect of electron–electron and electron–phonon interaction on the PE spectra. Data on superconductors and surface states of noble metals are presented. Similar

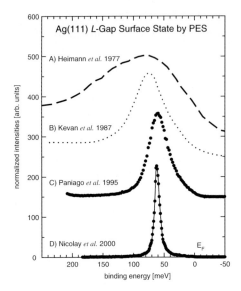

Fig. 1.8. Development of instrumental resolution, demonstrated on the example of the (111) surface state in the L-gap of the band structure of Ag. From the data of Nicolay et al. the intrinsic linewidth can be obtained with high accuracy [10, 11]

data on many-body effects are presented in the chapter by Johnson and Valla and in part in that of Shimada. The chapter by Claessen et al. describes high-resolution PE investigation of quasi one-dimensional solids. The chapter by Ortega and Himpsel treats low-dimensional artificial systems, namely chains of atoms on surfaces.

The chapter of Yokoya et al. shows how by using lasers as light sources the energy resolution in PES on solids can be brought to below 1 meV. The investigation of conventional superconductors with this technique reveals results on the gap which could hitherto be obtained only with tunnel spectroscopy.

This is followed by the only chapter on free molecules by Kimura. For the investigation of molecules by PES special very high-resolution techniques have been developed. They allow to reach an energy resolution of 7 µeV, much better than what has been obtained in instruments developed for the investigation of solids.

Quite intentionally a number of chapters on the investigation of high-temperature superconductors by PES have been incorporated into this volume. PES investigations on those materials have helped to understand their electronic structure. However at this point in time, after about 20 years of intensive research, our understanding of the superconducting state and – perhaps even more importantly – of the normal state of these systems is still missing. This is in part due to the fact that the coupling mechanism is still unknown (electron–phonon or electron–magnon or both). This however can possibly be revealed by PES and therefore many PES investigations on high-temperature superconductors address the question of the nature of the self-energy effects in the electron dispersion curves, which may shed light on the superconducting state.

The chapters by Shen, Takahashi, and Fink and the one by Johnson and Valla demonstrate indeed that high-resolution PES has contributed considerably to our understanding of the electronic structure of high-temperature superconductors. One of the main topics addressed in all these contributions is that of the dispersion of the bands near the Fermi energy. In that energy regime notable renormalization effects have been observed which may eventually help to understand the superconducting state in these fascinating materials. Johnson and Valla and Fink et al. discuss the question whether phonon or magnetic interactions (or both) couple to the electron to create the renormalization of the electronic bands near the Fermi energy. Shen and Shen show that at least for some system the broadening of the quasi-particle signal is produced by Frank–Condon broadening. Takahashi et al. deal with the renormalization in hole-doped materials and also in detail with electron-doped materials showing that the anisotropy in the gap function for these latter systems is more complicated than for the hole-doped systems. An additional chapter in this section by Wadati et al. deals with the investigation of non-superconducting oxide systems with the perovskite structure.

Finally the chapters of Sekiyama et al. and Takata (and in part that mentioned earlier by Shimada) deal with high-resolution high-energy PES. This allows, e.g., clearly to distinguish bulk from surface features in the spectra. Data on rare-earth compounds and semiconductors demonstrate that low-photon energy PES experiments can be distorted by strong signals from surface contributions.

References

1. Stefan Hüfner: *Photoelectron Spectroscopy, Principles and Applications*, Springer-Verlag, Berlin–Heidelberg–New York, 3rd edition (2003)
2. G. A. Sawatzky: Nature **342**, 176 (1989)
3. D. W. Turner et al: *Molecular Photoelectron Spectroscopy* (Wiley, New York 1970)
4. F. Reinert et al: Phys. Rev. Lett. **85**, 3930 (2000)
5. P. Heimann et al: Solid State Commun. **39**, 219 (1981)
6. S. Hüfner and G. K. Wertheim: Phys. Lett. **51**A, 299 (1975)
7. F. Reinert et al: Physica B **351**, 229–234 (2004)
8. T. Kiss et al: Phys. Rev. Lett. **94**, 057001 (2005)
9. U. Hollenstein et al: J. Chem. Phys. **115**, 5461 (2001)
10. F. Reinert et al: Phys. Rev. B **63**, 115415 (2001)
11. G. Nicolay et al: Phys. Rev. B **62**, 1631 (2000)

Part I

Many-Body Effects

2

Photoemission Spectroscopy with Very High Energy Resolution: Studying the Influence of Electronic Correlations on the Millielectronvolt Scale

F. Reinert[1] and S. Hüfner[2]

[1] Universität Würzburg, Experimentelle Physik II, Am Hubland, D-97074 Würzburg, Germany
reinert@physik.uni-wuerzburg.de
[2] Universität des Saarlandes, FR 7.2 – Experimentalphysik, Postfach 151150, D-66041 Saarbrücken, Germany
huefner@mx.uni-saarland.de

Abstract. A short review of the instrumentation and the theory of photoemission spectroscopy (PES) is presented, with special emphasis on high energy resolution. Experimental results on the electron-phonon interaction in a two-dimensional metallic system ($\overline{\mathrm{L}}$-gap surface state of Cu(111)) and a three-dimensional conventional superconductor (Pb(110)) are provided. The influence of overlayers (rare gases, Ag) on the $\overline{\mathrm{L}}$-gap surface states on Cu(111), Ag(111), and Au(111) is described. Finally the investigation of the Kondo resonance in several Ce-based heavy-Fermion compounds is presented.

2.1 Introduction

Photoemission spectroscopy (PES) has been established as one of the most important methods to study the electronic structure of molecules, solids and surfaces [1, 2]. Furthermore, PES has widespread practical implications in various fields [3] like surface chemistry or material science, and has significantly contributed to the understanding of fundamental principles in solid state physics.

In one of the four famous publications in 1905 – Einstein's *annus mirabilis* – he introduced [4] the concept of the *photon* and deduced the relation between the photon energy $h\nu$ and the maximum kinetic energy E_{kin}^{max} of the emitted electrons, the fundamental photoelectric equation:

$$E_{kin}^{max} = h\nu - \Phi_0 \ . \tag{2.1}$$

Φ_0 is a characteristic constant of the sample surface and is known today as the *work function*.

At that time, the maximum kinetic energy of the photoelectrons could be determined under vacuum conditions by the retarding-field technique. However, it took several years until Einstein's formula was experimentally confirmed by confirming a strictly linear dependence between maximum kinetic energy and the frequency of the light, i.e. the photon energy $h\nu$ (a review about the history of photoemission can be found in [5]).

The fundamental principle of the photoemission process is sketched in Fig. 2.1 [6]. This simplified picture shows the attractiveness of photoemission spectroscopy, because in that view, which will be refined later, the properties of the photoelectrons basically reflect the electronic eigenstates of the investigated system. Basically one distinguishes between ultraviolet photoemission (UPS), mainly for the (angular resolved) investigation of valence-band states

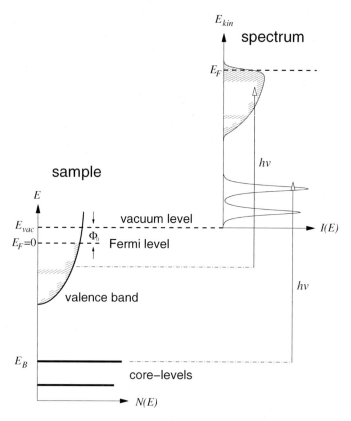

Fig. 2.1. Schematic view of the photoemission process in the single-particle picture. Electrons with binding energy E_B can be excited into free-electron states above the vacuum level E_{vac} by photons with energy $h\nu > E_B + \Phi_0$. The photoelectron distribution $I(E_{kin})$ can be measured by the analyser and is – in first order – an image of the occupied density of electronic states $N(E_B)$ of the sample

(ARUPS), and x-ray photoemission spectroscopy (XPS), providing the investigation of core-level states at higher binding energies.

2.2 Experimental Considerations

Today, a photoemission experiment for spectroscopy is basically performed in the same way as more than hundred years ago (see Fig. 2.2). Photons from a monochromatised light source, which can be a laboratory source – for vacuum ultraviolet (VUV) or soft x-ray radiation – or a synchrotron device, are directed towards a sample and the photoelectrons, liberated by the photoelectric effect, are analysed with respect to emission angle and kinetic energy by an electrostatic analyser (generally of the hemispherical type).

In the laboratory, gas discharge lamps and soft x-ray sources have been used for the photoexcitation with photon energies ranging from 11.8 eV (Ar I) to 1486.6 eV (Al-K_α). A larger energy range is covered by synchrotron radiation, which has become increasingly important because it allows measurements that cannot be performed with usual VUV or x-ray sources in the laboratory. The main difference compared to laboratory sources is that the photon energy can be selected by use of a monochromator from a continuous spectrum over a wide energy range. Other important advantages of the synchrotron light are e.g. very high intensity and brightness, variable polarisation, small photon spots, or the possibility of time resolution. However, for

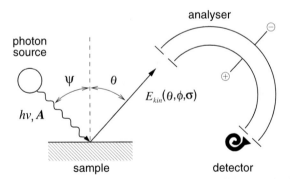

Fig. 2.2. Principle of a modern photoemission spectrometer. Monochromatic photons with energy $h\nu$ and polarisation ($\boldsymbol{A}$ is the vector potential of the electromagnetic field) are produced by a light source, e.g. an Al-K_α x-ray anode for XPS or a helium discharge lamp for UPS, and hit the sample surface under an angle Ψ with respect to the surface normal. The kinetic energy E_{kin} of the photoelectrons can be determined by use of electrostatic analysers (usually added by a retarding field) as a function of the experimental parameters, e.g. emission angle (θ, ϕ), the electron spin orientation σ, or the photon energy or polarisation. The whole setup is evacuated to ultra high vacuum (UHV, typically $p \lesssim 10^{-10}$ mbar)

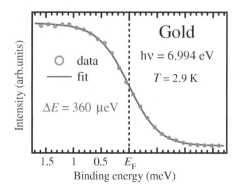

Fig. 2.3. Ultra-high resolution photoemission spectrum on a polycrystalline gold sample (evaporated Au film) for the determination of the energy resolution. The Fermi edge was measured at $T=2.9$ K using a frequency tripled ($KBe_2BO_3F_2$ crystal, KBBF) YVO_3 laser for the photoexcitation ($h\nu = 6.994$ eV) [15]

many applications the use of the –comparatively simple and cheap– laboratory sources is still advantageous.

Using standard laboratory sources, the kinetic energies of the produced photoelectrons lie in the energy range of the order of 10 eV up to ≈ 1.5 keV (cf. Eq. (2.1)). The escape depth of photoelectrons in this energy range is rather small, typically of the order of a few Å. Therefore, a characteristic property of PES experiments is their high *surface sensitivity*. The consequently small information depth of PES can be a drawback if one is interested in studying bulk properties of solids. However it seems that for many materials investigated so far, the observed electron spectra reflect predominantly the bulk properties. On the other hand, there are notable exceptions like e.g. $La_{1-x}Sr_xMnO_3$ [7,8] or $YbInCu_4$ [9–12], where recent high photon-energy PES has shown that even at 1486.6 eV the surface contributions which deviate significantly from the bulk contribution may be large and consequently can influence the results. A straight-forward way to reduce the surface contributions is the use of high-energy photons (HAXPS) for the photoexcitation, i.e. energies above 2 keV, preferably in the range of 5 keV or even more [13].

On the other hand, the information depth increases also for very low kinetic energies of a few electron volts. Suitable sources are e.g. frequency multiplied lasers, which allow – due to their small linewidth – an extremely high resolution of the photoemission spectra in energy. First examples of this rather new technique have been published in the literature [14–16] and can be found later in this volume (see contribution by Yokoya et al. in this volume). A typical setup is shown in Fig. 2.4 [15].

Whereas for band-structure measurements or Fermi-surface maps an energy resolution of the order of 50 meV is sufficient, the influence of many-body effects on the photoemission spectra close to the Fermi level is often on the scale of a few meV and requires an energy resolution of at least one

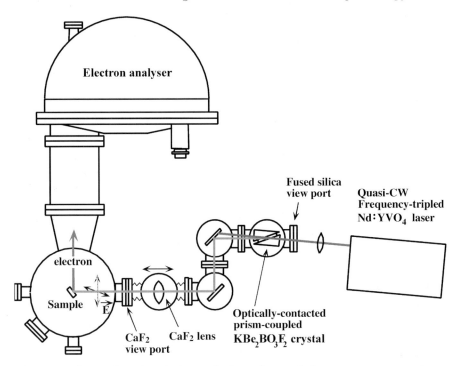

Fig. 2.4. Setup for ultra-high resolution photoemission spectroscopy using a frequency-multiplied laser source [15]

order of magnitude better. In this context, the technical improvement in the field of high-resolution spectrometers during the last few years is remarkable. Figure 2.3 shows an ultra-high resolution spectrum of a Fermi edge at low temperature, which can be measured with present-day commercially available instruments combined with a frequency-multiplied laser source.

The energy resolution can be determined from such a measurement by numerical least-squares fitting methods, using a convolution of the Fermi–Dirac distribution (FDD) (at the experimental temperature) with a transmission function describing the energy broadening of the instrument; usually the latter can be reasonably well approximated by a Gaussian. For the given spectrum (see Fig. 2.3) the experimental resolution, i.e. the full width at half maximum (FWHM) of the Gaussian, was determined to $\Delta E = 0.36$ meV, including contributions from the finite linewidth of the light source and the finite resolution of the analyser. This is the best energy resolution obtained in a photoemission spectrum on a solid up to now.

The heart of nearly all the present high-resolution PES instruments (except those used for studies of atoms and molecules – see contribution of Kimura in this volume, who uses a time-of-flight spectrometer) are Gammadata-Scienta electron analysers (see [17–19] for a detailed description of this system). Their design concept is based on the classical hemispheric analyser with Herzog plate termination and an electrostatic focusing lens system.

The high performance of modern spectrometers is closely connected with the high efficiency of the photoelectron detection, e.g. by use of a 2D MCP-CCD (two-dimensional micro-channel plate plus charge-coupled device) detector that allow a simultaneous measurement of photoelectrons with different kinetic energies and angle or spacial origin from the sample surface. The efficient parallel detection is important because the improved resolution in energy and angle reduces the phase space volume represented by one data point and therewith the photoelectron count – even more if other parameters of the photoemission process are additionally resolved, as e.g. the sample position, the photoelectron spin, or the temporal evolution.

The principle of a modern electron analyser system is shown in Fig. 2.5. The electrostatic lens system focuses the electrons emitted from the sample onto the entrance slit plane of the analyser. This allows parallel detection of

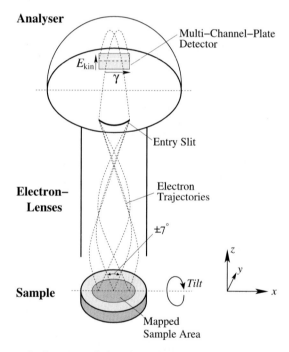

Fig. 2.5. Schematic drawing of the photoelectron trajectories in angular-resolved measurements using the a modern hemispherical analyser with parallel detection. Over a certain range, the two-dimensional detector (multi-channel-plate plus CCD) maps simultaneously the photoelectron kinetic energy E_{kin} in the radial direction, and perpendicular to it along the entrance slit, either the emission angle γ or different sample positions along the slit direction (not shown here), depending on the lens parameter settings [20]

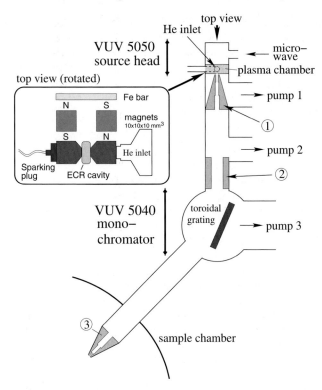

Fig. 2.6. Schematic drawing of the VUV excitation of the photoelectron spectrometer by a microwave-driven discharge lamp and a toroidal grating monochromator. To obtain a low pressure in the sample chamber during measurements (typically $p_\mathrm{meas} \leq 1.2 \times 10^{-9}$ mbar) the cross-sections (①–③) are reduced by apertures, and one has to introduce several differential pumping stages (turbo pump 1–3) [20]

the electrons over a certain solid angle. For a high resolution, the analyser needs a very good magnetic shielding and power supplies with high stability and accuracy.

Because of the reduction of the phase space volume, the photoemission experiments with a high-resolution analyser need intense light from either a synchrotron or a high-power – usually monochromatized – discharge lamp (or less frequently a laser, see Fig. 2.4) as is shown in Fig. 2.6. In the case of the VUV sources in the laboratory, alternative designs of discharge lamps have been employed, as e.g. the duoplasmatron, known from ion beam sources, or the microwave resonator as sketched in Fig. 2.6.

The discharge in the He gas is driven by microwaves, and various pumping stages reduce the flow of the He into the measuring chamber. A typical reference spectrum for the determination of the energy resolution, measured on a cooled ($T = 8$ K) polycrystalline silver sample, is shown in Fig. 2.7. The circles represent the experimental data, the dashed curve gives the Fermi-Dirac

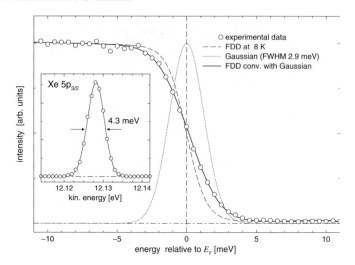

Fig. 2.7. Calibration of the energy resolution of the photoelectron spectrometer using a Ag Fermi edge at a sample temperature of $T = 8\,\mathrm{K}$. The convolution (*full line*) of a Fermi–Dirac distribution (*long-dashed line*) with a Gaussian (*dotted line*) of full width at half maximum (FWHM) of 2.9 meV full line represents the experimental (*circles*) data very well. The inset shows the gas-phase calibration using the Xe $5p_{3/2}$ line [21]

distribution (FDD) at the measuring temperature $T = 8$ K. The convolution of a Gaussian (dots) with 2.9 meV full width at half maximum with the FDD curve results in the full curve which gives an excellent representation of the measured data points and therefore allows to accurately determine the energy resolution of the instrument (lamp plus analyser) as 2.9 meV. An alternative characterization of the energy resolution can be done by measuring the photoemission spectrum of Xe gas in an appropriate gas cell (see inset of Fig. 2.7). Here the contribution of the lamp (1.2 meV) plus the Doppler broadening of the Xe atoms amounts to 3.4 meV, leading to an analyser resolution of 2.7 meV, in good agreement with the 2.6 meV obtained from the Ag Fermi edge calibration.

Finally Fig. 2.8 shows a labelled photograph of a typical laboratory instrument – consisting of various vacuum components, analyser, x-ray and VUV-monochromator – indicating the complexity of the system.

Even much better energy resolutions can be obtained in a particular photoemission experiment, namely PFI-ZEKE (pulsed-field-ionisation, zero-kinetic-energy) employing pulsed lasers as light sources and a time-of-flight spectrometer for the electron detection. For such an arrangement the energy resolution is several orders of magnitude better [40] than in an instrument with a hemispherical electron energy analyser. Figure 2.9 gives the the time-of-flight spectrum of the $\mathrm{Ar}^+\,^2P_{3/2} \leftarrow \mathrm{Ar}\,^1S_0$ transition on gaseous argon [23]. The experimental linewidth (FWHM) of 0.06 cm^{-1} is equal to an energy broadening

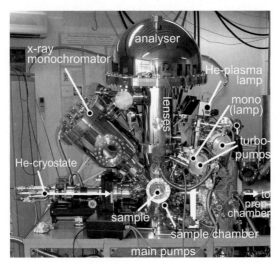

Fig. 2.8. Typical UHV (ultra high vacuum) setup of a modern photoelectron spectrometer in the laboratory (SCIENTA SES200), including sample cooling and manipulation, excitation sources, *in situ* surface preparation, and other vacuum components [22]

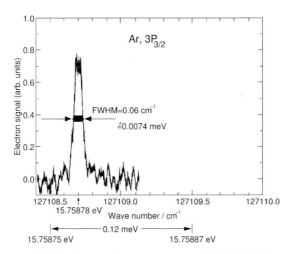

Fig. 2.9. Time-of-flight photoemission spectrum (PFI-ZEKE) on $Ar^{+\,2}P_{3/2} \leftarrow Ar\,^1S_0$ from a gas phase measurement [23]. The net energy width of the measured photoelectron distribution (FWHM) is only 0.0074 meV (FWHM)

of 0.0074 meV, meaning almost three orders of magnitude better than in the best standard photoelectron spectroscopy experiment. It has to be seen how this kind of technology can be transferred to classical PES experiments on solids.

2.3 Theory of the Photoemission Spectrum

The excitation of a photoelectron is actually a much more complicated process than illustrated by the simple picture above (Fig. 2.1). The sample represents always a many-body system that is involved as a whole in the photoemission process. However, the simplified single-particle picture is a good starting point for the understanding of many photoemission applications, as long as the spectrum is not significantly influenced by electronic correlation effects or one is interested in features like line shapes, satellites, or details in the band dispersion on the meV scale.

Over the decades in which photoemission has been applied for the spectroscopic investigation of matter, there have been many important theoretical studies to describe and analyse the spectral features in photoemission data [24–26]. The most general and widely applied theoretical description of the photoemission spectrum is based on using *Fermi's Golden Rule* as a result of perturbation theory in first order. In this approach, the photocurrent J is the result of a photon induced excitation of a system in the ground state $|\Psi_i\rangle$ into a final state $|\Psi_f\rangle = |\Psi_{\boldsymbol{\kappa},s}\rangle$, resulting in a photoelectron with momentum $\boldsymbol{\kappa}$ and kinetic energy $E_{kin} = \epsilon_{\boldsymbol{\kappa}} = \hbar^2 \kappa^2 / 2m_e$ and the remaining $(N{-}1)$-electron system:

$$J_{\boldsymbol{\kappa}}(h\nu) = \frac{2\pi}{\hbar} \sum_s |\langle \Psi_{\boldsymbol{\kappa},s} | H_{\mathrm{PE}} | \Psi_i \rangle|^2 \delta(\epsilon_{\boldsymbol{\kappa}} - \epsilon_s - h\nu) \,. \tag{2.2}$$

The index s refers to a set of quantum numbers that contains all possible excitations in the final state of the system, including phonons, plasmons, electron–hole pairs, and multiple excitations.

The perturbation operator H_{PE} describes the interaction of a (spin-less) electron in the system with the electromagnetic field $\boldsymbol{A}$ and is given by the transformation $\boldsymbol{p} \to \boldsymbol{p} - \frac{e}{c}\boldsymbol{A}$ of the generalised momentum operator $\boldsymbol{p} = -i\hbar\nabla$ in the unperturbed Hamiltonian $H_0 = p^2/2m_e + eV(\boldsymbol{r})$ (the scalar potential can be omitted by choosing an appropriate gauge of the electromagnetic field):

$$H = \frac{1}{2m_e}\left[\boldsymbol{p} - \frac{e}{c}\boldsymbol{A}\right]^2 + eV(\boldsymbol{r}) \tag{2.3}$$

$$= \frac{p^2}{2m_e} + \frac{e}{2m_e c}(\boldsymbol{A}\cdot\boldsymbol{p} + \boldsymbol{p}\cdot\boldsymbol{A}) + \frac{e^2}{2m_e c^2}A^2 + eV(\boldsymbol{r})$$

$$= H_0 + H_{PE}$$

with the photoemission perturbation operator

$$H_{PE} = \frac{e}{2m_e c}(\boldsymbol{A}\cdot\boldsymbol{p} + \boldsymbol{p}\cdot\boldsymbol{A}) + \frac{e^2}{2m_e c^2}A^2 \,. \tag{2.4}$$

The quadratic term in $\boldsymbol{A}$ becomes relevant only for extremely high photon intensities that are usually not produced by standard light sources in the laboratory. Furthermore, Eq. (2.4) can be simplified when particular surface effects (i.e. *surface photoemission*, see e.g. [27–31]) are neglected and one gets

$$H_{\text{PE}}^{vol} = \frac{e}{m_e c} \boldsymbol{A} \cdot \boldsymbol{p} , \tag{2.5}$$

which is an appropriate basis for the theoretical description of most photoemission studies.

However, for the calculation of the spectrum, a very central simplification has to be made, which is known as the *sudden approximation* (S.A.). It decouples the photoelectron in the final state $|\Psi_f\rangle = |\Psi_{\boldsymbol{\kappa},s}\rangle$ from the remaining solid so that all *extrinsic* interactions are neglected. In other words, the final state in Eq. (2.2) is replaced by

$$|\Psi_{\boldsymbol{\kappa},s}\rangle = |\boldsymbol{\kappa}; N-1, s\rangle \xrightarrow{\text{S.A.}} c_{\boldsymbol{\kappa},s}^{\dagger} |N-1, s\rangle , \tag{2.6}$$

with the creation operator for the photoelectron $c_{\boldsymbol{\kappa},s}^{\dagger}$. With this simplification one can easily transform Eq. (2.2) and gets

$$J_{\boldsymbol{\kappa}}(h\nu) = \frac{2\pi}{\hbar} \sum_{k} |\Delta_{\boldsymbol{\kappa}k}|^2 A_{\boldsymbol{k}}^{<}(\epsilon_{\boldsymbol{\kappa}} - h\nu) , \tag{2.7}$$

with the one-electron spectral function $A_{\boldsymbol{k}}^{<}(E) = \sum_s |\langle N-1, s|c_k|N\rangle|^2 \cdot \delta(E - \epsilon_s)$ and the photoemission matrix element $\Delta_{\boldsymbol{\kappa}k} = \langle \Psi_{\boldsymbol{\kappa}}|H_{PE}|\Psi_k\rangle$, describing the transition probability of a single electron from state $|\Psi_k\rangle$ into the final state $|\Psi_{\boldsymbol{\kappa}}\rangle$ (the spectral function $A^{<}$ must not be confused with the vector potential $\boldsymbol{A}$). For most photoemission applications, this matrix element is assumed to be constant over the investigated energy range [2].

By definition, the spectral function for the occupied electronic states $A_{\boldsymbol{k}}^{<}(E)$ is connected to the one-particle Green's function by

$$A_{\boldsymbol{k}}^{<}(E) = -\frac{1}{\pi} \Im\text{m} \left\{ G_{\boldsymbol{k}}(E - i0^+) \right\} \cdot f(E, T) . \tag{2.8}$$

with the Fermi–Dirac distribution $f(E, T)$ and the Green's function

$$G_{\boldsymbol{k}}(E) = \frac{1}{E - \epsilon_{\boldsymbol{k}} - \Sigma_{\boldsymbol{k}}(E)} , \tag{2.9}$$

with the complex self-energy $\Sigma_{\boldsymbol{k}}(E) = \Re\text{e}\Sigma_{\boldsymbol{k}}(E) + i\Im\text{m}\Sigma_{\boldsymbol{k}}(E)$, that contains all contributions from many-body processes like electron–electron, electron–phonon, or electron–impurity interaction that determine the *intrinsic* quasiparticle spectrum [32] or photoemission line shape

$$\Sigma_{\boldsymbol{k}}(E) = \Sigma_{\boldsymbol{k}}^{el-el}(E) + \Sigma_{\boldsymbol{k}}^{el-ph}(E) + \Sigma_{\boldsymbol{k}}^{el-imp}(E) + \cdots . \tag{2.10}$$

Thus, we get the spectral function from inserting Eq. (2.9) in Eq. (2.8)

$$A_{\mathbf{k}}^{<}(E) = \frac{1}{\pi} \frac{|\Im m \Sigma_{\mathbf{k}}(E)|}{[E - \epsilon_{\mathbf{k}} - \Re e \Sigma_{\mathbf{k}}]^2 + [\Im m \Sigma_{\mathbf{k}}(E)]^2} \cdot f(E,T) . \quad (2.11)$$

A particularly simple case is when the self-energy is constant in energy, $\Sigma_{\mathbf{k}} = \Re e \Sigma_{\mathbf{k}} + i \Im m \Sigma_{\mathbf{k}}$. Then, the spectrum yields a Lorentzian centered at $\epsilon_{\mathbf{k}} + \Re e \Sigma_{\mathbf{k}}$ with a full width at half maximum (FWHM) of $\Gamma_{\mathbf{k}} = 2 \Im m \Sigma_{\mathbf{k}}$ (with the wave vector $\mathbf{k}$ and the single-particle dispersion $\epsilon_{\mathbf{k}}$). Other examples are discussed in Sect. 2.4.

In general, the sudden approximation is at high kinetic energies a well suited *ansatz* for the description of finite systems such as free or adsorbed atoms or molecules, but for solids one has always to take extrinsic losses into account [33]. In the three-step model by Berglund and Spicer [34], this is accomplished simply by a convolution of the spectral function with a loss function [35]. However, the exact treatment of extrinsic losses – e.g. for a detailed investigation of lifetime effects [36, 37] – requires more complicated theories because there is a quantum-mechanical interference between the intrinsic spectral function and the extrinsic losses.

More detailed information about the theory of photoemission can be found for example in [24–26, 38].

2.3.1 Final State Effects

Without having it mentioned explicitly, we used the Green's function – and consequently the spectral function – of the photohole, which describes the initial-state properties of the system. However, in general the contribution of the photoelectron final state to the photoemission spectrum is large and must not be neglected. In principle, the result is a convolution of the spectral function of the photohole, given by the definition above, and the spectral function of the photoelectron. The exact treatment is very complicated, but the influence of the photoelectron properties to the spectrum can be illustrated by kinematic considerations for the lifetime width, as found in [36, 39]. Here one assumes that the contributions from photohole (initial state) and photoelectron (final state) can be described by Lorentzians with the linewidths Γ_i and Γ_f, respectively. The resulting linewidth Γ_m is a linear combination of these two, given by

$$\Gamma_m = \frac{\Gamma_i / |v_{i\perp}| + \Gamma_f / |v_{f\perp}|}{\left| \frac{1}{v_{i\perp}} \left(1 - \frac{m v_{i\|} \sin^2 \theta}{\hbar k_\|} \right) - \frac{1}{v_{f\perp}} \left(1 - \frac{m v_{f\|} \sin^2 \theta}{\hbar k_\|} \right) \right|} , \quad (2.12)$$

with the group velocities $\hbar v_{i\perp} = \partial E_i / \partial k_\perp$ and so forth. This relation simplifies in normal emission, i.e. $\theta = 0$. If furthermore $|v_{i\perp}| \ll |v_{f\perp}|$ one gets

$$\Gamma_m \approx \Gamma_i + \left| \frac{v_{i\perp}}{v_{f\perp}} \right| \Gamma_f . \quad (2.13)$$

In a perfectly two-dimensional system, e.g. the Shockley states discussed further below, the group velocity of the photohole perpendicular to the surface vanishes, i.e. $v_{i\perp} = 0$, and the measured linewidths in Eq. (2.12) is given by $\Gamma_m = \Gamma_i/(1 - mv_{i\parallel}\sin^2\theta/(\hbar k_\parallel))$. Consequently, in quasi two-dimensional systems the spectra at or close to normal emission represent the initial state properties of the photohole. On the other hand, at angles far from normal emission the measured linewidth can even be smaller than Γ_i, an effect known as *kinematic compression* [29].

2.4 Scattering at Phonons, Electrons, and Impurities

Among the different contributions to the electron self-energy in Eq. (2.10), the most important for the description of a non-magnetic metal are the electron–electron, the electron–phonon, and the electron–impurity interaction. The latter is usually regarded as independent of temperature and energy and is not studied in detail by theory up to know. Therefore, the experimentalist tries to keep his data as free as possible from the influence of contaminations. However, if there is a considerable contribution to the intrinsic linewidth from impurity scattering, one can consider a certain defect concentration by including the electron–impurity term $\Sigma_k^{el-imp}(E) = i(\frac{1}{2}\Gamma^{el-imp})$, which does not depend on energy and temperature. Usually, there are additional consequences from the existence of impurities on the spectra, e.g. an increase of the background intensity.

For many cases, the electron–electron contribution can be described by the 3D Fermi-liquid scenario, where on gets for excitations near the Fermi level the temperature and energy dependent self-energy [32, 41, 42]

$$\Sigma^{el-el}(E,T) = \alpha \cdot E + i\beta\left[E^2 + (\pi k_B T)^2\right] . \tag{2.14}$$

Whereas Σ^{el-el} is the dominant contribution at higher binding energies [43–46], the energy dependence close to the Fermi level is mainly determined by the electron–phonon contribution Σ^{el-ph} that increases rapidly on the scale of the Debye energy $\hbar\omega_D$ even if the electron–phonon coupling is small [47]. In these cases, Σ^{el-el} can be incorporated to a large extent [48] in the single-particle band-structure energies ϵ_k.

2.4.1 The Eliashberg Function

The central property in the description of electron–phonon coupling is the Eliashberg function $\alpha^2 F(\hbar\omega)$. It contains the density of phonon states $F(\hbar\omega)$ and the coupling between the phonons and the electrons, expressed by $\alpha^2 F(\hbar\omega)$. Important for the connection to superconductivity, as discussed later, the information about the phonon spectrum can be taken from the inversion of the gap equations in an analysis of tunneling data for superconductors.

If one averages over the k-dependence in $\alpha^2 F$ one gets the electron–phonon self-energy [49]

$$\Sigma^{el-ph}(E,T) = \int dE' \int_0^{\hbar\omega_{max}} d(\hbar\omega)\, \alpha^2 F(\hbar\omega)$$
$$\times \left[\frac{1 - f(E',T) + n(\hbar\omega,T)}{E - E' - \hbar\omega} + \frac{f(E',T) + n(\hbar\omega,T)}{E - E' + \hbar\omega} \right], \quad (2.15)$$

with the Bose–Einstein distribution $n(\hbar\omega,T)$, the Fermi–Dirac distribution $f(E,T)$, and the maximum energy of the phonons $\hbar\omega_{max}$. In the particular example of the Debye model which we discuss in more detail below, the maximum phonon energy is given by the Debye energy $\hbar\omega_D$.

For the imaginary part of Σ_k^{el-ph}, which is equivalent half of the full Lorentzian width $\frac{1}{2}\Gamma_k^{el-ph}$, one gets

$$\Im m \Sigma_k^{el-ph}(E,T) = \pi \int_0^{\hbar\omega_m} \alpha^2 F_k(E')\, [1 + 2n(E') + \cdots$$
$$+ f(E+E') - f(E-E')]\, dE'. \quad (2.16)$$

2.4.2 Mass-Enhancement Factor

A very important parameter in the context of electron–phonon coupling and superconductivity is the *mass-enhancement factor* λ [48]. It is defined by the change of the electronic group velocity $v_k = (1/\hbar)\, \partial E/\partial k$ which changes by a factor of $1/(1+\lambda)$ close to the Fermi level [50]. Consequently the electronic density of states and the band mass increase by a factor of $(1+\lambda)$. From the change of the group velocity we get the connection to the real part of the electron–phonon self-energy:

$$\lambda_k = -\left. \frac{\partial \Re e \Sigma_k^{el-ph}(E)}{\partial E} \right|_{E=E_F} \quad (2.17)$$

In general, the mass-enhancement factor is anisotropic and temperature dependent; it decreases with T and vanishes at high temperatures (for $T \gtrsim \Theta_D/2$, with the Debye temperature $\Theta_D = \hbar\omega_D/k_B$). However, in most cases, one is interested only in the low-temperature limit of λ and, therefore, the temperature dependence is suppressed in the following.

The effect of mass renormalisation by electron–phonon coupling can be observed for various physical properties as e.g. the electronic heat capacity at low temperatures, the cyclotron effective mass m_c in the de Haas-van Alphen effect, or the Fermi velocity. The shape of the Fermi surface, however, is not renormalised (Luttinger theorem [51]).

Very important is the relation of λ to the Eliashberg function: at $T=0$ the coupling parameter is equal to the first momentum of the Eliashberg coupling function

$$\lambda = \int_0^{\hbar\omega_{max}} \frac{\alpha^2 F(\hbar\omega)}{\hbar\omega}\, \mathrm{d}(\hbar\omega)\,, \tag{2.18}$$

which is the central function in the description of the electron–phonon interaction induced properties of a solid.

Equivalent to λ is another renormalisation parameter $Z_{\bm{k}} = 1/(1+\lambda_{\bm{k}})$, which is called the wave-function renormalisation constant. At $T=0$ it describes the discontinuity in the momentum distribution function $n_{\bm{k}}$

$$n_{\bm{k}} = \int_{-\infty}^{+\infty} A_{\bm{k}}^{<}(E)\, \mathrm{d}E = \int_{-\infty}^{E_F} A_{\bm{k}}(E)\, \mathrm{d}E\,, \tag{2.19}$$

exactly at $k=k_F$. In the following, we omit the index $\bm{k}$ in λ and Z, which is actually only justified in an isotropic system. For the Shockley-type surface states, which we shall discuss below, this is rather well fulfilled in the two dimensions of the surface plane.

2.4.3 Debye Model

In the Debye model in three dimensions the dispersion of the phonon modes is linear [50], $\hbar\omega \propto |\bm{k}|$, and one assumes that only the longitudinal modes interact with the electrons with $\alpha(\omega) = const$. Thus, the Eliashberg function is proportional to the density of phonon levels and is simply given by

$$\alpha^2 F(E) = \begin{cases} \lambda \left(\dfrac{E}{\hbar\omega_D}\right)^2 & \text{for } E \leq \hbar\omega_D \\ 0 & \text{for } E > \hbar\omega_D \end{cases} \tag{2.20}$$

In the limit of high temperatures, one gets a linear dependence of the imaginary part and the photoemission linewidth on the temperature by putting Eq. (2.20) in Eq. (2.16):

$$\Im m \Sigma^{el-ph}(T) = \frac{1}{2}\Gamma^{el-ph}(T) = \pi\lambda k_B T \quad \text{for} \quad k_B T \gg \hbar\omega_D\,. \tag{2.21}$$

Furthermore, in the limit of $T=0$ one gets an analytical description for the energy dependence of the real and imaginary part of the self-energy, namely

$$\Im m \Sigma^{el-ph}(E) = \Gamma^{el-ph}(E)/2 = \begin{cases} \dfrac{\pi}{3}\lambda\dfrac{E^3}{(\hbar\omega_D)^2} & \text{for } E \leq \hbar\omega_D \\ \dfrac{\pi}{3}\lambda\,\hbar\omega_D & \text{for } E > \hbar\omega_D \end{cases} \tag{2.22}$$

and

$$\Re e \Sigma^{el-ph}(E) = -\lambda \frac{\hbar\omega_D}{3}$$

$$\times \left[\left(\frac{E}{\hbar\omega_D}\right) + \left(\frac{E}{\hbar\omega_D}\right)^3 \ln\left|1-\left(\frac{\hbar\omega_D}{E}\right)^2\right| + \ln\left|\frac{1+\left(\frac{E}{\hbar\omega_D}\right)}{1-\left(\frac{E}{\hbar\omega_D}\right)}\right| \right]. \tag{2.23}$$

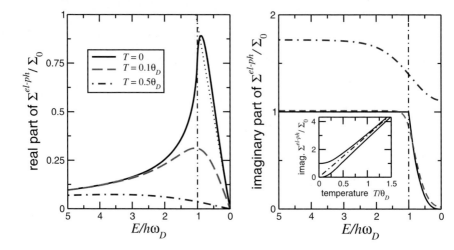

Fig. 2.10. Plots of the energy dependence of the self-energy in the Debye model as given by Eqs. (2.15) and 2.16, using $\lambda = 1$. The *left panel* gives the real part $\Re\Sigma^{el-ph}(E,T)$, the *right panel* gives the imaginary part $\Im\Sigma^{el-ph}(E,T)$, the *inset* shows the temperature dependence at $E=0$ (*lower curve*) and at $E=5\times\hbar\omega_D$ (*upper curve*), both approaching asymptotically the linear behaviour given in Eq. (2.21) (*dotted line*). All values are given on a reduced scale, with the Debye temperature θ_D, the Debye energy $\hbar\omega_D$, and the normalisation constant $\Sigma_0 = (\pi/3)\lambda\hbar\omega_D$

The self-energy of the Debye model, as given by Eqs. (2.15)–(2.23) is plotted in Fig. 2.10; the parameters are given on reduced scales. The real part of the self-energy $\Re\Sigma^{el-ph}$ at $T=0$ increases linearly from $E=0$ and shows a maximum close to the Debye energy $\hbar\omega_D$. With increasing temperature, the maximum shifts to higher energies and $\Re\Sigma^{el-ph}$ decreases. Consequently, the coupling parameter λ, as defined in Eq. (2.18), decreases significantly with the temperature. The imaginary part $\Im\Sigma^{el-ph}$ shows also a pronounced temperature dependence: The sharp edge at $\hbar\omega_D$ smears out and the whole curve shifts to larger values.

2.4.4 Modelled Spectra – Two or Three Feature Dispersion and Lineshape

Usually, the photoemission spectra resulting from the Debye model are approximated as Lorentzians, which can be appropriate for the description of many systems, in particular when the electron–phonon coupling is weak. However, the details of the spectra bear interesting information that we would like to discuss in more detail in the following. Therefore, we modelled the spectral function in Eq. (2.11) in one $\mathbf{k}$-direction by inserting the Debye self-energy from Eqs. (2.22) and (2.23). For simplicity we omit the energy and momentum units in the following and calculate the spectra at $T=0$.

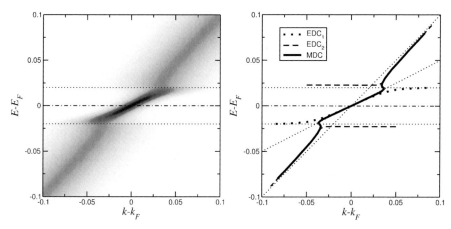

Fig. 2.11. Result of a 1D model calculation for the intensity distribution and dispersion in the spectral function $A_k(E)$ in the Debye model at $T=0$ as a function of the momentum k and the energy E. The model parameters are $\lambda=1$, $\hbar\omega_D=0.020$, the unrenormalised dispersion was set to $\partial\epsilon_k/\partial k=1$. The *left panel* gives the intensity on a grey scale (white represents low intensity, black is maximum intensity). The *right panel* displays the result of the peak positions in horizontal (MDC) and vertical cuts (EDC$_1$ and EDC$_2$), as given in Fig. 2.12. The *thin dotted lines* indicate the unrenormalized dispersion ϵ_k and the extrapolated slope of the renormalized dispersion at (E_F, k_F) which is $1/(1+\lambda)$

As model parameters we used $\lambda = 1$, i.e. a comparatively strong coupling, and $\hbar\omega_D=0.020$. The dispersion of the single-particle band was set to $\partial\epsilon_k/\partial k = 1$. Note that qualitatively, the results do not depend on the choice of parameters. In Fig. 2.11, left panel, the result of the spectral intensity $A_k(E)$ above and below the Fermi level is given in a grey scale as a function of energy and momentum. One can see that the band disperses diagonally for $|E| \gg \hbar\omega_D$ with constant linewidth and intensity. Close to the Fermi level, i.e. for $|E| \ll \hbar\omega_D$, the slope of the dispersion is obviously reduced, the linewidth decreases and the intensity reaches its maximum exactly at $(E_F, k_F) \equiv (0,0)$.

To analyze the dispersion and the spectral shape in more detail, one has to cut the intensity distribution in the same way as one does with the photoemission intensity in the experiment. The energy distribution curve (EDC) is the intensity versus energy E at constant k, the momentum distribution curve (MDC) is the dependence of the intensity on k at constant binding energy. The result of such vertical and horizontal cuts through the intensity distribution in the left panel of Fig. 2.11 are given in Fig. 2.12. The general shape of the EDC and MDC curves differ substantially: Whereas the MDCs are perfect Lorentzians, as one would expect from Eq. (2.11) with constant $\Im m\Sigma$, the EDCs show a complicated three-peak structure. For large k, the EDC is dominated by a Lorentzian contribution with a full width at half maximum (FWHM) of ≈ 0.042, which is identical with the value for $2\Im m\Sigma^{el-ph}$

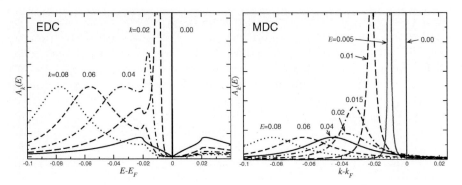

Fig. 2.12. Cuts through the intensity distribution in the *left panel* of Fig. 2.11. *Left panel:* energy distribution curves (EDC), i.e. vertical cuts for $k \leq k_\mathrm{F}$; *right panel:* momentum distribution curves (MDC), i.e. horizontal cuts. The EDC show three peaks (for $k < k_\mathrm{F}$ two below and one above E_F), the MDC cuts at constant energy are single Lorentzians. For the description of realistic photoemission data one has to include the Fermi–Dirac distribution, the influence of finite temperature in $\Sigma_{\boldsymbol{k}}(E)$, and the experimental broadening due to finite angle and energy resolution

for $E > \hbar\omega_\mathrm{D}$ in Eq. (2.22). In addition, there are two weaker spectral features at approximately ± 0.02, i.e. at $\pm\hbar\omega_\mathrm{D}$. For k-values closer to k_F, the EDC spectra get even more complicated. There appears a sharp line close to the Fermi level that gains intensity when k approaches k_F. For $k = k_\mathrm{F}$ the EDC is symmetric in energy, consisting of a δ-function at E_F and two weak features with a maximum around $\pm\hbar\omega_\mathrm{D}$. Note that except for the latter case, the spectral weight in the EDCs vanishes exactly at E_F.

The Electron–Phonon "Kink"

If one analyses the dispersion of the peak maxima from these cuts, one gets obviously different results for EDC and MDC. The right panel of Fig. 2.11 shows the result of the dispersion analysis from the spectra in Fig. 2.12. Since there is only one peak maximum in each MDC, the dispersion can be described by one single curve (solid line), displaying the famous "kink" [50], which has been extensively discussed in the context of the high-temperature superconductors (HTSC) as a key feature for the understanding of the coupling mechanism in these compounds [52–56]. The slope of this curve is reduced for energies close to E_F, exactly by the mass-enhancement factor $1/(1+\lambda)$. Many examples for the experimental observation of a quasi-particle renormalization in metallic systems, i.e. the kink in the band dispersion, can be found in the literature [57, 57–62] and are presented later in this volume.

Quasi-Particle and Satellites

If one analyses the energy position of the spectral features in the EDC spectra instead, one obtains three curves in the dispersion plots. As displayed in the right panel of Fig. 2.11, the two equivalent dispersions for states $k \gg k_F$ are related to the two Lorentzians with constant $\Gamma = (2\pi/3)\lambda\hbar\omega_D$ (EDC$_2$), dashed line). The dotted line (EDC$_1$) describes the position of the sharp feature close to E_F. Very close to the Fermi surface and far away from k_F the spectra are dominated by one spectral feature – the so-called *quasi-particle peak* – whose line shape can be well approximated by a Lorentzian. The width and the dispersion of this quasi-particle peak is immediately described by the real and the imaginary part of the self-energy, as given in Eqs. (2.22) and 2.23. Here, the analysis of the MDC and the EDC give identical results. However, in the the energy range around $\pm\hbar\omega_D$, where two neighboring structures get comparable spectral intensity, one cannot speak about a quasi-particle peak anymore and one should have this always in mind when line shapes and dispersions are discussed. In this cross-over range, one has the transition from one, lets call it phonon mode with constant energy $\hbar\omega_D$, to another, the electron mode, which is dispersing with the unrenormalized ϵ_k. For states far away from the cross-over range, the spectral weight of the "satellites" with their maxima at $E \approx \hbar\omega_D$ is small, however not vanishing completely.

The Momentum Distribution Function

The existence of these satellites has an immediate influence on the momentum distribution function n_k at $T = 0$, which we evaluated from the data in Fig. 2.11 when the intensity is normalized to $\int A_k(E)dE = 1$. Figure 2.13 shows the influence of the electron–phonon coupling as the deviation from a step function, i.e. as a finite n_k above k_F and a finite $(1-n_k)$ below. In comparison to the interaction-free electron gas, the discontinuity Z at k_F is reduced (to a value of $Z=0.5$ for the present example), as theoretically defined

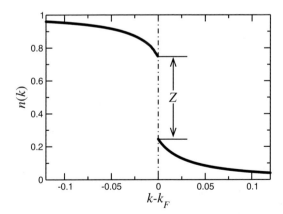

Fig. 2.13. Result for the momentum distribution function n_k at $T = 0$ from an analysis of the numerical data in Figs. 2.11 and 2.12, with $n_k = \int_{-\infty}^{E_F} A_k(E)\,dE$. Since we have $\lambda=1$, the discontinuity in n_k at k_F amounts to $Z=1/(1+\lambda)=1/2$. In the interaction free case, one would get at $T=0$ a step function with $Z=1$

by $Z=1/(1+\lambda)$. It is equivalent with the spectral weight of the δ-function dispersing through E_F; or more general, Z is the strength of the quasi-particle pole at $k = k_\mathrm{F}$.

A very detailed and precise discussion about the meaning of the self-energy and the spectral function can be found in [25]. For a more general treatment of Green's functions and many-particle physics, the reader is referred to the excellent textbooks in [63–66].

2.4.5 Electron–Phonon Coupling in Shockley-Type Surface States

Shockley-type surface states [67, 68], observed on noble metal faces in (111) orientation, have been a classical example for extremely narrow photoemission features since the first observation more than 20 years ago [69, 70]. These particular quasi two-dimensional electronic states appear on many metal surfaces [57–60, 71] and are suitable as model systems for the experimental investigation of fundamental solid state and surface properties, including band dispersion, lifetime effects, the influence of surface modifications, and many-body band renormalisation, essentially by angular resolved photoemission spectroscopy [43, 72–81]. In addition, the interest on surface states has been increased recently by the development of spectroscopy with the scanning tunneling microscope (STM), which produce spectral information on surface electronic states above and below the Fermi level with additional local information [45, 82–86].

Experimental Results

Figure 2.14 shows a typical experimental ARUPS data set on the Shockley state on Cu(111) [44]. The left panel gives the photoemission intensity on a grey scale as a function of binding energy E and the parallel momentum $k_\parallel$ along $\overline{\Gamma}\,\overline{\mathrm{K}}$ (high intensity appears dark). The lighter area, in which the Shockley state is imbedded, marks the gap in the bulk bands projected to the (111) surface, the so-called $\overline{\mathrm{L}}$-gap. The right panel shows the analysis of the Shockley-state band-dispersion by fitting the $k_\parallel$-dependence of the peak maxima positions in the EDC (at maximum binding energy) and MDC cuts by a parabola. The resulting fit parameters are given in the figure captions.

Electron–Phonon Contribution to the Self-Energy of Shockley States

As in the model data for the Debye model in Fig. 2.11, the investigation of the influence of many-body effects requires a detailed analysis of the line shape in the EDC or MDC spectra. Since the coupling for the Shockley state on the noble metal surfaces is small (λ is of the order of 0.1) there is only one structure resolvable in the EDC cuts, which can be well approximated by a Lorentzian. For states close to the Fermi level, however, an analysis of the linewidth from MDC is favorable because the contributions from the finite experimental resolution can be separated more easily.

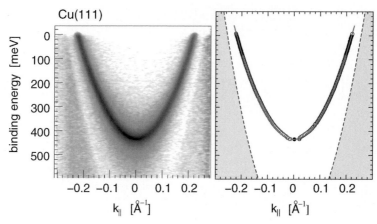

Fig. 2.14. High-resolution ARUPS on the Shockley state of Cu(111) along the $\overline{\Gamma K}$ direction. The *left panel* gives the raw photoemission data (high intensity appears dark); the *right panel* shows the result of fitting the dispersion (i.e. position of peak maxima in MDC, from EDC at band minimum). From this least-squares fit one can extract mainly three parameters: the maximum binding energy $E_0 = 435 \pm 1\,\mathrm{meV}$, the effective band mass $m^*/m_e = 0.412$, and the Fermi vector $k_F = \pm 0.215\,\text{Å}^{-1}$ (from [44])

Figure 2.16 shows the result of a detailed analysis of the dispersion and the linewidth of the Shockley state on Cu(111) as a function of binding energy. As demonstrated for the model spectra above, the deviation from the single-electron dispersion (here taken from the parabolic high-energy behavior) represents the real part of the self-energy, with a slope of λ in the limit of $E \rightarrow E_F$ and a maximum close to the Debye energy $\hbar\omega_D = 27\,\mathrm{meV}$. The linewidth, i.e. the imaginary part of Σ, shows an increase on the scale of $\hbar\omega$ with an offset of $\approx 5\,\mathrm{meV}$ due to impurity scattering. The Debye model, using the parameters given in the figure captions, describes the experimentally observed behavior already rather well. However, first-principle calculations [46,88] can explain the influence of many-body effects on the Shockley states very precisely today [45,75,89]

Electron–Electron Contribution to the Linewidth of Shockley States

For energies $E \gg \hbar\omega_D$ the contribution of the electron–phonon interaction remains constant in energy, a further broadening is due to electron–electron scattering. As displayed in the inset of Fig. 2.16, the linewidth of the Cu(111) Shockley state increases roughly linearly towards is maximum value at the band minimum at 435 meV [20], the simple one-band Landau scenario of a 3D Fermi liquid, leading to the energy dependence of Eq. (2.14), is not sufficient to describe the energy dependence quantitatively. For the surface states, more elaborated theories are required [46,88]. Table 2.1 gives a summary of calculated and experimental linewidths for the Shockley states on

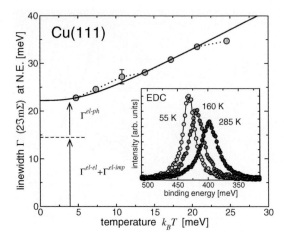

Fig. 2.15. Temperature dependence of the Shockley state on Cu(111) near the $\overline{\Gamma}$-point (at band minimum in normal emission, $E = 435$ meV). The *circles* give the experimental results, the *solid line* gives the result of the Debye model ($\lambda = 0.137$, $\hbar\omega_D = 27$ meV fixed). In addition to the zero-temperature value of $\Gamma^{el-ph}(E > \hbar\omega_D) = 8$ meV, there is an offset of $\Gamma^{off} \approx 14.5$ meV which is the sum of the electron–electron contribution $\Gamma^{el-el} \approx 9.5$ meV at this energy and impurity scattering $\Gamma^{el-imp} \approx 5$ meV (for the decomposition of the individual contributions see Fig. 2.16). The inset gives the EDCs at the band minimum for three different temperatures (from [87])

Table 2.1. Intrinsic linewidth $\Gamma = 2\Im m\Sigma$ of the Shockley states on noble metal (111)-faces at the band minimum. The values for theory and STM are taken from [45], the photoemission results can be found in [44]. More information about the calculations can be found in [88]

	Γ_{ARUPS} [meV]	Γ_{theo} [meV]	Γ_{STM} [meV]
Ag(111)	6 ± 0.5	7.2	6
Cu(111)	23 ± 1	21.7	24
Au(111)	21 ± 1	18.9	18

Cu(111), Ag(111), and Au(111) at the band minimum, where the contribution of electron–electron scattering is dominant.

It should be noted that it was assumed that the intrinsic linewidth of surface states is concealed in PES data, because the existence of surface defects – in particular step edges – is unavoidable over the large area defined by the sampling area (spot size) in photoemission experiments. Indeed, high-resolution spectroscopy with an STM [90], taken on large defect-free noble-metal terraces, showed a discrepancy [45] to the PES results published at that time. Meanwhile, it has been demonstrated that this discrepancy can be removed by high-resolution photoemission and a careful consideration of the

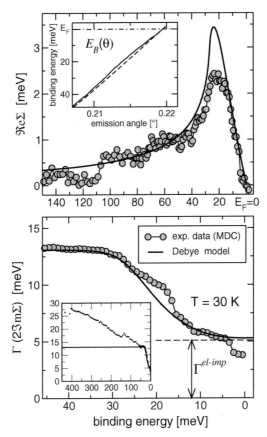

Fig. 2.16. Energy dependence of the real and imaginary part of the self-energy Σ^{el-ph} for the Cu(111) Shockley-state, extracted from the band renormalization (see inset) and the energy dependence of the linewidth, respectively. The *solid lines* give the results of the Debye model for $\lambda = 0.137$ and $\hbar\omega_D = 27$ meV (real part $T = 0$, imaginary part $T = 30$ K) The inset in the lower panel shows the linewidth over the whole energy range (from [20, 87])

surface quality [91, 92]: STM, PES, and theoretical results have converged to the values given in Table 2.1 [44], indicating that a further improvement of the energy resolution will not yield other values for the intrinsic linewidths of these Shockley states.

2.4.6 Electron–Phonon Coupling in a Three-Dimensional Solid

In a three-dimensional solid the observation of small renormalization effects and a detailed analysis of the photoemission lineshape can be thwarted by the contribution of the photoelectron lifetime [36] as discussed above. This effect dominates when the hole-velocity $v_\perp$ perpendicular to the surface is large and, consequently, the resulting linewidth does not represent the intrinsic hole properties anymore (see discussion on p. 24). In the case of quasi-low dimensional solids, i.e. a crystal with a pronounced anisotropy in the electronic properties, the hole velocity $v_\perp$ can be reduced significantly, e.g. by a factor of 1000, and the experimental results do reflect the photo hole, which is the matter of interest. Since these systems are bulk crystals – in contrast to surface or interface related states as e.g. the Shockley states discussed above – there exist results from bulk sensitive methods which can be compared quantitatively with the spectroscopy results. There are many examples for quasi low-dimensional solid systems in the literature, in particular the high-temperature superconductors (HTSC) or the transition metal dichalcogenides. Among the latter there is one compound, namely $TiTe_2$, which has been regarded as the prototype for a 2D Fermi-liquid system for a long time [93, 94].

$TiTe_2$, a Fermi Liquid?

Early photoemission data on this compound were analysed in the frame of the Fermi-liquid scenario and allowed to determine model parameters which are in reasonable agreement with parameters extracted from transport measurements [93]. Since these first PES measurements, an improvement of the energy resolution by a factor of 10 has been achieved, and a reinvestigation of $TiTe_2$ by high-resolution ARUPS shows that the intrinsic line shape can not sufficiently be described by the Fermi-liquid scenario, i.e. the electron–electron coupling, alone. Figure 2.17 shows a high-resolution k_F-spectrum of

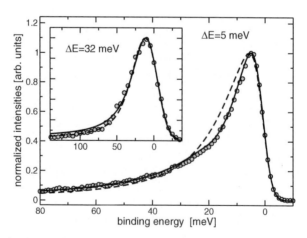

Fig. 2.17. Analysis of the photoemission spectra of $TiTe_2$ at k_F (*circles*) [47]. The experimental broadening amounts to $\Delta E = 5$ meV and $\Delta\theta = 0.3°$ (HeI$_\alpha$, $T = 10$ K). The curves give the modelled spectra without (*dashed*) and with (*solid*) electron–phonon contribution (cf. Eqs. (2.10), (2.14), and (2.22) with $\beta = 0.005$ eV^{-1}, $\Im m\Sigma^{el-imp} = 14$ meV, $\hbar\omega_D = 20$ meV, $\lambda = 0.2$). The inset shows the same model spectrum with an increased experimental broadening of $\Delta E = 32$ meV in comparison with data from [93] (normalized to the same maximum intensity)

TiTe$_2$ at a temperature of $T = 10\,\mathrm{K}$ [95]. A detailed look at the lineshape reveals a blip around a binding energy of 20 meV, which is not described by the Fermi-liquid scenario, neither in the largely simplified form of Eq. (2.14) (Taylor expansion) nor in more elaborated theories [96]. The numerical fit of the data can be significantly improved if both the electron–phonon interaction and the defect scattering in the system is considered as well. With Debye model parameters from the literature (see [47] and references therein) and a constant impurity contribution of $\Im m \Sigma^{el-imp} = 14\,\mathrm{meV}$ we get the solid curve in Fig. 2.17 which fits significantly better to the data then the result without the electron–phonon contribution (dashed line).

In the presented energy range, the model result changes only slightly by changing the electron–electron parameter β, because the total self-energy in Eq. (2.10) is dominated by the energy dependence of the electron–phonon contribution Σ^{el-ph}. Changes of β modify mainly the background intensity at higher binding energies. The inset shows the comparison of a data set measured with an energy resolution of 32 meV, in comparison with two normalized model curves with the same parameters as in the high resolution plot, except an increased Gaussian broadening to consider the larger experimental linewidth. One can observe that the presence of the electron–phonon contribution does not significantly alter the result, except the apparent background intensity. In other words, only the high-resolution data allow to analyse the lineshape for the individual contributions.

Real Three-Dimensional Solids

Even for a three-dimensional solid one can assume that the temperature and energy dependence of the photoelectron contribution is small, even if the photoelectron contribution dominates the net linewidth. In the case of lead one can observe, in HeI$_\alpha$-ARUPS spectra on the bulk states, lines with a total width of approximately 100 meV at low temperatures [87]. Increasing the temperature leads to an additional increase of the linewidth, as shown in Fig. 2.18. Analogously to the temperature dependence of the surface state in Fig. 2.15 the linewidths increases slowly with increasing temperature and approaches the linear asymptotical behaviour at high temperatures. Apart from the offset, the temperature dependence is nicely described by Eq. (2.16) using the Eliashberg function $\alpha^2 F$ as given in the literature (see inset of Fig. 2.18). In contrast to the two-dimensional case, where the offset is only due to Γ^{el-imp} and Γ^{el-el}, one has to take here the photoelectron contribution into account.

However, high-resolution photoemission allows a detailed investigation of the relative changes of the linewidth, e.g. with temperature and energy, and a precise determination of the peak maxima dispersion. Figure 2.19 shows the experimentally determined real and imaginary part of $\Sigma^{el-ph}(E)$ in comparison with the calculated energy dependence from the literature [97, 99]. Although one has to cope with the final state effects of the photoelectron, which make the experimental linewidth much broader than the intrinsic photohole

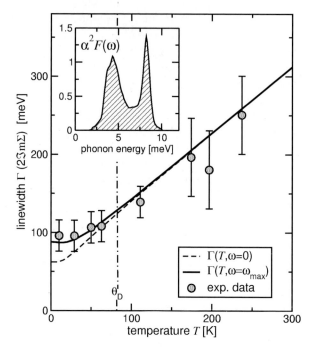

Fig. 2.18. Electron–phonon contribution to the photoemission linewidth Γ of Pb(110): Temperature dependence of Γ for energies $\omega \gtrsim \omega_D$ (*circles*). The *dashed* and the *solid line* represent the calculated electron–phonon contribution at $E_F = 0$ and $E = \hbar\omega_{\max} = 12\,\mathrm{meV}$, shifted vertically to match the experimental data. The inset shows $\alpha^2 F(\omega)$ as used for the calculation [49]. The vertical offset is due to impurity scattering and final state effects; the electron–electron Γ^{el-el} contribution vanishes at E_F (from [97])

contribution, the agreement is suprisingly good. However, one should note that the used photon energy of $h\nu = 21.23\,\mathrm{eV}$ results in a comparatively small experimental linewidth, as it was determined by photon-energy dependent investigations with synchrotron radiation at the considered Pb band [100]. In general, one could use additional information about the photoemission final states, as e.g. obtained by VLEED (very low energy electron diffraction) [101], to avoid problems with the photoelectron final state, if this is necessary.

2.4.7 Consistency with Superconductivity

Bardeen, Cooper, and Schrieffer (BCS) have demonstrated [103] that already a weak attractive coupling of two electronic states by a phonon leads to a superconducting ground state below a certain transition temperature T_c. The electronic density of states $N(E)$ in the superconducting state shows a temperature dependent band gap of the size $2\Delta(T)$ confined by two singularities at $\pm\Delta(T)$. In more detail, it is given by $N(E)/N_0(E) = |E|/\sqrt{E^2 - \Delta^2(T)}$,

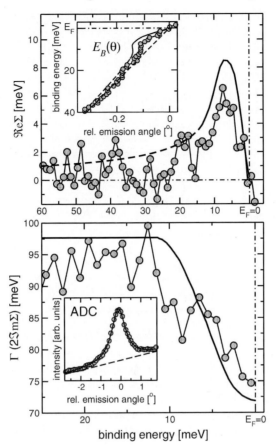

Fig. 2.19. Real and imaginary part of the electron–phonon self-energy Σ^{el-ph} for lead; experimental photoemission (He I$_\alpha$) data on the Pb(110) surface (*circles*) and calculated from [98] (*full lines*). *Upper inset:* measured band dispersion close to E_F in the normal phase ($T \approx 8\,\mathrm{K}$); the *dashed line* gives the result of a linear fit to the data points at higher energies ($E_B > 20\,\mathrm{meV}$). The experimental data points have been determined from the maximum position of a Lorentzian fit to the angular distribution curves (ADC); see lower inset for an ADC at $E_B = 10.5\,\mathrm{meV}$ (from [97, 99])

with N_0 being the density of states in the normal conducting phase. However, if the coupling is strong, the BCS theory fails to describe the superconducting properties quantitatively. For strongly coupled superconductors, the *Eliashberg theory* gives an important extension of the BCS theory by introducing a complex and energy dependent gap function $\Delta(E,T)$, which is immediately connected to the Eliasberg function discussed above in the context of the metallic systems. Due to this energy dependent gap function, the electronic density of states shows a spectral signature of the Eliashberg coupling function αF, which can be investigated by photoemission spectroscopy.

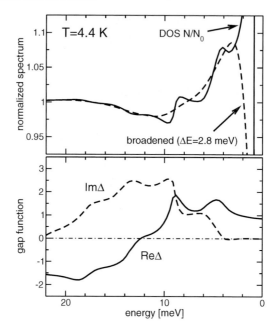

Fig. 2.20. *Lower panel:* real and imaginary part of the gap function of lead, interpolated at $T=4.4\,\text{K}$ from the calculated gap functions at $T=0$ and $0.98T_c$ given in [102]. *Upper panel:* resulting spectrum, raw normalized density of states N/N_0 (*solid line*) and broadened by a convolution with a Gaussian (FWHM $\Delta E=2.8\,\text{meV}$) to simulate the resolution function

For this purpose we have chosen lead again, for which we discussed above already the spectral properties in the normal state. Pb is a classical strongly coupled superconductor with a transition temperature of $T_c = 7.19\,\text{K}$. For a strongly coupled superconductor, the imaginary part of the complex gap function $\Delta(E)$ shows usually a pronounced structure around $E = \hbar\omega_D + \Re\Delta(E, T=0)$ which leads to a dip in the superconducting density of states N/N_0. Figure 2.20 shows the gap function and the resulting density of states, based on the Eliashberg coupling function given in Fig. 2.18 (see [102]). The calculated curves show that phonon fine structure will be smeared out in a photoemission measurements with an energy resolution of $\Delta E = 2.8\,\text{meV}$. However, the peak-dip structure with a 10% intensity difference between peak maximum and dip minimum should be observable. In fact, as demonstrated in Fig. 2.21, the calculated curve, including the finite energy resolution of $\Delta E = 2.8\,\text{meV}$, describes nicely the normalized experimental data without a further fitting of any parameters [97].

This example demonstrates that high-resolution photoemission spectroscopy provides the combination of information about both the normal and superconducting phase. In the case of lead, one single model is able to describe consistently the observations in both phases, namely the Eliashberg

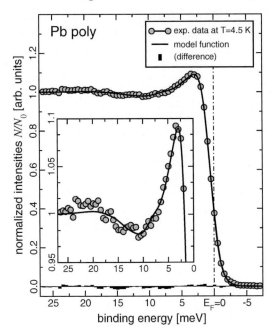

Fig. 2.21. Polycrystalline lead in the superconducting phase: Comparison of the modelled function from Fig. 2.20 (*solid line*) with the experimental data (*circles*). The *black bars* at the x-axis indicate the difference between experiment and theory. The inset shows a blow-up of the peak and dip structure

theory based on a strong electron–phonon coupling. In general, one can infer from the existence or absence of such a consistency about the coupling mechanism and the applicability of a certain model. A more detailed discussion of photoemission spectroscopy on low-T_c superconductors can be found in the article by Yokoya et al. later in this volume.

2.5 Surface Modification and the Influence on Shockley States

In the following we come back to the Shockley-type surface states introduced already in Subsect. 2.4.5. Because of their localization to the surface of a crystal, Shockley states are extremely sensitive to surface modifications, like coverages, reconstructions, and disorder. On the other hand, surface states participate in surface processes, as e.g. adsorption or catalysis [104]. Here, we give a few examples for this interplay between Shockley-states and surface modification, with particular respect to recent results by high-resolution ARUPS.

Rare-Gas Adsorbates and Energy Shifts

Already the weak interaction of a physisorbed rare-gas monolayer with the noble metal substrate leads to a considerable shift of the Shockley-state band towards the Fermi level [105]. The three panels of Fig. 2.22 show the Shockley state of the Cu(111) surface covered with 1 ML of Kr, Ar, and Xe, respectively. Depending on the nobleness of the rare-gas, the Shockley state shifts considerably towards the Fermi level. On the other hand, a systematic study with different substrates and noble gases shows that this shift is – in first order – independent from the substrate [80]. The rare-gas atoms are bonded by van der Waals forces and there is only little charge transfer between adsorbate and substrate. The energy shift can be explained by the Pauli repulsion between the closed shell atoms and the surface-state electrons of the substrate [105], which are particularly sensitive to this interaction.

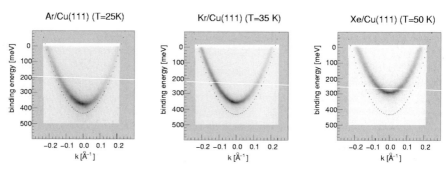

Fig. 2.22. Modification of the Shockley-state on Cu(111) by rare-gase adsorbates. The grey scale plot presents experimental ARUPS data (He I_α) on a surface covered with one closed monolayer of Ar, Kr, and Xe, respectively. The *points* indicate the dispersion of the Shockley state of the clean Cu(111) surface (cf. Fig. 2.14) [77]

In the case of the rare-gas adsorbates, the main change in the Shockley-state properties is in the binding energy, and therewith the Fermi surface and the number of surface state electrons. The effective mass (i.e. band curvature) remains constant. Only in the case of the Au(111) surface state, which is known for its characteristic spin-orbit splitting [106, 107], the size of the splitting increases for the covered surfaces which can be explained as a direct consequence of the Pauli repulsion as well [80]. For other adsorbate systems, as e.g. thin epitaxial noble metal films or sub-monolayer alkali coverages, one has to cope with other effects that influence the Shockley state significantly or destroy it completely [108].

Superlattices and Surface Structure

Among rare-gas monolayer systems on noble metal (111) surfaces, there is only the combination of Xe on Cu(111) where one observes a commensurate

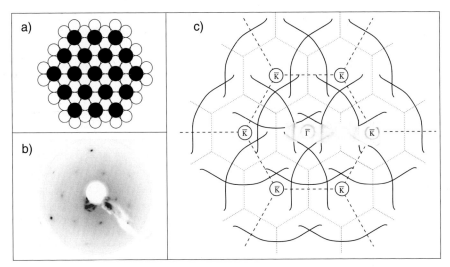

Fig. 2.23. Superlattice of the Xe monolayer on Cu(111). Panel (**a**) $\sqrt{3} \times \sqrt{3}R30°$ ordering in real space. The Xe adatoms (*black*) on-top of the substrate atoms (*white*); (**b**) low energy electron diffraction (LEED) image ($E_0 = 120\,\text{eV}$) of the monolayer system; (**c**) Fermi surface map (FSM)

superstructure. The $\sqrt{3} \times \sqrt{3}R30°$ reduces the area of the surface Brillouin zone (SBZ) by a factor of 3, leading to a backfolding of the surface state bands. In the Fermi surface maps (FSM) given in Fig. 2.23 the Shockley state appears in the Xe covered data also at the former $\overline{K}$ points which form $\overline{\Gamma}$ points in the reconstructed SBZ (R-SBZ) [80]. However, a further significant modification of the Shockley state – except the binding energy shift discussed above – can not be observed.

Backfolding – the Appearance of Band Gaps

This is in contrast to cases where the backfolding leads to overlapping Shockley-state bands of neighboring R-SBZ. At the zone boundaries, this overlap leads to the formation of band gaps, as it can most easily described by the nearly-free electron model [50]. A recent example of the appearance of a band gap [110] in a Shockley-state band is displayed in Fig. 2.24. One monolayer of Ag on Cu(111) can form different surface structures [111, 112], e.g. a commensurate triangular structure including the formation of local dislocation loops in the Cu(111) substrate [113]. In this case, the backfolding-induced modification of the Shockley-state band dispersion – in particular the opening of the band gaps – has been discussed as a possible driving mechanism for the formation of the commensurate superstructure [110, 114, 115]. Similar band gaps in Shockley states have been observed e.g. for the herringbone reconstructed ($23 \times \sqrt{3}$) Au(111) surface [79], for the Si(111)-Ag,Au [116] or on vicinal noble metal surfaces [117–119].

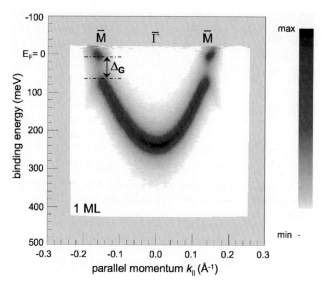

Fig. 2.24. Dispersion of the Shockley-state on Cu(111) covered by 1 ML of Ag [109]. The grey-scale plot gives the photoemission intensity $I(E_B, \mathbf{k}_\parallel)$ as a function of the binding energy and the parallel momentum in the direction of $\overline{\Gamma}$–$\overline{\mathrm{M}}$. The Ag forms a superlattice close to (9×9), inducing new zone boundaries at which one can observe the formation of a surface-state band-gap Δ_G

2.6 Another High-resolution Paradigm: The Kondo Resonance

Another typical many-body effect in metallic solids is the *Kondo effect* that appears as a characteristic minimum in the temperature dependence of the electrical resistivity in systems with magnetic impurities, e.g. 3d or 4f elements [121]. This effect is a consequence of scattering processes between the conduction electrons and the impurities, which lead to the creation of a characteristic many-body feature in the electronic density of states close to the Fermi-level [122–124]. This energetically very narrow feature is called the *Abrikosov-Suhl* or *Kondo resonance*.

The Single-Impurity Anderson Model

The simplest theoretical description of the Kondo resonance is given by the single-impurity Anderson model (SIAM) [125] that we will use for the discussion here. The SIAM model contains the single-particle energies of the conduction electrons $\varepsilon_{\mathbf{k}}$ and the localised f (or d) states ε_{fm}, the hybridisation $V_{\mathbf{k}m}$ between these states, and – finally – the Coulomb correlation energy U between two electrons in the f states of the impurity:

$$H_{\text{SIAM}} = \sum_{k\sigma} \varepsilon_k c^\dagger_{k\sigma} c_{k\sigma} + \sum_{m\sigma} \varepsilon_{fm} f^\dagger_{m\sigma} f_{m\sigma} + \sum_{km\sigma} \left(V_{km} c^\dagger_{k\sigma} f_{m\sigma} + \text{h.c.} \right)$$
$$+ \frac{U}{2} \sum_{(m\sigma) \neq (m'\sigma')} f^\dagger_{m\sigma} f_{m\sigma} f^\dagger_{m'\sigma'} f_{m'\sigma'} . \qquad (2.24)$$

Although all energies are typically at least of the order of several 100 meV or more (for many calculations U is even set to infinity), the resulting energy eigenvalues are characterised by a small energy scale, expressed by the *Kondo temperature* T_K, that gives the temperature range for the observation of correlation effects in thermodynamic, magnetic, or transport measurements (e.g. Kondo minimum in the resistivity). Typical Kondo temperatures found for Ce systems reach from a few Kelvin (γ-Ce like systems) up to several 1000 K (α-Ce like) [121]. Spectroscopically, the Kondo temperature characterises the width and the position of the Kondo resonance at low temperatures.

Numerical methods, like the non-crossing approximation (NCA) [126], allow to calculate the SIAM spectrum at finite temperatures, as shown in the inset of Fig. 2.25). In addition to the Kondo resonance with a maximum slightly above E_F, the energy range close to E_F shows several narrow features

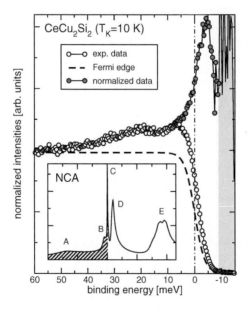

Fig. 2.25. Photoemission spectrum of polycrystalline CeCu$_2$Si$_2$ close to the Fermi level before and after normalizing to the Fermi distribution (He I, $T = 10$ K). The normalization restores the spectral information up to an energy of $5k_B T$ above E_F [120]. The inset shows the typical SIAM-NCA spectrum for a γ-Ce like compound, displaying the Kondo resonance (C), the crystal-field features (B, D), and the spin-orbit satellites (A, E)

that are due to the spin-orbit splitting (A and E) and the crystal-field splitting (B and D) of the f-states.

Figure 2.25 shows a high-resolution, angle-integrated photoemission spectrum on the prototype heavy Fermion compound $CeCu_2Si_2$ ($T_K = 11$ K) for which the Kondo resonance appears slightly above the Fermi level, where the photoemission intensity is suppressed by the Fermi–Dirac distribution (FDD) [127]. However, the intensity above E_F is not completely zero, and there are numerical methods by which the spectral information can be restored up to approximately $5k_B T$ above E_F [120, 128, 129], provided that the energy resolution is of the order of the thermal broadening or better. We demonstrate here the *normalisation method*, in which the experimental spectrum is divided by – i.e. normalised to – the Fermi–Dirac distribution $f(E, T_{eff})$ at an appropriate effective temperature $T_{eff} \gtrsim T_{sample}$ to take the finite experimental broadening into account [120]. Whereas the raw photoemission spectrum near E_F shows only the occupied tail of the Kondo resonance and a weak crystal-field satellite, the Kondo resonance in the normalised data appears as a narrow line with a peak maximum at only a few meV above E_F, from which it can be analysed now in more detail.

Since the accessible energy range above E_F depends linearly on the sample temperature, one can reach spectral structures farther above E_F when the sample temperature is increased. However, the intrinsic spectrum itself has also a temperature dependence on the scale given by the Kondo temperature T_K. The lower panel of Fig. 2.26 shows the normalised spectra at different temperatures. One can see that the Kondo resonance becomes broader and merges with the neighbouring structures, which are due to crystal-field excitations. At 200 K, i.e. approximately $10 \times T_K$, there is only one broad structure left, but still with intensity at E_F. The same behaviour can be observed in the numerically calculated spectra of the SIAM, for which the information about crystal-field energies, spin-orbit splitting, and one-particle $4f$ energy of $CeCu_2Si_2$ were determined independently [127, 130, 131, 133].

Quantitative Analysis – Extraction of the Kondo Temperature

For a quantitative comparison one has to consider that the restored experimental spectra are still influenced by the the finite experimental resolution, what can lead to an artificial shift and a broadening of the intrinsic spectrum – in particular at low temperatures. However, this can be numerically considered in a iterative fitting procedure [134] from which the model parameters and the Kondo temperature can be obtained. Figure 2.27 shows the result of such a fitting, applied to the temperature dependent photoemission spectra of $CeCu_6$ and $CeSi_2$ [130, 131]; equivalent analyses on other Ce systems can be found in the literature [133, 134]. The resulting Kondo temperatures and crystal electric field (CEF) splittings, compared to values from other experimental bulk-sensitive methods (from references cited in [130, 131]), are given in Table 2.2.

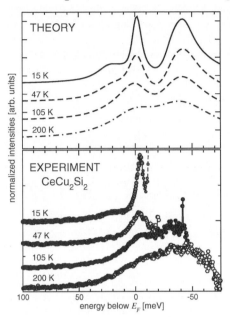

Fig. 2.26. Temperature dependence of the photoemission spectra of CeCu$_2$Si$_2$. The experimental data (*lower panel*) are compared with the theoretical NCA spectra (*upper panel*) for different temperatures. The NCA spectra are calculated with one fixed parameter set, only the temperature was varied (from [127])

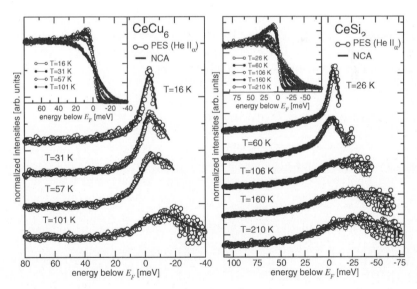

Fig. 2.27. Fit of the NCA spectra (*solid lines*) to the temperature dependence of the normalised experimental data (*circles*), applied to two Ce systems with different Kondo temperatures. *Left panel:* CeCu$_6$ with $T_K \approx 5$ K [130]; *right panel:* CeSi$_2$ with $T_K \approx 40$ K [131]. The NCA spectra are calculated with one fixed parameter set, only the temperature was varied. The insets show the raw experimental data

Table 2.2. Summary of the Kondo temperatures T_K and the energies for the two dominant crystal electric field transitions $\Delta_{1\to 2/3}$ determined from PES/NCA compared to values from the literature, determined by other experimental methods, namely from neutron scattering studies, Raman scattering, and specific heat measurements

	from other experiments			from PES/NCA		
	T_K [K]	CEF [meV]		T_K [K]	CEF [meV]	
		$\Delta_{1\to 2}$	$\Delta_{1\to 3}$		$\Delta_{1\to 2}$	$\Delta_{1\to 3}$
$CeCu_6$	5.0 ± 0.5	7.0	13.8	4.6	7.2	13.9
$CeCu_2Si_2$	$4.5\cdots 10$	$30\cdots 36$	–	6	32	37
$CeRu_2Si_2$	16	19	34	16.5	18	33
$CeNi_2Ge_2$	29	(4)	34	29.5	26	39
$CeSi_2$	22/41	25	48	35	25	48

Obviously, already the simplified picture of a single magnetic impurity reproduces the main features of the $4f$ density of states, with surprisingly good quantitative agreement. This again is an example where the comparison of the experimentally observed structures with theoretical results allows, on the one hand, the identification of the experimental structures and, on the other hand, the verification of the applicability of the model. However, one should note that the SIAM is – of course – inherently not able to describe lattice effects, that are responsible e.g. for the existence of Kondo insulators or that can renormalise the Fermi surface of a metallic Kondo system. Other theoretical approaches, as e.g. the periodic Anderson model (PAM), have to be used for the description of these phenomena.

2.7 Summary and Conclusions

In this introductory chapter a short outline of the theory of photoemission spectroscopy and some examples for state-of-the-art experiments were given. In brief, photoemission spectroscopy measures the spectral function of the system under investigation which contains – via the self-energy – the many-body effects of the system. These are in particular the electron–electron, the electron–phonon, and the electron–impurity interactions. In comparison to an interaction-free system, these interaction processes give rise to a finite linewidth and a renormalisation of the band energies with distinct temperature and energy dependencies. The predictions of selected theoretical models have been tested with simple two-dimensional systems namely the $\overline{L}$-gap Shockley-type surface states of the (111)-surfaces of the noble metals Cu, Ag, and Au. The data show quantitative agreement with theory. However, even in the best prepared crystals there is a considerable contribution of defect scattering at impurities to the photoemission linewidth. This indicates, that with the resolution obtained now in photoemission spectroscopy sample perfection can become an important issue.

In another system, namely the quasi-two dimensional system $TiTe_2$, which has served as a prototype Fermi-liquid system, indeed the electron–impurity scattering term is still the dominant one (at low temperatures and energies) and the exact determination of the electron–electron interaction parameter, which is so important in Fermi-liquid theory, is still not possible from the photoemission data.

Electron–phonon coupling is the reason for the superconducting ground state in conventional superconductors. It was demonstrated that in the classical strongly-coupled superconductor lead, a consistent description of the photoemission spectra in both the normal state and the superconducting state was obtained.

Finally, high-resolution photoemission experiments on Ce-based Kondo systems, like e.g. $CeCu_2Si_2$, allow the observation and quantitative analysis of the Kondo resonance and its temperature dependence, which can be described reasonably by the single-impurity Anderson model.

In the first chapter of this volume it was tried to stay more general in order to give a useful introduction for the following topical chapters. In all it can be stated that photoemission spectroscopy with an energy resolution of the order of 1 meV and an angular resolution clearly below 1° opens a new domain for this type of electron spectroscopy. This will be demonstrated by the examples in the following chapters of this book.

Acknowledgements

The experimental work summarised in this chapter was performed in collaboration with Azzedine Bendounan, Dirk Ehm, Brigitte Eltner, Frank Forster, Georg Nicolay, and Stefan Schmidt at the universities in Würzburg and Saarbrücken. We would like to thank Johannes Kroha (Bonn), Peter Wölfle (Karlsruhe), and Eugeni Chulkov (San Sebastian) for theoretical support and helpful discussions. The work was supported by the Deutsche Forschungsgemeinschaft (grant nos. Hu-149, Re-1469, and SFB 277/TP B5).

References

1. M. Cardona and L. Ley, editors: *Photoemission in Solids I*, volume 26 of *Topics in Applied Physics*. Springer-Verlag, Berlin–Heidelberg–New York (1978)
2. Stefan Hüfner: *Photoelectron Spectroscopy. Principles and Applications.* Springer-Verlag, Berlin–Heidelberg–New York, 3rd. edition (2003)
3. Franz Himpsel and Per-Olof Nilsson, editors: *Focus on Photoemission and Electronic Structure*. Number 7 in New J. Phys. Deutsche Physikalische Gesellschaft & Institute of Physics (2005)

4. Albert Einstein: Ann. Physik **17**(1), 132–148 (1905)
5. H. P. Bonzel and Ch. Kleint: Phys. Stat. Sol. B **49**(2), 107–153 (1995)
6. F. Reinert and Stefan Hüfner: New J. Phys. **7**, 97, April 2005. Focus Issue on 'Photoemission and Electronic Structure'.
7. M. Imada et al: Rev. Mod. Phys. **70**, 1039–1263 (1998)
8. K. Horiba et al: Phys. Rev. Lett. **93**, 236401 (2004)
9. S. Ogawa et al: Solid State Commun. **67**(11), 1093–1097 (1988)
10. F. Reinert et al: Phys. Rev. B **58**, 12808–12816 (1998)
11. F. Reinert et al: Phys. Rev. B **63**, 197102 (2001)
12. H. Sato et al: Phys. Rev. Lett. **93**, 246404 (2004)
13. *Proceedings of the E-MRS Fall Meeting 2004, Symposium D – Applications of Linear and Area Detectors for X-Ray and Neutron Diffraction and Spectroscopy*, volume **547** of Nuclear Instruments & Methods in Physics Research A, (2005)
14. T. Kiss et al: J. Electron Spectrosc. Relat. Phenom. **144–147**, 953–956 (2005)
15. T. Kiss et al: Phys. Rev. Lett. **94**, 057001 (2005)
16. J. D. Koralek et al: Phys. Rev. Lett. **96**, 017005 (2006)
17. U. Gelius et al: J. Electron Spectrosc. Relat. Phenom. **52**, 747–785 (1990)
18. M. Mårtensson et al: J. Electron Spectrosc. Relat. Phenom. **70**, 117–128 (1994)
19. G. Beamson and D. Biggs: *High Resolution XPS of Organic Polymers*, John Wiley & Sons, Chichester–New York–Brisbane–Toronto–Singapore (1992)
20. Georg Nicolay: PhD thesis, Universität des Saarlandes, Saarbrücken, Germany, October (2002)
21. F. Reinert et al: J. Electron Spectrosc. Relat. Phenom. **114–116**, 615–622 (2001)
22. S. Schmidt, Universität des Saarlandes, Saarbrücken, unpublished (2001)
23. U. Hollenstein et al: J. Chem. Phys **115**(12), 5461–5469 (2001)
24. B. Feuerbacher et al: *Photoemission and the electronic properties of surfaces*. John Wiley & Sons, Chichester–New York–Brisbane–Toronto (1978)
25. Lars Hedin and Stig Lundqvist: *Effects of Electron-Electron and Electron-Phonon Interactions on the On-Electron States of Solids*, volume **23** of *Solid State Physics*. Academic Press, New York & London (1970)
26. Carl-Olof Almbladh and Lars Hedin: *Beyond the one-electron model / many-body effects in atoms, molecules and solids*, volume **1** of *Handbook on Synchrotron Radiation*, chapter 8, pages 607–904. North-Holland (1983)
27. T. Miller et al: Phys. Rev. Lett. **77**(6), 1167–1170 (1996)
28. E. D. Hansen et al: Phys. Rev. B **55**(3), 1871–1875 (1997)
29. E. D. Hansen et al: Phys. Rev. Lett. **80**(8), 11766–1769 (1998)
30. T. Michalke et al: Phys. Rev. B **62**(15), 10544–10547 (2000)
31. F. Pforte et al: Phys. Rev. B **63**, 115405 (2001)
32. David Pines and Philippe Nozières: *The Theory of Quantum Liquids – Normal Fermi Liquids*, volume 1. Addison-Weseley, New York–Amsterdam–Tokyo (1966)
33. L. Hedin et al: Phys. Rev. B **58**(23), 15565–15582 (1998)
34. C. N. Berglund and W. E. Spicer: Phys. Rev. **136**(4A), 1030 (1964)
35. S. Tougaard and P. Sigmund: Phys. Rev. B **25**(7), 4452 (1982)
36. N. V. Smith et al: Phys. Rev. B **47**, 15476–15481 (1993)
37. S. Hüfner et al: J. Electron Spectrosc. Relat. Phenom. **100**(1–3), 191–213 (1999)

38. S. D. Kevan, editor: *Angle Resolved Photoemission – Theory and Current Applications*, volume **74** of *Studies in Surface Science and Catalysis*. Elsevier Science Publishers, Amsterdam–London–New York–Tokyo (1992)
39. T.-C. Chiang et al: Phys. Rev. B **21**(8), 3513–3522 (1980)
40. Ingo Fischer: Int. J. Mass. Spec. **216**, 131–153 (2002)
41. J. M. Luttinger and J. C. Ward: Phys. Rev. **118**(5), 1417–1427 (1960)
42. J. M. Luttinger: Phys. Rev. **121**(4), 942–949 (1961)
43. B. A. McDougall et al: Phys. Rev. B **51**(19), 13891 (1995)
44. F. Reinert et al: Phys. Rev. B **63**, 115415 (2001)
45. J. Kliewer et al: Science **288**, 1399–1402 (2000)
46. P. M. Echenique et al: Surf. Sci. Rep. **52**, 219–317 (2004)
47. G. Nicolay et al: Phys. Rev. B **73**, 045116 (2006)
48. Göran Grimvall: *The Electron–Phonon Interaction in Metals*, volume XVI of *Selected Topics in Solid State Physics*. North-Holland, Amsterdam–New York–Oxford (1981)
49. W. L. McMillan and J. M. Rowell: Phys. Rev. Lett. **14**(4), 108–112 (1965)
50. Neil W. Ashcroft and N. David Mermin: *Solid State Physics*. Saunders College, Philadelphia (1988)
51. J. M. Luttinger: Phys. Rev. **119**(4), 1153–1163 (1960)
52. J. C. Campuzano et al: Phys. Rev. Lett. **83**(18), 3709–3712 (1999)
53. P. D. Johnson et al: Phys. Rev. Lett. **87**(17), 177007 (2001)
54. A. Lanzara et al: Nature **412**, 510–514 (2001)
55. T. Sato et al: Phys. Rev. Lett. **91**(15), 157003 (2003)
56. T. Takahashi et al: New J. Phys. **7**, 105 (2005) Focus Issue on 'Photoemission and Electronic Structure'.
57. M. Hengsberger et al: Phys. Rev. B **60**(15), 10796–10802 (1999)
58. T. Valla et al: Phys. Rev. Lett. **83**(10), 2085–2088 (1999)
59. S. LaShell et al: Phys. Rev. B **61**(3), 2371–2374 (2000)
60. Eli Rotenberg et al: Phys. Rev. Lett. **84**(13), 2925–2928 (2000)
61. J. Schäfer et al: Phys. Rev. Lett. **92**(9), 097205 (2004)
62. C. Kierkegaard et al: New Journal of Physics **7**, 99 (2005)
63. G. Rickayzen: *Green's Functions and Condensed Matter*, volume **5** of *Techniques of Physics*. Academic Press Inc., London–San Diego–New York (1991)
64. Eleftherios N. Economou: *Green's Functions in Quantum Physics*, volume **7** of *Solid-State Science*. Springer, Berlin–Heidelberg–New York (1990)
65. Richard D. Mattuck: *A Guide to Feynman Diagrams in the Many-Body Problem*. McGraw-Hill, Inc., New York (1976)
66. Gerald D. Mahan: *Many-Particle Physics*. Plenum Press, New York (1981)
67. William Shockley: Phys. Rev. **56**, 317–323 (1939)
68. Sydney G. Davison and Maria Stęślicka: *Basic Theory of Surface States*. Oxford University Press, Oxford–New York (1992)
69. P. O. Gartland and B. J. Slagsvold: Phys. Rev. B **12**(10), 4047–4058 (1975)
70. P. Heimann et al: J. Phys., Condens. Matter **10**, L17–L21 (1977)
71. T. Balasubramanian et al: Phys. Rev. B **57**(12), R6866–R6869 (1998)
72. S. D. Kevan: Phys. Rev. Lett. **50**(7), 526–529 (1983)
73. S. D. Kevan and R. H. Gaylord: Phys. Rev. B **36**(11), 5809–5818 (1987)
74. R. Paniago et al: Surf. Sci. **331–333**, 1233–1237 (1995)
75. Asier Eiguren et al. Phys. Rev. Lett. **88**(6), 066805 (2002)
76. F. Reinert: J. Phys., Condens. Matter **15**, S693–S705 (2003)

77. F. Forster et al: Surf. Sci. **532–535**, 160–165 (2003)
78. H. Cercellier et al: Phys. Rev. B **70**, 193412 (2004)
79. F. Reinert and G. Nicolay: Appl. Phys. A **78**, 817–821 (2004)
80. F. Forster et al: J. Chem. Phys. B **108**(38), 14692–14698 (2004)
81. H. Cercellier et al: Phys. Rev. B **73**, 195413 (2006)
82. J. Li et al: Phys. Rev. B **56**(12), 7656–7659 (1997)
83. J. Li et al: Phys. Rev. Lett. **81**(20), 4464–4467 (1998)
84. R. Berndt et al: Appl. Phys. A **69**, 503–506 (1999)
85. L. Bürgi et al: Surf. Sci. **447**, L157–L161 (2000)
86. L. Limot et al: Phys. Rev. Lett. **91**(19), 196801 (2003)
87. F. Reinert et al: Physica B **351**, 229–234 (2004)
88. P. M. Echenique et al: Chem. Phys. **251**, 1–35 (2000)
89. B. Hellsing et al: J. Electron Spectrosc. Relat. Phenom. **129**, 97–104 (2003)
90. J. Li et al: Surf. Sci. **422**, 95–106 (1999)
91. G. Nicolay et al: Phys. Rev. B **62**(3), 1631–1634 (2000)
92. G. Nicolay et al: Surf. Sci. **543**, 47–56 (2003)
93. R. Claessen et al: Phys. Rev. Lett. **69**(5), 808–811 (1992)
94. L. Perfetti et al: Phys. Rev. B **64**, 115102 (2001)
95. G. Nicolay: High-resolution photoemission measurements on $TiTe_2$. unpublished (2001)
96. K. Matho: Physica B **206&207**, 86–88 (1995)
97. F. Reinert et al: Phys. Rev. Lett. **91**(18), 186406 (2003)
98. Göran Grimvall: Phys. kond. Mat. **9**, 283–299 (1969)
99. F. Reinert et al: Phys. Rev. Lett. **92**(8), 089904 (2004)
100. F. Reinert et al: High-resolution photoemission spectroscopy on lead at HiSOR. unpublished (2003)
101. V. N. Strocov et al: Phys. Rev. B **63**, 205108 (2001)
102. D. J. Scalapino et al: Phys. Rev. Lett. **14**(4), 102–105 (1965)
103. J. Bardeen et al: Phys. Rev. **108**, 1175–1204 (1957)
104. E. Bertel and M. Donath, editors: *Electronic surface and interface states on metallic systems*, Singapore, 1994. World Scientific.
105. E. Bertel and N. Memmel: Appl. Phys., A **63**, 523–531 (1996)
106. S. LaShell et al: Phys. Rev. Lett. **77**(16), 3419–3422 (1996)
107. G. Nicolay et al: Phys. Rev. B **65**, 033407 (2002)
108. F. Forster et al: Surf. Sci. (2006) ECOSS23, in print.
109. A. Bendounan et al: Phys. Rev. B **72**, 075407 (2005)
110. F. Schiller et al: Phys. Rev. Lett. **94**, 016103 (2005)
111. W. E. McMahon et al: Surf. Sci. Lett. **279**, L231–235 (1992)
112. B. Aufray et al: Microsc. Microanal. Microstruct. **8**, 167–174 (1997)
113. A. Bendounan et al: Phys. Rev. B **67**, 165412 (2003)
114. A. Bendounan et al: Phys. Rev. Lett. **96**, 029701 (2006)
115. A. Bendounan et al: Surf. Sci. (2006) ECOSS23, in print.
116. J. N. Crain et al: Phys. Rev. B **66**, 205302 (2002)
117. A. Mugarza et al: Phys. Rev. B **66**, 245419 (2002)
118. F. Baumberger et al: Phys. Rev. Lett. **92**(1), 016803 (2004)
119. C. Didiot et al: Phys. Rev. B (2006) accepted.
120. T. Greber et al: Phys. Rev. Lett. **79**(22), 4465–4468 (1997)
121. Karl A. Gschneidner, Jr. and Le Roy Eyring, editors: *Handbook on the physics and chemistry of rare earths*. North-Holland, Amsterdam–New York–Oxford (1982–2002)

122. O. Gunnarssonand K. Schönhammer: Phys. Rev. Lett. **50**(8), 604–607 (1983)
123. O. Gunnarsson and K. Schönhammer: Phys. Rev. B **28**(8), 4315–4341 (1983)
124. O. Gunnarsson and K. Schönhammer: Phys. Rev. B **31**(8), 4815–4834 (1985)
125. P. W. Anderson: Phys. Rev. **124**(1), 41–53 (1961)
126. N. E. Bickers et al: Phys. Rev. B **36**(4), 2036–2079 (1987)
127. F. Reinert et al: Phys. Rev. Lett. **87**(10), 106401 (2001)
128. W. von der Linden et al: Phys. Rev. Lett. **71**(6), 899–902 (1993)
129. Felix Schmitt: Master's thesis, Universität Würzburg, Würzburg, July 2005.
130. D. Ehm et al: Acta Phys. Pol. B **34**, 951–954 (2003)
131. J. Kroha et al: Physica E **18**(1–3), 69–72 (2003)
132. G. Nicolay, Universität des Saarlandes, Saarbrücken, unpublished (2002)
133. D. Ehm et al: Phys. Rev. B (2006) in preparation.
134. D. Ehm et al: Physica B **312–313**, 663–665 (2002)

3

Photoemission as a Probe of the Collective Excitations in Condensed Matter Systems

P. D. Johnson and T. Valla

Condensed Matter and Materials Science Department, Brookhaven National Laboratory, Upton, NY, 11973, USA

Abstract. Recent advances in photoemission are allowing detailed studies of the role of collective many-body excitations in the decay of a photohole. These collective excitations include phonons, charge density waves and magnetic or spin excitations. With these developments angle resolved photoemission with its momentum resolving capabilities has become a powerful probe of the transport properties in condensed matter systems. We review these advances and examine the application of high resolution photoemission to studies of both metallic systems and the new high-T_c superconductors.

3.1 Introduction

New developments in instrumentation have recently allowed photoemission measurements to be performed with very high energy and momentum resolution [1]. This has allowed detailed studies of the self-energy corrections to the lifetime and mass renormalization of excitations in the vicinity of the Fermi level. These developments come at an opportune time. Indeed the discovery of high-temperature superconductivity in the cuprates and related systems is presenting a range of challenges for condensed matter physics [2]. Does the mechanism of high-T_c superconductivity represent new physics? Do we need to go beyond Landau's concept of the Fermi liquid [3]? What, if any, is the evidence for the presence or absence of quasi-particles in the excitation spectra of these complex oxides? The energy resolution of the new instruments is comparable to or better than the energy or temperature scale of superconductivity and the energy of many collective excitations. As such, photoemission has again become recognized as an important probe of condensed matter.

Studies of the high-T_c superconductors and related materials are aided by the observation that they are two-dimensional. To understand this, we note that the photoemission process results in both an excited photoelectron and a photohole in the final state. Thus the experimentally measured photoemission peak is broadened to a width reflecting contributions from both the finite lifetime of the photohole and the momentum broadening of the outgoing photoelectron. The total width Γ is given by [4]

$$\Gamma = \left(\Gamma_h + \frac{v_h}{v_e}\Gamma_e\right)\left(\left|1 - \frac{v_h}{v_e}\right|\right)^{-1} \tag{3.1}$$

where Γ_h is the width of the hole state, Γ_e the width of the electron state, and v_h and v_e the respective perpendicular velocities. In a two-dimensional system with $v_h = 0$, the width of the photoemission peak is therefore determined entirely by the inverse lifetime or scattering rate of the photohole, Γ_h. This observation offers the possibility that the technique may be useful as a probe of the related scattering mechanisms contributing to the electrical transport in different materials. Unlike other probes of these transport properties, photoemission has the advantage that it is momentum resolving. In drawing conclusions from such studies, it is important to remember that the single-particle scattering rate measured in photoemission is not identical to the scattering rate measured in transport studies, τ_{tr}. However with certain assumptions, the two are approximately related and the transport scattering rate can be written $\hbar/\tau_{tr} = \hbar/\tau \left(1 - \langle\cos\vartheta\rangle\right)$ where $\hbar/\tau$ represents the single-particle scattering rate and $\langle\cos\vartheta\rangle$ represents the average value of $\cos\vartheta$ with ϑ the scattering angle [5,6].

In the following sections we first review the photoemission process with particular reference to the role of coupling to many-body excitations. We then examine in more detail the coupling to a variety of excitations including phonons, charge density waves (CDW) and magnetic or spin excitations. Finally we review studies of the high-T_c materials with an emphasis on measurements of self-energy effects. We note that our discussion is heavily concentrated around our own work but recognize the many important studies that have been reported by other groups.

3.2 The Photoemission Process

In photoelectron spectroscopy, a photon of known energy, $h\nu$, is absorbed and the outgoing electron's energy ($h\nu - \phi - \varepsilon_k$) and angle are measured. These properties determine the binding energy ε_k and parallel momentum $k_{||}$ of the hole left in the occupied valence bands [7]. Interaction effects, including for instance Coulomb and electron–phonon, cause the sharp line spectrum of independent electron theory, $A_0(\mathbf{k},\omega) = \text{Im } G_0(\mathbf{k},\omega) = \text{Im } 1/(\omega - \epsilon_{\mathbf{k}0} - i\eta)$, where $\epsilon_{\mathbf{k}0}$ represents a bare band dispersion, to evolve into $\text{Im } 1/[\omega - \epsilon_{\mathbf{k}0} - \Sigma(\mathbf{k},\omega)]$ where the complex self-energy $\Sigma(\mathbf{k},\omega)$ contains the effects of the many-body interactions. The single-particle spectral function of the hole-state, $A(\mathbf{k},\omega)$, then takes the form

$$A(\mathbf{k},\omega) \propto \frac{\Sigma_2(\mathbf{k},\omega)}{[\omega - \epsilon_{\mathbf{k}0} - \Sigma_1(\mathbf{k},\omega)]^2 + (\Sigma_2(\mathbf{k},\omega))^2} \tag{3.2}$$

Thus the real part, $\Sigma_1(\mathbf{k},\omega)$, gives a shift in energy and associated mass enhancement, while the imaginary part $\Sigma_2(\mathbf{k},\omega)$ gives the lifetime broadening $h/\tau_\mathbf{k}$. Here $\tau_\mathbf{k}$ is the typical time before the hole state $(\omega,\mathbf{k})$ scatters into other states, $(\omega',\mathbf{k}')$. In the limit of $\omega \to 0$, the real part of the self-energy may be written as $\Sigma_1(\mathbf{k},\omega) \sim -\omega\lambda_\mathbf{k}$ with $\lambda_\mathbf{k}$ representing a coupling constant describing the coupling to excitations that scatter the hole from $(\omega,\mathbf{k})$ to other states. The process of coupling is illustrated in Fig. 3.1 where we consider coupling to a mode described by an Eliashberg function, $\alpha^2 F$. Here $\alpha^2 F$ represents the product of the density of states of the relevant excitation and a matrix element reflecting the coupling strength [8]. For the present purposes, $\alpha^2 F$ in Fig. 3.1(a) is simply represented by a single Gaussian peak at energy ω_0. Coupling to such a mode (at $T = 0$) will result in a broadened step function in the scattering rate or imaginary part of the self-energy, Σ_2. The step function reflects the observation that when the photohole has enough energy to create the mode ($\omega \geq \omega_0$), scattering from the mode opens up a new decay channel, thereby shortening the lifetime. The real and imaginary parts of the self-energy are related via causality through a Kramers–Kronig transform. Thus the step function in Σ_2 results in a cusp function for Σ_1 (panel (c)). Such an energy dependence of the self-energy affects the measured spectra in two ways. Above and below the mode energy there will be a noticeable change in the spectral function as illustrated in panel (d). Secondly, as noted above,

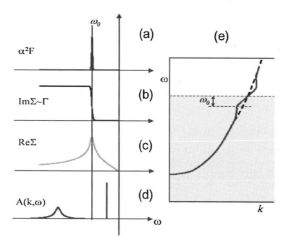

Fig. 3.1. An electron scattering from a mode with $\alpha^2 F$ as in (a) will experience a step function at the mode energy in the imaginary component of its self-energy, ImΣ or Σ_2, as in panel (b). A Kramers–Kronig transform of Σ_2 will produce a cusp function in the real part of the self-energy, ReΣ or Σ_1 as shown in panel (c). The Σ_2 shown in panel (b) results in a spectral function having the form shown in panel (d) above and below the mode energy. Panel 1(e) shows the mass renormalization in the immediate vicinity of E_F.

the measured dispersion will be given by $\epsilon_{k0} + \Sigma_1(\mathbf{k}, \omega)$. Thus with Σ_1 taking the form shown in panel (c), the dispersion will display the mass enhancement observed immediately below the Fermi level as presented in Fig. 3.1(e).

The intensity $I(\mathbf{k}, \omega)$ of photoelectrons measured as a result of the photoemission process is given by

$$I(\mathbf{k}, \omega) = |M|^2 A(\mathbf{k}, \omega) f(\omega) \tag{3.3}$$

where M represents the matrix element linking the initial and final states in the photoemission process, $A(\mathbf{k}, \omega)$ is the single-particle spectral function given in Eq. (3.2) and $f(\omega)$ is the Fermi function which enters because the photoemission process is restricted to excitation from occupied states. Modern photoelectron spectrometers allow the simultaneous measurement of photoelectron intensities from a finite range in both energy and momentum space. A typical image is shown in Fig. 3.2 [9]. The ability to obtain such images has led to the development of new methodologies for the extraction of self-energies. As such, the spectral response in Fig. 3.2 may be analysed by taking an intensity cut at constant angle or momentum, the so called energy distribution curve (EDC) or by taking an intensity cut at constant energy, a momentum distribution curve (MDC). The former has been the traditional

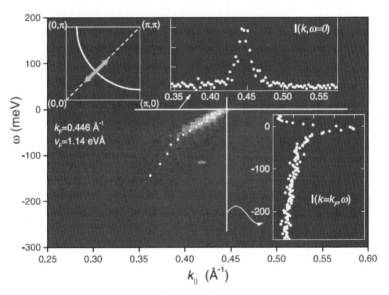

Fig. 3.2. Two-dimensional spectral plot showing the intensity of emission as a function of ω, the binding energy, and $k_\parallel$, the parallel momentum. The photon energy is 21.2 eV and the sample temperature is 48 K. *Clockwise from upper left*, the insets show the region of the Brillouin zone sampled in the experiment, a cross section through the intensity at constant energy ($\omega = 0$) as a function of momentum (an MDC), and a cross section through the intensity at constant angle or momentum ($k = k_F$) as a function of ω (an EDC)

method, the latter is a new method enabled by the new instrumentation. Let us consider the MDC method. If the binding energy is fixed and in the limit of a momentum independent self-energy, the spectral function (MDC) takes the simple form:

$$A(\mathbf{k}, \omega_0) = \frac{\Sigma_2(\omega_0)}{[\omega_0 - \epsilon_{\mathbf{k}0} - \Sigma_1(\omega_0)]^2 + [\Sigma_2(\omega_0)]^2} \quad (3.4)$$

In the vicinity of the Fermi level we may approximate the bare dispersion with a linear form such that $\epsilon_{k0} = v_0(k - k_F)$ with v_0 the bare velocity. As has been discussed in several papers [10–13], the MDC is then a simple Lorentzian, centered at $k_m = k_F + [\omega_0 - \Sigma_1(\omega_0)]/v_0$ and with the full width at half maximum $\Delta k = 2\Sigma_2(\omega_0)/v_0$. The self-energy can thus be simply extracted from MDC peaks at any binding energy.

Noting that the measured or renormalized velocity $v = v_0/(1+\lambda)$, where, as before, λ represents the coupling constant, the equivalent EDC has a width ΔE such that

$$\Delta E = v \Delta k = \frac{2\Sigma_2}{(1+\lambda)} = \frac{2\Sigma_2}{1 - \partial \Sigma_1/\partial \omega} \quad (3.5)$$

If the real part of the self-energy displays no frequency dependence, the width ΔE is directly related to the scattering rate. Both EDCs and MDCs will have a Lorentzian line shape. However this is no longer true if the real part of the self-energy is frequency dependent and particularly in the vicinity of a mode, the width of the EDC, ΔE, will be strongly dependent on the renormalization of the velocity. This can result in the EDC having a complex two peaked structure that is more difficult to interpret.

3.3 Electron–Phonon Coupling in Metallic Systems

In this section we focus on photoemission studies of electron–phonon coupling in metallic systems. The electron–phonon coupling contribution, Γ_{e-ph}, to the total scattering rate may be calculated via the Eliashberg equation such that [5]

$$\Gamma_{e-ph}(\omega, T) = 2\pi\hbar \int_0^\infty d\omega' \alpha^2 F(\omega') [2n(\omega') + f(\omega' + \omega) + f(\omega' - \omega)] \quad (3.6)$$

where again $\alpha^2 F$ is the Eliashberg coupling constant and $f(\omega)$ and $n(\omega)$ are the Fermi and Bose–Einstein functions, respectively. Γ_{e-ph} increases monotonically with energy up to some cut-off defined by the Debye energy. At $T = 0$ the electron–phonon coupling constant is given by [8]

$$\lambda = 2 \int_0^\infty \frac{\alpha^2 F(\omega')}{\omega'} d\omega' \quad (3.7)$$

Early photoemission studies focused on the observation that at higher temperatures, above approximately one third the Debye energy, Eq. (3.6) reduces to $\Gamma_{e-ph} = \pi \lambda k_B T$. Thus a measurement of the width of a photoemission peak as a function of temperature provides direct access to the coupling constant, λ. This approach has been used in several studies including a study of the electron–phonon contribution to quasi-particle lifetimes of surface states on the Cu(111) [14] and Be(0001) [15] surfaces. In the former case the electron–phonon coupling constant for the surface, $\lambda = 0.14$ was close to that measured for bulk copper, $\lambda = 0.15$. In the case of Be, the surface was found to have a dramatically enhanced value of $\lambda = 1.15$, which is to be compared with the bulk value of $\lambda = 0.24$. Hengsberger et al. found a similar value, $\lambda = 1.18$, for the electron–phonon coupling parameter in the surface region of Be by measuring the velocity renormalization in the surface band [16]. However the most recent study of the same surface reduced the value λ to 0.7, a value obtained from a determination of the rate of change of the real part of the self-energy, $-(\partial \Sigma_1/\partial \omega)$, in the vicinity of E_F. Enhanced values of λ have led to speculation on the possibility of enhanced superconducting transition temperatures in the surface region [15].

The introduction of the new instrumentation in the nineties allowed the first direct imaging of the mass renormalization due to electron–phonon coupling. Figure 3.3 shows an image of the spectral intensity excited from a two-dimensional surface resonance in the Γ–N azimuth of a Mo(110) crystal with the sample held at 70 K [17]. The state shown in the figure corresponds to a surface resonance which closes an elliptical hole Fermi orbit around the

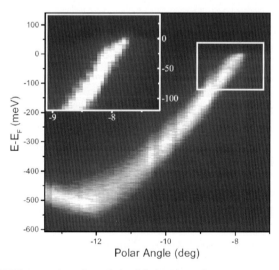

Fig. 3.3. ARPES intensity plot of the Mo(110) surface resonance recorded along the Γ–N line of the surface Brillouin zone at 70 K. Shown in the inset is the spectrum of the region around k_F taken with special attention to the surface cleanliness

center of the zone, $\bar{\Gamma}$ [18]. In the vicinity of the Fermi level there is a notable change in the rate of dispersion, or mass enhancement, and a rapid change in the width of the band. The self-energy corrections resulting in these changes reflect three principal contributions, electron–electron scattering, electron–phonon scattering and electron–impurity scattering. These different contributions all add linearly to give the total scattering rate Γ such that

$$\Gamma = \Gamma_{e-e} + \Gamma_{e-ph} + \Gamma_{imp} \quad (3.8)$$

In a Fermi liquid the electron–electron scattering term is given by $\Gamma_{e-e}(\omega, T) = 2\beta\left[(\pi k_B T)^2 + \omega^2\right]$ where, within the Born approximation, $2\beta = (\pi U^2)/(2W^3)$, with U the on-site Coulomb repulsion and W the bandwidth of the state. As noted earlier the electron–phonon contribution may be calculated via the Eliashberg equation, Eq. (3.6). The final contribution in Eq. (3.8), impurity scattering, is elastic in that the impurity atoms are considered to have no internal excitations. Thus the scattering-rate, Γ_{imp}, is proportional to the impurity concentration, but independent of energy and temperature. At sufficiently low temperature, impurity scattering represents the dominant decay mechanism for a hole close to E_F.

Figure 3.4 shows the measured Σ_2 of the Mo surface state as a function of binding energy. The data points are extracted from the image of Fig. 3.3 in

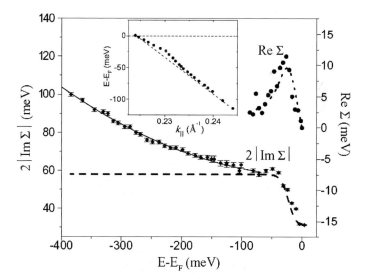

Fig. 3.4. The photohole self-energy as a function of binding energy at 70 K of the Mo(110) surface along the $\bar{\Gamma}$–$\bar{N}$ line. The real part is obtained from the MDC-derived dispersion shown in the inset. The imaginary part is obtained from the width of the quasi-particle peak. The *solid line* is a quadratic fit to the high-binding energy data ($\omega < 80$ meV). The *dashed (dotted)* line shows the calculated electron–phonon contribution to the imaginary (real) part of the self-energy. The *dashed line* is shifted up by 26 meV

two ways, either from EDCs or from MDCs. The calculated electron–phonon contribution to the self-energy is indicated in the figure. In the vicinity of the Fermi level, the agreement between the calculation using the theoretical $\alpha^2 F$ of bulk molybdenum [19] and the experimentally measured widths is excellent. There is a rapid change in the scattering rate up to the Debye energy at ~ 30 meV. At binding energies greater than this, the electron–phonon contribution saturates. However, also shown in the figure is a quadratic fit to the measured widths at the higher binding energies. The quadratic dependence is an indication that electron–electron scattering, as in a Fermi liquid, plays an important role. In a purely two dimensional system there should be a logarithmic correction to the quadratic term [20]. Thus Γ_{e-e} will be proportional to $\omega^2 \ln \omega$. However the simple quadratic fit works well as indicated in the figure because the surface state shown in Fig. 3.3 is, as previously noted, a surface resonance with good coupling to bulk states [18]. The quadratic fit is consistent with the prefactor in the expression for Γ_{e-e} having $U \sim 0.6$ eV, as predicted for molybdenum [21], and $W \sim 1.3$ eV the approximate bandwidth of the surface state. The measured widths also have an energy-independent contribution due to scattering from hydrogen impurity centers [17].

The calculated real component of the self-energy, Σ_1, derived through a Kramers–Kronig transform of Σ_2 is also shown in Fig. 3.4 where it is compared with the experimentally derived values. From Σ_1 it is possible to determine a value for the electron–phonon coupling constant of 0.4 to be compared with the bulk value of 0.42. As we have already noted the coupling constant can also be

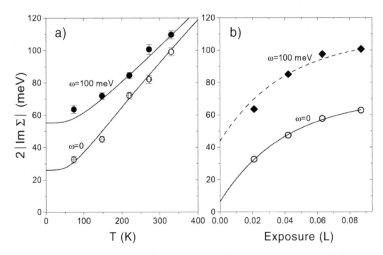

Fig. 3.5. The peak width as a function of (**a**) temperature and (**b**) exposure to background hydrogen, measured for two binding energies. For the exposure dependence the sample was held at 70 K. Lines in (a) are calculated electron–phonon contributions, shifted up by 26 meV to match the data. Lines in (b) represent fits (see text for details)

derived from the temperature dependence of the peak widths. This is shown in Fig. 3.5(a) for two different binding energies, $\omega = 0$ and $\omega = 100\,\text{meV}$. By doing linear fits to the experimental data points, values for the coupling constant of 0.52 and 0.35 are obtained respectively. These values are again close to the bulk value.

The observation that the width of the quasi-particle peak always has a significant constant term indicates the presence of impurity scattering. It is known that this surface state is very sensitive to hydrogen adsorption. Figure 3.5(b) shows how the width changes with the exposure to residual hydrogen. Note that it saturates with exposure θ. If the scattering rate is proportional to the concentration of adsorbed particles, the experimental points become a measure of the concentration. Since the number of free adsorption sites decays exponentially with exposure, the concentration of adsorbed atoms as a function of exposure should change as $c(\theta) = c_0 + c_{sat}(1 - e^{-p\theta})$, where p is the adsorption probability and c_0 (c_{sat}) is the initial (saturation) concentration. The width of the quasi-particle peak can be fitted with the same dependence (lines). It is notable that extrapolation to zero exposure results in a residual width of $6 \pm 5\,\text{meV}$ at $\omega = 0$. electron–phonon coupling contributes with $\approx 5\,\text{meV}$ for $T = 70\,\text{K}$. However, we should also note some uncertainty in the initial coverage due to the change in adsorption conditions between flashing the sample and the measurement.

3.4 Studies of the Dichalcogenides

The family of layered dichalcogenides provide a range of interesting phenomena for study. These systems exhibit both charge-density wave (CDW) for-

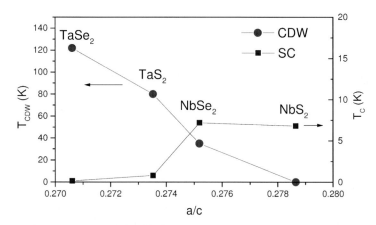

Fig. 3.6. A comparison of the CDW transition temperature T_{CDW} and the superconducting transition temperature T_c for the dichalcogenides plotted as a function of the ratio of lattice parameters a/c (reproduced from [23])

mation and superconductivity [22, 23]. As shown in Fig. 3.6, the fact that the CDW transition temperature decreases while the superconducting critical temperature, T_c, increases on going from TaSe$_2$ through TaS$_2$ and NbSe$_2$ to NbS$_2$ suggests that the two order parameters represent competing ground states. Indeed, it has been found that in TaS$_2$ and NbSe$_2$, T_c increases under pressure while T_{CDW} decreases [24, 25]. While these studies suggested that after CDW order disappears, T_c remains approximately constant, a more recent study of NbSe$_2$ indicates that as a function of pressure, T_c at first increases up to some maximum value and then decreases [26]. In NbS$_2$, the system with no CDW order, T_c is insensitive to pressure. Although various anomalies, including an apparent anisotropy of the superconducting gap, have been observed [27–30], it is generally believed that superconductivity in the dichalcogenides is of the conventional BCS character, mediated by strong electron–phonon coupling [29]. However, consensus on the exact mechanism that drives the system into the CDW state has still not been reached. Some authors [22, 23, 31, 32] argue, in analogy with a Peierls transition in one-dimensional systems, that the CDW transition is driven by a Fermi-surface instability or nesting, with some portions of the Fermi surface spanned by a CDW vector q_{CDW}. In another scenario, the CDW instability is induced by the nesting of van Hove singularities or saddle points in the band structure if they are within a few $k_B T_{CDW}$ of the Fermi energy [33].

The Fermi surface of the 2H-dichalcogenides is rather complicated, being dominated by several open (2D-like) sheets and one small 3D S(Se)-derived pancake-like Fermi surface [29]. In such a situation, one may anticipate anisotropic properties and in particular, an anisotropic electron–phonon coupling. The resistivity anisotropy, of the order of 10–50, is much smaller than in layered oxides, indicating a substantial inter-layer hopping [34]. Transport properties show relatively small anomalies at T_{CDW}, suggesting that only a small portion of the Fermi surface becomes gapped in the CDW state. In addition, the 2H-dichalcogenides become better conducting in the CDW state, indicating a higher degree of coherence.

Several ARPES studies of the dichalcogenides have measured the form or shape of the Fermi surface, the focus being on the identification of the appropriate nesting vector associated with the CDW. In Fig. 3.7 we reproduce the results of a recent study showing the measured Fermi surfaces of both TaSe$_2$ and NbSe$_2$ [35]. The figure also shows superimposed the results of a simple tight binding fit, by the authors of the study, to the electronic structure.

Our own study of TaSe$_2$ [12] found, as in earlier studies [36–38], that the hole pocket at the center of the zone remains ungapped even in the CDW state. However the studies of the electronic states forming this Fermi surface found strong evidence of the formation of the CDW state as shown in Fig. 3.8. The figure shows the photoemission intensity, recorded in the CDW state at $T = 34$ K, as a function of binding energy and momentum along the line through the two-dimensional Brillouin zone indicated in the inset of the figure. A band is observed crossing the Fermi level at a point on the hole-like

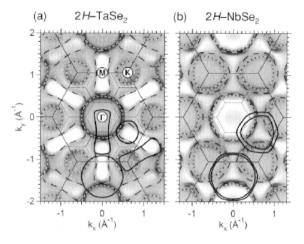

Fig. 3.7. E_F-ARPES intensity maps of (a) $2H$-TaSe$_2$ and (b) $2H$-NbSe$_2$, measured in the normal states at 125 and 65 K, respectively ($h\nu = 100$ eV). Raw data are shown in the *upper right* quadrants. The rest of the data was obtained by mirror symmetry operations. The intensity value at each k point was normalized by the intensity integrated over the occupied transition-metal d bandwidth (~ 400 meV). The *darker grayscale* indicates higher photoemission intensity. The *small-dotted* hexagons are the Brillouin-zone scheme for the 3×3 superlattice. *Short and long dashed lines*: Simulated Fermi contours of two transition-metal d-derived bands. *Solid lines* in the lower right quadrants: Umklapp-shifted Fermi contours (reproduced from [35])

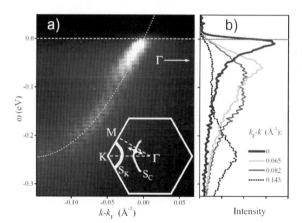

Fig. 3.8. The photoemission intensity of TaSe$_2$ in the CDW state ($T = 34$ K) as a function of binding energy and momentum along the line indicated in the inset by the *double-headed arrow*. The intensity is represented by a *grayscale* map, with *white* representing the highest intensity. The dispersing state is a part of the hole like Fermi surface S_C, centered at Γ. This Fermi surface is not gapped in the CDW state. (b) EDCs, measured for several momenta as discussed in the text

Fermi surface S_C, centered at Γ. The Fermi surface of the dichalcogenides is double walled, suggesting that every band should be split into two [36, 37]. Indeed, in our own study [12] we detected both bands and observed that the splitting, as well as the relative intensity of the two bands, is strongly dependent on momentum, photon energy, polarization and surface quality. In some circumstances, only one band can be observed. However, when both bands are well resolved, they show similar behavior in the vicinity of the Fermi level and similar self-energy corrections. Therefore the presence (or absence) of the second band would appear irrelevant. The most remarkable feature in Fig. 3.8 is the "kink" in the band's dispersion, accompanied by a sharpening in the vicinity of the Fermi level. The figure also shows EDC cuts through the intensity at constant momenta. In this energy range, the EDCs show a two-peaked structure, behavior that is again characteristic of the interaction of the photohole with some excitation of the system with energy range limited approximately to the energy scale of the kink. As discussed earlier and presented in Eq. (3.4), the real and imaginary components of the self-energy, $\Sigma_1(\omega)$ and $\Sigma_2(\omega)$, can be extracted directly from a momentum-distribution curve. The fitting is possible without imposing any particular model for the interaction. The non- interacting dispersion in Fig. 3.8 may be approximated with a second-order polynomial that coincides with the measured dispersion at $k = k_F$ and at higher binding energies, close to the bottom of the band: thus $\Sigma_1 = 0$ for $\omega = 0$ and for $\omega < -200\,\text{meV}$. Figure 3.9 shows several MDCs with corresponding fits. In contrast to the lineshapes in Fig. 3.8(b) for EDCs, the lineshapes in Fig. 3.9 are approximately Lorentzian at low binding energies developing an asymmetry at higher binding energies. The latter asymmetry mostly reflects the quadratic term in the non-interacting dispersion. The advantage of using MDCs in the analysis is obvious in that the self-energies are more dependent on energy than on momentum.

The results of the fitting procedure, which produces pairs of Σ_1 and Σ_2 for every MDC are shown in Fig. 3.10 for several temperatures. Σ_2 as obtained by fitting EDCs when the latter have a Lorentzian lineshape are also included. The real part of the self-energy is concentrated in the region of binding energies less than 150 meV. At the lowest temperature, it has a maximum at a binding energy of ~ 65 meV, approximately coincident with the value corresponding to the sharp drop in Σ_2. Such behavior is indicative of the scattering of the photohole from some collective excitation or "mode" of the system. The striking similarity with the behavior observed in ARPES studies of a photohole interacting with phonons [16, 17] would point to the electron–phonon coupling as the source of this behavior. However this would imply the presence of ~ 70 meV phonons in the CDW state where the highest calculated and measured phonon frequency is ~ 40 meV [39]. The measured temperature dependence of the self-energy also argues against phonons. With increasing temperature, the peak in Σ_1 loses its magnitude and the structure shifts to lower energies. At a temperature of 111 K, only a small peak is left at a binding energy of ~ 30 meV and this survives in the normal state to at

3 Photoemission as a Probe of Collective Excitations   67

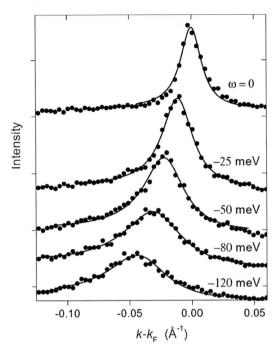

Fig. 3.9. MDCs of TaSe$_2$, measured at different binding energies (*symbols*), fitted with a momentum-independent spectral function (*solid lines*) as discussed in the text

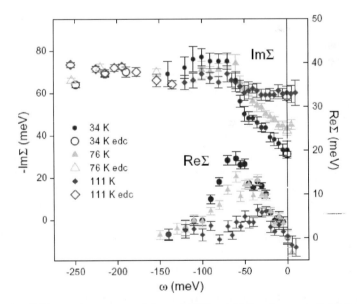

Fig. 3.10. Self-energies extracted from MDCs of TaSe$_2$ for several temperatures (*solid symbols*). Results for Σ_2 obtained from EDCs are shown as *open symbols*

least 160 K. The latter peak may be of the same CDW origin, but may also be caused by conventional electron–phonon coupling, since it is within the range of the phonon spectrum. At low temperatures the imaginary part of the self-energy or scattering rate shows a sharp reduction for binding energies lower than 70 meV. As the temperature increases, this reduction becomes less pronounced.

In a more recent photoemission study [35], it has been suggested that the higher energy "kink" observed in TaSe$_2$ is associated with a band folding associated with the CDW transition. However it is important to note that the study of Rossnagel et al. [35] was at a lower energy resolution and the authors reported a lack of observation of causality, i.e. no defined relationship between the measured real and imaginary parts of the self-energy. This differs from the results of the studies discussed here and shown in Fig. 3.10. We believe that the high-energy kink is closely related to the CDW gap, either in a conventional way, where the kink will shift from phonon frequency ω to $\omega + \Delta_q$ where Δ_q represents the (CDW) gap that opens in the final state at scattering vector q, or in a more exotic way, where a new excitation, or a fluctuation of magnitude of the CDW order parameter, couples to holes and produces the mass enhancement.

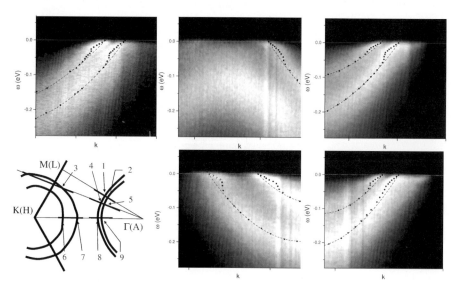

Fig. 3.11. The photoemission intensities of NbSe$_2$ in the CDW state at $T = 10$ K for several momentum lines indicated in the schematic view of the Brillouin zone (*lower left panel*) by the *dark-gray lines*. The *light-gray lines* represent Nb-derived Fermi sheets. The nine Fermi points are numbered. The MDC derived dispersions are represented by *filled circles*. The high-energy part of the dispersions is fitted with a second-order polynomial (*dashed lines*), and the low energy part is fitted with *straight lines*

Studies of the related system NbSe$_2$ show somewhat different behavior [40]. Figure 3.11 shows the photoemission intensity, recorded at $T = 10$ K, as a function of binding energy and momentum along three different momentum lines in the Brillouin zone. Nine Fermi crossings are included: three pairs on the double-layer split Fermi sheets centered around Γ and three crossings on the split sheets centered at the K point. A characteristic change in the quasi-particle velocity ("kink") can easily be identified in all crossings. The kinks are also accompanied by a sharp change in the quasi-particle widths at the "kink" energy. These observations are again indicative of (bosonic) excitations interacting with the quasi-particles. Compared to TaSe$_2$, the excitation spectrum in the CDW state is limited to significantly lower energies. It also appears that the kink is not unique; its strength and energy depend on k, being different for different crossings. The band dispersions in the figure are determined from the peak intensities of MDCs fitted to the spectra. As shown in the figure, the high-energy part of the extracted dispersion can be fitted with a parabola that crosses through k_F, whereas the low energy part ($\omega < 15$–20 meV) is fitted with a straight line. Assuming that the parabola represents the "non-interacting" dispersion, then the slopes of these two lines at $\omega = 0$ may be used to directly extract the coupling constant, again using the expression $\lambda = v_0/v_F - 1$ with v_0 the "non-interacting" or bare Fermi velocity and v_F the renormalized one. The "non-interacting" parabolas are subtracted from the measured dispersions to extract $\Sigma_1(\omega)$. The results are shown in Fig. 3.12(a) for several crossings from Fig. 3.11. $\Sigma_1(\omega)$ gives the same coupling constant $\lambda = -(\partial \Sigma_1/\partial \omega)_0$, but also provides additional information about the spectrum of excitations interacting with the quasi-particles.

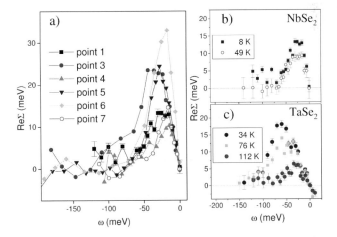

Fig. 3.12. (a) Real parts of self-energies of NbSe$_2$, ReΣ, extracted from measured dispersions from Fig. 3.11 for several Fermi points. Temperature dependence of $\Sigma_1(\omega)$ for (b) NbSe$_2$ for point 1 from Fig. 3.11 and for (c) TaSe$_2$ near the same region on the Γ-centered Fermi surface reproduced from Fig. 3.10 for comparison

It is obvious from Fig. 3.12(a) that not only is the magnitude of $\Sigma_1(\omega)$ different for different states, but also the peaks are at different energies, ranging from ~ 13 meV to ~ 35 meV. Various experimental and theoretical studies have shown that the phonon spectrum is fully consistent with these energies, with acoustic phonon branches lying below $\omega \sim 12$ meV, and optical branches spanning the region $15 < \omega < 40$ meV [41]. Shifts of the $\Sigma_1(\omega)$ maxima would further suggest that some electronic states are coupled predominantly to acoustic modes while others couple more strongly to the optical modes, even though the states are sometimes very close in momentum (compare points 4 and 5, for example). A strong k-dependence of Σ would complicate the MDC line-shape in the energy region where the momentum dependence exists. It is interesting that in spite of these differences in Σ, the resulting coupling constant does not vary much, $\lambda \sim 0.85 \pm 0.15$, within the experimental uncertainty. The only exception is the inner K-centered sheet (point 6 in Fig. 3.11), where $\lambda \sim 1.9 \pm 0.2$. This seemingly too large coupling constant is however in good agreement with the large measured value of linear specific heat coefficient, $\gamma \sim 18.5$ mJmol^{-1}K^{-2}, [27, 42] which is proportional to the renormalized density of states (DOS) at the Fermi level, $N(0)(1+\lambda)$, through $\gamma = (1/3)\pi^2 k_B^2 N(0)(1+\lambda)$. Band structure calculations give the "bare" DOS $N(0) \sim 2.8$ states eV^{-1} unit cell^{-1} [29], suggesting $\lambda \sim 1.8$. However even this might be an underestimate as γ measures an average over the Fermi surface, weighted by the DOS. A similar value for λ is obtained from c-axis optical conductivity [43] suggesting that the c-axis transport is probably dominated by the K–H-centered cylinders with largest warping.

It is instructive that in TaSe$_2$ the CDW gap opens up in the same region of the Fermi surface [35, 37, 44], while the Γ–A-centered Fermi cylinders remain ungapped, and gain coherence in the CDW state [12]. It therefore seems plausible that both the superconductivity and the CDW state originate from the inner K sheet and are driven by strong electron–phonon coupling. This seems to be in line with the original suggestion of Wilson [31] that the self-nesting of the inner K sheet drives the CDW in the 2H-dichalcogenides. A lack of CDW gap on the Γ-centered sheets in all the 2H-dichalcogenides studied in ARPES suggests that these sheets support neither the self-nesting nor the nesting which would mix them with the K-centered sheets. In particular, a proposed f-wave symmetry for the CDW gap [23] may be ruled out. The relative strength of the CDW and superconducting ordering is determined by the nesting properties of the inner K cylinder, while the upper limit for T_c (when the CDW is destroyed by applying pressure, for example) is given by λ. Nesting weakens with increasing 3D character (increased warping with k_z) under pressure and on moving from TaSe$_2$ to NbS$_2$. The coupling constant, λ increases from TaSe$_2$ to NbSe$_2$ and is only weakly pressure dependent. ARPES studies of NbSe$_2$ [30, 45, 46], have shown no evidence of a CDW gap suggesting that the nested portion of the Fermi surface is small and not sampled. However as there is a non-trivial k_z-dispersion or warping in this material, it is possible that the in-plane k_F might be tuned into a nesting configuration and that the gap opens only near certain k_z. Note that the energy splitting between the double walled sheets is larger for K-centered sheets. A similar k-dependence is also expected for the interlayer hopping, $t_\perp$, that produces the warping. Additionally, as the Fermi velocities are larger for Γ-centered sheets, it is reasonable to expect that the in-plane k_F varies with k_z much less on the Γ-cylinders than on the K-cylinders (the change in the in-plane Fermi momentum is approximately given by $\Delta k_F \propto t_\perp / v_F$). The measured Fermi surfaces centered at Γ are too large at the sampled k_z, and are therefore not expected to ever reach the self-nesting condition $2k_F = q_{CDW}$. On the other hand, the inner K-centered sheet seems to be very close to producing the required nesting. It is interesting to note that according to STM studies [47], the CDW gap is large ($\Delta_{CDW} \sim 35$ meV) and should be easily measurable in ARPES. The overall electronic properties in NbSe$_2$ are much less sensitive to the CDW transition than in TaSe$_2$. Even the CDW induced structure in the self-energy that existed in TaSe$_2$ is absent in NbSe$_2$. Both the "kink" and the scattering rate are remarkably insensitive to the CDW (see Fig. 3.12(b)), an observation that is consistent with the relative positions of NbSe$_2$ and TaSe$_2$ in Fig. 3.6.

3.5 Magnetic Systems

In magnetic systems, aside from phonon scattering, the possibility also exists for scattering from spin excitations. Such effects have been found in photoemission studies of gadolinium [48, 49] and of iron [50]. The spin dependent electronic structure of these materials has been studied with spin-resolved photoemission [51]. However there has only been one such study with sufficiently high energy resolution to examine in detail the spin-resolved self-energy effects. That is a study of gadolinium [49].

The ground state of gadolinium is ferromagnetic with a Curie temperature T_C of 293 K. The (0001) surface of this material has been shown both theoretically [52] and experimentally [53] to support a surface state derived from the Gd 5d orbitals. The state, which is spin polarized through an exchange interaction with the localized 4f orbitals has an important history and indeed it was spin-resolved photoemission studies of the surface state that finally confirmed that the surface moments were ferromagnetically aligned with the bulk of the material [54].

Figure 3.13 shows spectral-density maps recorded from the clean Gd(0001) surface in the ΓX azimuth at two different temperatures [48]. The EDC width of the surface state at a binding energy of ∼170 meV increases as the temperature is raised from 82 K to 300 K. The increase reflects a reduction in the lifetime of the photohole as a result of increased electron–phonon and electron–magnon scattering at the higher temperature. In the low-temperature plot, the state has a width approximately constant until the angle of emission exceeds 5°. At this point, according to calculated band structures [52],

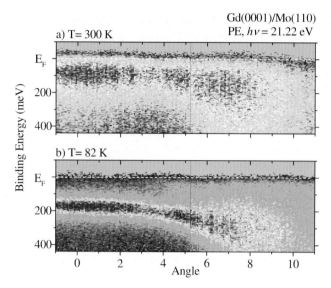

Fig. 3.13. *Upper panel:* Spin-integrated spectral response for the Gd(0001) surface as a function of binding energy and angle of emission measured from the surface normal. The sample T is 300 K and the incident photon energy is 21.2 eV. *Lower panel:* As above but now the sample T is 82 K

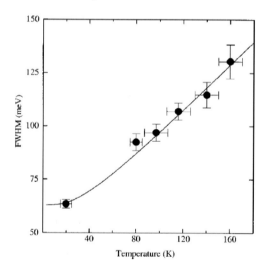

Fig. 3.14. The full width-half maximum (FWHM) of the majority spin peak of the Gd(0001) surface state as a function of T. The *solid line* indicates a fit to the data using Eq. (3.6) as given in the text

the surface state leaves the bulk band gap and begins to resonate with bulk bands. This accounts for the increased broadening or reduced lifetime. Figure 3.14 shows the width of the peak as a function of temperature and also shows a fitting to the data points using the expression given in Eq. (3.6). The latter results in a value of $\lambda \sim 1.0$ for the electron–phonon coupling constant, which may be compared with a value of 1.2 (bulk, spin averaged), extracted from the measured specific heat [55], using the calculated density of states and assuming only electron–phonon renormalization, and a theoretical value of 0.4 (also bulk and spin-averaged) obtained in a spin-polarized calculation of the electron–phonon coupling constant [56]. At the low temperatures indicated in Fig. 3.13 the state is predominantly majority spin. The electron–phonon coupling parameter may be written as $\lambda = N_S \langle I_S^2 \rangle / M \langle \omega^2 \rangle$ where N_S represents the spin-projected density of states at the hole binding energy, $\langle I_S^2 \rangle$ is the Fermi surface average of the electron–phonon matrix element, M is the atomic mass and $\langle \omega^2 \rangle$ is an average phonon frequency. Wu et al. have calculated an enhanced magnetic moment in the Gd surface layer [52]. Using their calculated majority and minority spin densities in the surface layer, one obtains $\lambda \sim 1.15$ and 0.25 for the surface majority and minority spin electron–phonon coupling, close to the value $\lambda = 1$ derived from the plot of Fig. 3.14 and again assuming that the latter is dominated by the majority spin channel.

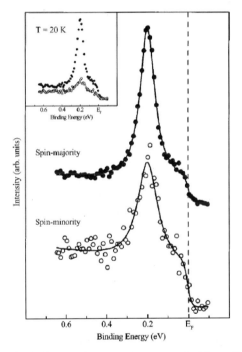

Fig. 3.15. Spin-resolved photoemission spectra recorded from the Gd(0001) surface at 20 K. The *upper* and *lower* spectra represent the emission in the majority- and minority-spin channels, respectively. The *lines* indicate Lorentzian fits to the spectra superimposed on appropriate backgrounds. The inset shows the relative intensities in the two spin channels

The results of a spin-resolved photoemission study of the same surface state held at $T = 20\,\text{K}$ are illustrated in Fig. 3.15 [49] As noted earlier, both experiment [54, 57] and theoretical calculations [52] indicate that the surface state should be 100% majority spin, reflecting parallel alignment of the surface and bulk moments. The coexistence of both spin components at the same energy in Fig. 3.15 is therefore an intrinsic property of the surface state arising from a combination of spin-orbit and spin-exchange processes. A simple model yields a polarization $P = \Delta/\sqrt{\Delta^2 + \xi^2}$ for each quasi-particle state. With a spin-orbit parameter $\xi = 0.3\,\text{eV}$ and an exchange splitting $\Delta = 0.7\,\text{eV}$ at $0\,\text{K}$, we get a spin-orbit induced mixing $R = (n^+/n^-) = (1\text{-}P)/(1\text{+}P) \sim 5\%$. Here n^+ and n^- represent the number of electrons with spin-up and spin-down, respectively. R increases to 8% at $T = 150\,\text{K}$ as the exchange splitting between the occupied and unoccupied surface states gets smaller [58].

Fitting the spectra in Fig. 3.15 with Lorentzian line shapes shows that the minority spin peak has a larger width than its majority spin counterpart, 116 meV as opposed to 86 meV. Removing the contribution from the experimental resolution, these widths become approximately 105 meV in the minority spin channel and 70 meV in the majority channel. electron–phonon, electron–magnon and electron–electron scattering each give distinct spin dependent contributions to the scattering rate. electron–electron scattering by exchange processes favors the two holes in the final state being of opposite spin [59]. From consideration of the total density of states in the spin channels, we estimate the scattering rate from this process to be equal for a majority spin hole and a minority spin hole. The electron–phonon and impurity scattering rate are proportional to the density of states at the hole binding energy for the same spin while the electron–magnon rate is proportional to the density of states for the opposite spin. Since the majority-spin density of states is large while the minority-spin part is small, impurity and electron–phonon scattering should be more important in the majority spin channel. The observation that the minority spin channel is broader suggests electron–magnon scattering is the dominant decay mechanism. At $T = 0\,\text{K}$, the minority-spin component of a photohole can scatter to the majority spin component of a hole state higher in the surface band by emitting a spin wave (tilting the spins of the localized f-electrons). The corresponding spin-flip process is not available to the majority-spin component of the photohole at $T = 0$ because the localized f-spins have saturated magnetization and are not able to tilt upwards when the hole tilts down. At higher temperatures, inelastic scattering can occur back and forth between the two spin channels mediated by the emission or absorption of magnons, but the minority-spin component always has the higher density of final states to scatter into. An approximate treatment [60] using the "$s - f$" Hamiltonian [61] found the result

$$\hbar/\tau(\downarrow) = \frac{\sqrt{3}}{4}\frac{P'(\uparrow)m^*}{S}\left(\frac{2JSa}{\hbar}\right)^2 \tag{3.9}$$

for the decay of the minority ($\downarrow$) spin component due to spin flip scattering with magnon emission. Here J is the $s - f$ exchange parameter giving the exchange splitting $2J = 0.65\,\text{eV}$ measured for the surface state [58], $m^* = 1.21 m_e$ is the effective mass measured for the surface band, and $P'(\uparrow) = 0.87$ is the experimentally measured majority component of the band. With S = 7/2 and $a = 3.6\,\text{Å}$, $\hbar/\tau(\downarrow) \approx 0.095\,\text{eV}$. Conversely, replacement of $P'(\uparrow)$ by $P'(\downarrow) = 1 - P'(\uparrow)$ gives $\hbar/\tau(\downarrow) \approx 0.014\,\text{eV}$ for the majority spin component. Thus at low T, the majority spin channel is dominated by electron–phonon scattering whereas the minority spin channel is dominated by electron–magnon scattering. Based on the relative spin-dependent densities of states it is possible to provide estimates of the contribution of phonon scattering in the two spin channels. These would be 46 meV in the majority spin channel and 10 meV in the minority spin channel, leaving approximately 10 meV in each channel due to impurity scattering, probably from hydrogen as in the case of molybdenum discussed earlier.

It is interesting to note that when looking at unoccupied states the converse should be true [62]. At low temperatures, an electron added to an unoccupied minority spin band should decay preferentially via phonon scattering and an additional excited electron in a majority spin band should decay preferentially via magnon scattering.

Although not spin-resolved, another study has examined the possibility of scattering from spin excitations in the ferromagnetic material, iron [50]. In studies of the Fe(001) surface, Schäfer et al. identified a mass renormalization up to a binding energy of 120 meV. The latter energy was too large to be associated with phonons (Debye energy, $\theta_\text{D} \sim 39\,\text{meV}$) and thus the authors identified the self-energy corrections with scattering from spin excitations.

3.6 Studies of the High-T_C Superconductors

In this final section we discuss studies of the high-T_C superconductors and related compounds (see also Part V of this volume). As we have already noted, these materials discovered in 1986 [2] have presented and continue to present some of the biggest challenges in materials science today. ARPES with high energy and momentum resolution has emerged as one of the leading techniques for the study of such materials. Indeed it was the drive to understand the high-T_C superconductors that led to a renaissance in the use of ARPES. The technique has made many important contributions to our understanding of these materials including measurements of the anisotropy of both the superconducting gap [63] [64] and the normal state "pseudogap" [65, 66]. More recently, the discovery of a mass renormalization [9], evident in the dispersion in the vicinity of the Fermi level of the cuprate, $Bi_2Sr_2CaCu_2O_{8+\delta}$, has led to renewed speculation about the origin of high-temperature superconductivity and the possibility that the observed renormalization reflects coupling to some boson involved in the pairing.

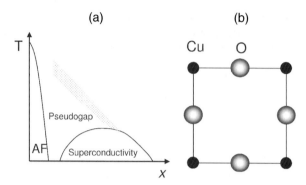

Fig. 3.16. (a) A schematic phase diagram showing the different ground states encountered in the cuprates as a function of temperature and doping; (b) the atomic layout of the copper oxygen planes that are thought to be responsible for the superconductivity in the cuprates

Before discussing the renormalization effects in more detail we first review some aspects of the high-T_C superconductors. It is generally accepted that the superconductivity in the cuprates evolves from a parent insulating state by doping carriers into the two-dimensional CuO_2 planes. With half-filled band, the ground state of the parent compound is an antiferromagnetic Mott insulator. With doping, the systems move from the antiferromagnetic state through to a regime where superconductivity is possible. The commonly accepted phase diagram for the cuprates is shown in Fig. 3.16(a). The materials exhibit superconductivity in the region under the dome. However in the underdoped region a gap or "pseudogap" is observed in the normal state at temperatures well above the superconducting transition temperature, T_c. At optimal doping corresponding to the maximum T_c the materials are considered non-Fermi liquids in the normal state. The structure of the Cu-O plane is shown in Fig. 3.16(b). In the superconducting state the order parameter has d-wave symmetry. In terms of the Cu-O plane, the d-wave symmetry is reflected in the gap being maximum in the copper–oxygen bond direction and non-existent in the direction along the diagonal or copper–copper direction. The latter corresponding to the (π, π) direction of the Brillouin zone is commonly referred to as the nodal direction and the former in the $(\pi, 0)$ direction of the Brillouin zone as the anti-nodal direction.

The first photoemission studies of the high-T_C superconductors [67, 68] identified the copper d-bands and in the case of $YBa_2Cu_3O_{6+x}$, a Fermi level [68]. With improved crystals the superconducting gap was identified [69] followed by measurements of the anisotropy of the gap in the a-b plane associated with the d-wave symmetry [63, 64]. These studies were extended to similar measurements of the anisotropy of a pseudogap observed in the normal state in the underdoped materials [65, 66]. There have also been a number of studies of the spectral function in the vicinity of the $(\pi, 0)$ direction. In the superconducting state this is characterized by a "peak dip hump" structure similar to that in the vicinity of the gap in a BCS like superconductor. As such, the observation has promoted considerable discussion along the lines of the BCS mechanism. Reviews of much of this and previous work have been presented elsewhere [70, 71].

In the present discussion we focus our attention on studies of the nodal region, primarily because a mass renormalization observed in spectra recorded in that direction has all the hallmarks of the mass renormalizations that we have discussed in earlier sections in this chapter. However, while we restrict our discussion to this region, observations in the nodal direction clearly have implications for observations throughout the zone.

The first high-resolution study of the electronic structure in the nodal direction revealed a new feature, a mass enhancement of the low energy excitations immediately below the Fermi level [9]. The relevant spectral intensity plot has been shown earlier in Fig. 3.2. With certain assumptions about the non-interacting dispersion, the authors reported an increased effective mass m^* such that $m^*/m_b \sim 1.6$ where m_b represents the observed mass at higher binding energies. This observation has potentially important implications for the mechanism driving high-T_c superconductivity and an obvious question is whether or not it points to a BCS-like mechanism whereby the electrons or renormalization and associated "kink" have become central issues in subsequent ARPES work with considerable controversy regarding their source [72–76]. Are they related to the presence of spin excitations or do they reflect an interaction with phonons or indeed any other collective mode? In the case of the cuprates, this is not an easy issue to resolve as the various energy scales are nearly identical. However, there is broad agreement on the experimental observations.

All studies agree that the "coupling" is largest in the underdoped regime as is evident in the spectra of Fig. 3.17 [75]. By coupling we mean, as discussed above, that the measured velocity is decreased compared to the bare velocity in the absence of coupling. However it has also been noted that the measured Fermi velocity shows little variation as a function of doping [75, 77]. Thus the biggest change is not in the measured velocity, rather it is in the assumed bare velocity, i.e. the "bare" velocity is largest in the underdoped regime. This is counterintuitive in that the underdoped regime is more insulating-like and one would naively anticipate that the velocity would be less. Experimentally the observation of constant Fermi velocity is evident in spectra obtained from both the $Bi_2Sr_2CaCu_2O_{8+\delta}$ [75] and $La_{2-x}Sr_xCuO_4$ [77] families. It is also reproduced in certain theoretical calculations [78].

In examining the mass enhancement some groups have focused more closely on the associated "kink" in the dispersion and suggested that its presence at a similar energy in studies of all of the different cuprates is an indication of coupling to a phonon mode [74]. Indeed neutron studies do indicate the presence of phonon modes at similar energies [79]. The authors of these studies have also suggested that an "unconventional isotope effect" is an indication of

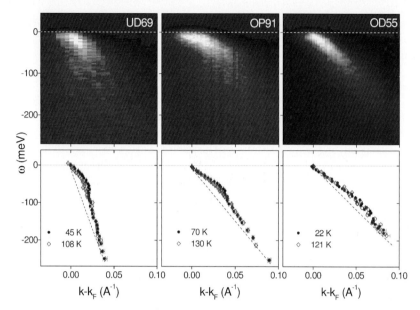

Fig. 3.17. *Upper panels:* two dimensional photoemission intensities for $Bi_2Sr_2CaCu_2O_{8+\delta}$ observed from (**a**) underdoped (UD), (**b**) optimally doped (OP), and (**c**) overdoped (OD) samples. The superconducting transition temperatures are indicated. *Lower panels:* the *dotted lines* indicate the MDC deduced dispersions for both the superconducting (*filled dots*) and normal states (*open diamonds*) corresponding to the different samples in the panels above

the role of phonons in that the substitution of O^{18} for O^{16} results in a change in the velocity of the higher energy electrons as opposed to the lower energy electrons in the vicinity of the Fermi level [80]. If the phonons play any role in the superconductivity, this observation is again counterintuitive. However we note that subsequent attempts to reproduce this effect have failed [81]. More recently, proponents of the phonon scenario have used the maximum entropy method (MEM) to extract an Eliashberg function, $\alpha^2 F$ [82]. The results of this study suggest a multimode structure for the phonon spectrum. While this is a distinct possibility we note that the analysis is also controversial at the present time [83,84].

There are several observations that argue against phonons as the source of kink. Firstly, we note that certainly in the optimally doped materials the resistivity is perfectly linear down to the superconducting transition temperature. A linear resistivity can be associated with phonon scattering. However, as was noted earlier, with respect to Eq. (3.6) this linearity extends down to approximately one third the Debye energy. In the case of $La_{2-x}Sr_xCuO_4$, T_c is approximately 40 K. Multiplying by a factor of three would correspond to a Debye energy of 10 meV, which is certainly too low to give the observed "kink" in the photoemission spectrum at 70 meV. As such, if phonons are involved,

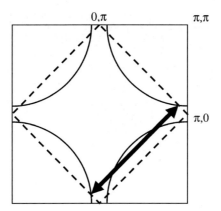

Fig. 3.18. Fermi surface of a doped cuprate system (*solid arcs*) and the antiferromagnetic Brillouin zone of an undoped insulator (*dashed line*). Regions on the Fermi surface ("Hot spots") that can be joined by the double ended arrow representing the antiferromagnetic wave-vector (π,π) can be strongly coupled to antiferromagnetic fluctuations

the mechanism is certainly not describable using the standard Eliashberg approach. Further, the measured scattering rates do not saturate at energies above the kink energy as would be expected on the basis of the Eliashberg equation if the kink reflected the Debye energy. Rather they show a continuous variation to higher binding energies suggesting that some form of electron–electron scattering plays an important role.

An alternative scenario would suggest that the mass enhancement reflects coupling to the spin excitations in the system. Such a coupling is expected to be strongest in the anti-nodal or ($\pi,0$) region reflecting the observation that the spin excitations are described primarily by the scattering vector $Q = (\pi,\pi)$, which couples the antinodal regions as shown in Fig. 3.18. However certain behavior in the nodal region would also appear to carry the hallmarks of such an interaction. Examining Fig. 3.17 it is possible, with certain assumptions about the bare velocities, to extract representative doping-dependent real components of the self-energy, Σ_1. These are reproduced in Fig. 3.19 where for each doping level the Σ_1 corresponding to the superconducting state is compared to the Σ_1 corresponding to the normal state [75]. Certainly in the underdoped and optimally doped regimes there is a marked difference in the spectra on entering the superconducting state. Similar behavior has been observed elsewhere both in ARPES studies [12,85] and in optical conductivity studies [86]. The changes in Σ_1 can be measured as a function of temperature as indicated in Fig. 3.20. From the latter figure we see a reasonably sharp onset around the superconducting transition temperature. Again similar data has been obtained in a more recent study reported by Terashima et al. [87]. Many properties of the high-T_c superconductors show a similar temperature dependence including the development of a sharp coherent peak

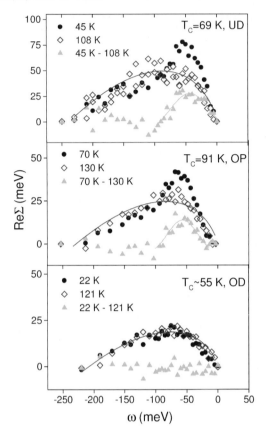

Fig. 3.19. Σ_1 as a function of binding energy in $Bi_2Sr_2CaCu_2O_{8+\delta}$ for the superconducting (*filled dots*) and normal states (*open diamonds*) for the UD69, OP91, and OD55 samples. The solid lines through the normal state data represent MFL fits to the data. The difference between the superconducting and the normal Σ_1 for each level of doping is also plotted (*filled triangles*). The lines through the latter are Gaussian fits to extract the peak energy ω_0^{sc}

in the $(\pi, 0)$ direction [88], the development of the superconducting gap at the Fermi surface and the rearrangement of the spin susceptibility associated with the formation of a magnetic resonance mode in the superconducting state [89]. All of these changes are identified with the electron channel. As we have noted above, it is less clear that such a marked temperature dependence exists in the phonon spectrum. Several authors have therefore associated the changes observed in Σ_1 with changes observed in the spin susceptibility, pointing to scattering from spin excitations as the source of mass renormalization [11, 75, 87, 90]. This is consistent with the observation that the coupling appears much stronger in the underdoped regime, a region where the spin excitations are more pervasive. Further the effects become more pronounced

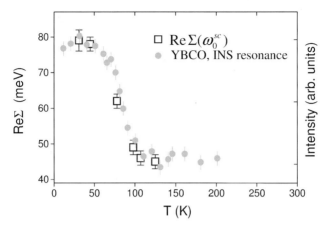

Fig. 3.20. Temperature dependence of $\Sigma_1(\omega_0^{sc})$ from the nodal line for an underdoped sample with $T_c = 69\,\text{K}$ (*open squares*) compared with the temperature dependence of the intensity of the resonance mode observed in INS studies of underdoped $YBa_2Cu_3O_{6+x}$, $T_c = 74\,\text{K}$ (*filled circles*) [89]

on moving away from the nodal direction towards the $(\pi, 0)$ direction. This is evident in the measured momentum dependence of the change in the Fermi velocity on entering the superconducting state [91] and is consistent with the observation that the spin excitations are described primarily by the scattering vector $Q = (\pi, \pi)$ coupling the antinodal regions.

3.7 Summary and Outlook

The new experimental developments combined with new analysis methods have allowed photoemission to become a powerful probe of the collective excitations in condensed matter systems. We can anticipate that such studies will continue and be extended to an ever larger array of new materials. We can also anticipate that the experimental capabilities will be improved. However this will not be easy. The total energy resolution in any experiment is influenced by the energy spread in the incident light beam and the resolving power of the electron spectrometer. These each present a challenge but not an insurmountable challenge. The temperature of the sample and also the quality of the sample surface will also be reflected in the measured peak widths. These contributions are intrinsic and represent more of a challenge. It will be challenge to get the sample much below 1 K but getting to low temperatures is worth the effort. It is a simple matter to show from equation (3.6) above that in the limit of 0 K the Eliashberg function, $\alpha^2 F$, is simply related to $\frac{d\Delta k(\omega)}{d\omega}$ where $\Delta k(\omega)$ is the width of an MDC at binding energy ω. The problems associated with sample surface quality will be somewhat alleviated in experiments that are less surface sensitive such as the new laser based techniques [92, 93].

Acknowledgments

We would like to acknowledge the many contributions of our collaborators on the work described here. These include A.V. Fedorov, S.L. Hulbert, P.-A. Glans, C. McGuinness, K.E. Smith. E.Y. Andrei, H. Berger, Q. Li, G.D. Gu, N. Koshizka, G. Reisfeld, J. Xue, F.J. DiSalvo, Z. Yusof, B.O. Wells, A.R. Moodenbaugh, C. Kendziora, S. Jian, D.G. Hinks, F. Liu, M. Weinert, T.E. Kidd and P.B. Allen. The research work described in this paper was supported by the Department of Energy under Contract No. DE-AC02-98CH10886.

References

1. N. Mårtensson et al: J. Electron Spectr. Relat. Phenom. **70**, 117, (1994)
2. J. G. Bednorz and K. A. Müller: Z. Phys. B: Condens. Matter **64**, 189 (1986)
3. D. Pines and P. Nozieres: *The Theory of Quantum Liquids* (Benjamin, New York, 1969)
4. N. V. Smith et al: Phys. Rev. B **47**, 15476 (1993)
5. G. D. Mahan: *Many Particle Physics* (Plenum Press, New York 1990)
6. N. V. Smith et al: Phys. Rev. B **64**, 155106 (2001)
7. *Angle-Resolved Photoemission*, Ed. S. Kevan (Elsevier, Amsterdam 1992)
8. G. Grimvall: *The electron–phonon Interaction in Metals* (North-Holland, New York, 1981)
9. T. Valla et al: Science **285**, 2110 (1999)
10. S. LaShell et al: Phys. Rev. B **61**, 2371 (2000)
11. A. Kaminski et al: Phys. Rev. Lett. **84**, 1788 (2000)
12. T. Valla et al: Phys. Rev. Lett. **85**, 4759 (2000)
13. A. A. Kordyuk et al: Phys. Rev. B, **71**, 214513 (2005)
14. B. A. McDougall et al: Phys. Rev. B **51**, 13891 (1995)
15. T. Balasubramanian et al: Phys. Rev. B **57**, R6866 (1998)
16. M. Hengsberger et al: Phys. Rev. Lett. **83**, 592 (1999)
17. T. Valla et al: Phys. Rev. Lett. **83**, 2085 (1999)
18. K. Jeong et al: Phys. Rev. B **38**, 10302 (1988); K. Jeong, R. H. Gaylord and S. D. Kevan, Phys. Rev. B 39, 2973 (1989)
19. S. Y. Savrasov and D. Y. Savrasov: Phys. Rev. B **54**, 16487 (1996)
20. C. Hodges et al: Phys. Rev. B **4**, 302 (1971)
21. W. A. Harrison: *Electronic Structure and the Properties of Solids* (W. H. Freeman & Co, San Francisco, 1980)
22. J. A. Wilson et al: Phys. Rev. Lett. **32**, 882 (1974)
23. A. H. Castro Neto: Phys. Rev. Lett. **86**, 4382 (2001)
24. F. Smith et al: J. Phys. C: Solid State Phys. **5**, L230 (1972); C. Berthier et al: Solid State Commun. **18**, 1393 (1976); D. W. Murphy et al: J. Chem. Phys. **62**, 967 (1975)
25. P. Moline et al: Phil. Mag. **30**, 1091 (1974)
26. H. Suderow et al: Phys. Rev. Lett. **95**, 117006 (2005)
27. P. Garoche et al: Solid State Commun. 19, 455 (1976); D. Sanchez et al: Physica B **204**, 167 (1995)
28. J. E. Graebner and M. Robbins: Phys. Rev. Lett. **36**, 422 (1976)

29. R. Corcoran et al: J. Phys. Condens. Matter **6**, 4479 (1994)
30. T. Yokoya et al: Science **294**, 2518 (2001); T. Kiss et al: Physica B **312-313**, 666 (2002)
31. J. A. Wilson: Phys. Rev. B **15**, 5748 (1977)
32. N. J. Doran et al: J. Phys. C **11**, 699 (1978)
33. T. M. Rice and G. K. Scott: Phys. Rev. Lett. **35**, 120 (1975)
34. B. Ruzicka et al: Phys. Rev. Lett. **86**, 4136 (2001)
35. K. Rossnagel et al: Phys. Rev. B **72**, 121103 (2005)
36. R. Liu et al: Phys. Rev. Lett. **80**, 5762 (1998)
37. R. Liu et al: Phys. Rev. B **61**, 5212 (2000)
38. Th. Straub et al: Phys. Rev. Lett. **82**, 4504 (1999)
39. G. Benedek et al: Europhys. Lett. **5**, 253 (1988); G. Brusdeylins et al: Phys. Rev. B **41**, 5707 (1990)
40. T. Valla et al: Phys. Rev. Lett. **92**, 086401 (2004)
41. J. L. Feldman: Phys. Rev. B **25**, 7132 (1982); G. Brusdeylins et al: Phys. Rev. B **41**, 5707 (1990); Y. Nishio: J. Phys. Soc. Jpn. **63**, 223 (1994)
42. J. M. E. Harper et al: Phys. Rev. B **15**, 2943 (1977); K. Noto et al: Nuovo Cimento **38**, 511 (1977)
43. S. V. Dordevic et al: Phys. Rev. B **64**, 161103 (2001)
44. R. Liu et al: Phys. Rev. B **61**, 5212 (2000); A. V. Fedorov et al: unpublished.
45. Th. Straub et al: Phys. Rev. Lett. **82**, 4504 (1999)
46. W. C. Tonjes et al: Phys. Rev. B **63**, 235101 (2001)
47. H. F. Hess et al: J. Vac. Sci. Technol. A **8**, 450 (1990)
48. A. V. Fedorov et al: J. Elect. Spectr. And Relat. Phenom. **92**, 19 (1998)
49. A. V. Fedorov et al: Phys. Rev. B 65, 212409 (2002)
50. J. Schäfer et al: Phys. Rev. Lett. **92**, 97205 (2004)
51. P. D. Johnson: Rep. Prog. Phys. **60**, 1217-1304 (1997)
52. R. Wu et al: Phys. Rev. B **44**, 9400 (1991)
53. D. Li et al: J. Magn. Magn. Mater. **99**, 85 (1991)
54. G. A. Mulhollan et al: Phys. Rev. Lett. **69**, 3240 (1992)
55. P. Wells et al: J. Phys. F **4**, 1729 (1974)
56. H. L. Skriver and I. Mertig: Phys. Rev. B **41**, 6553 (1990)
57. D. Li et al: Mat. Res. Socs Proc. **313**, 451 (1993)
58. A. V. Fedorov et al: Phys. Rev. B **50**, 2739 (1994); E. Weschke et al: Phys. Rev. Lett. **77**, 3415 (1996)
59. B. Sinkovic et al: Phys. Rev. B **52**, R15703 (1995)
60. P. B. Allen: Phys. Rev. B **63**, 214410 (2001)
61. C. Zener: Phys. Rev. **81**, 440 (1951); C. Zener: Phys. Rev. **82**, 403 (1951); C. Zener: Phys. Rev. **83**, 299 (1951)
62. A. Rehbein et al: Phys. Rev. B **67**, 033403 (2003)
63. Z.-X. Shen et al: Phys. Rev. Lett. **70**, 1553 (1993)
64. H. Ding et al: Phys Rev Lett. **74**, 2784 (1995)
65. A. G. Loeser et al: Science **273**, 325 (1996)
66. H. Ding et al: Nature **382**, 51 (1996)
67. B. Reihl et al: Phys. Rev. B **35**, 8804 (1987)
68. P. D. Johnson et al: Phys. Rev. B **35**, 8811 (1987)
69. C. G. Olsen et al: Science **245**, 731 (1989)
70. A. Damascelli et al: Rev. Mod. Phys. **75**, 473 (2003)
71. J. C. Campuzano et al in *Physics of Superconductors*, Vol.II, ed K. H. Bennemann and J.B. Ketterson, Springer Berlin, 2004 p. 167–272

72. P. V. Bogdanov et al: Phys. Rev. Lett. **85**, 2581 (2000)
73. A. Kaminski et al: Phys. Rev. Lett. **86**, 1070 (2001)
74. A. Lanzara et al: Nature **412**, 510 (2001)
75. P. D. Johnson et al: Phys. Rev. Lett. **87**, 177007 (2001)
76. T. K. Kim et al: Phys. Rev. Lett. **91**, 167002 (2003)
77. X.J. Zhou et al: Nature **423**, 398 (2003)
78. M. Randeria et al: Phys. Rev. B **69**, 144509 (2004)
79. R. J. McQueeney et al: Phys. Rev. lett. **82**, 628 (1999)
80. G.-H. Gweon et al: Nature **430**, 187 (2004)
81. F. Douglas et al: to be published
82. X. J. Zhou: Phys. Rev. Lett. **95**, 117001 (2005)
83. T. Valla: cond-mat/0501138 (2005)
84. X. J. Zhou et al: cond-mat/0502040 (2005)
85. T. Yamasaki et al: cond-mat/0603006 (2006)
86. J. Hwang et al: Nature **427**, 714, (2004)
87. K. Terashima et al: Nature Physics **2**, 27 (2006)
88. A. V. Fedorov et al: Phys. Rev. Lett. **82**, 2179 (1999)
89. P. Dai et al: Science **284**, 1346 (1999)
90. T. Valla: Proceedings SPIE – Volume 5932, *Strongly Correlated Electron Materials: Physics and Nanoengineering*, Ivan Bozovic, Davor Pavuna Editors, 593203 (2005)
91. T. Valla et al: Phys. Rev. Lett. **85**, 828 (2000)
92. T. Kiss et al: Phys Rev. Lett. **94**, 057001(2005)
93. J. D. Koralek et al: Phys. Rev. Lett. **96**, 017005 (2006)

High-resolution Photoemission Spectroscopy of Solids Using Synchrotron Radiation

K. Shimada

Hiroshima Synchrotron Radiation Center, Hiroshima University, Kagamiyama 2-313, Higashi-Hiroshima, Hiroshima 739-0046, Japan
kshimada@hiroshima-u.ac.jp

Abstract. We present high-resolution photoemission spectroscopy of Ni, Ce, and rare-earth compounds using tunable synchrotron radiation in a wide photon energy range from $h\nu = 6$ eV up to 6000 eV. In the ultraviolet and soft x-ray regime, such tunability enables us to clarify details of the electronic structures near the Fermi level at a specific point of the Brillouin zone. By means of a quantitative analysis of the spectral function, it is possible to clarify the many-body nature of the quasi-particles. The energy dependence of the inelastic mean free path can be utilized to decompose the spectra into the contributions originating from the surface and bulk electronic structure.

4.1 Introduction

High-energy-resolution photoemission spectroscopy using tunable synchrotron radiation (SR) has recently undergone rapid development [1–9]. Synchrotron radiation offers the advantages of tunability, polarization and cleanliness [10]. The wide energy range of its photons enables us to clarify valence-band and core-level photoemission spectra. With angle-resolved photoemission spectroscopy (ARPES), the electron-energy-band structure in the initial states, and the Fermi surface can be elucidated. New advances in high-energy and angular-resolution measurements enabled us to examine quantitatively fine spectral features near the Fermi level (E_F), which are directly related to the low-energy excitations of the materials [1–3, 8, 9]. For three-dimensional electron systems in particular, the final-state effect is significant, and for lineshape analyses, the tunability of the incident photon energy is critical [11, 12]. On the basis of the electron kinetic-energy dependence of the mean free path, one can study the surface, interface, and bulk electronic properties [7]. In this chapter, we will describe the high-resolution photoemission spectroscopy of Ni and Ce metals, and rare-earth compounds that employs SR over a wide photon-energy range from vacuum ultraviolet (VUV) and soft x-ray (SX) up to hard x-ray (HX), namely from $h\nu = 6$ eV up to $h\nu = 6000$ eV.

4.2 Inelastic Mean Free Path, Energy and Angular Resolution

Figure 4.1(a) shows the inelastic mean free path (IMFP) as a logarithmic function of electron kinetic energy in solids [13, 14]. The IMFP for the kinetic energies from 20 eV up to 1000 eV is about 5 to 20 Å; below 20 eV and above 200 eV, that value increases [13, 14].

Figure 4.1(b) schematically illustrates typical high-energy resolution (ΔE_{tot}) of photoemission spectroscopy of solids as a function of the electron kinetic energy. Here ΔE_{tot} includes contributions from an electron energy analyzer (ΔE_a) and a SR monochromator ($\Delta h\nu$): $\Delta E_{tot} = \sqrt{(\Delta E_a)^2 + (\Delta h\nu)^2}$. Now hemispherical electron-energy analyzers are commercially available with an energy resolving power $E_p/\Delta E_a \sim 4000$, where E_p is the pass energy of electron energy analyzer, and an angular resolution of $\Delta\theta \leq 0.3°$. In addition to the improved electron energy analyzers, a low-emittance electron storage ring, undulators, and a high-resolution monochromator have been rapidly developed. To date, high-resolution photoemission measurements using SR have been widely conducted, with a total energy resolving power of $E_K/\Delta E_{tot} \sim 10^3$–$10^4$, a level which becomes larger as the kinetic energy of the photoelectron E_K increases.

The examination of interesting physical properties of solids, such as superconductivity and magnetism, benefits greatly from a high absolute energy resolution depending on the energy scale. For example, if one wants to investigate phenomena of high-T_C superconductors with a superconducting gap of $\sim$20 meV, an energy resolution better than 20 meV is essentially needed. A

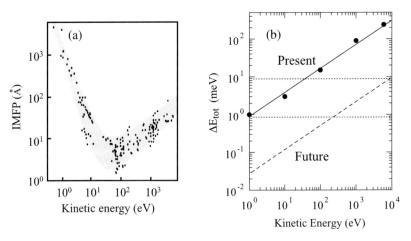

Fig. 4.1. (a) Inelastic mean free path (IMFP) as a function of electron kinetic energy in solids; (b) high energy resolution (ΔE_{tot}) of photoemission spectroscopy at present (*circles*) and anticipated future improvement (*dashed line*) as a function of electron kinetic energy. The *dotted lines* indicate the thermal energy corresponding to the temperatures, $T = 100$ K and 10 K [6, 14]

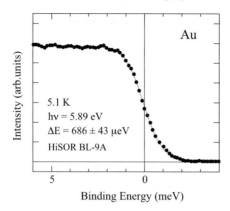

Fig. 4.2. High-resolution photoemission spectrum with SR of Au Fermi edge. Energy resolution $\Delta E \sim 0.7$ meV has been obtained at $h\nu = 5.89$ eV at a temperature of 5.1 K [17]

conventional superconductor such as Pb with a superconducting gap ~ 3 meV, requires an absolute energy resolution better than 3 meV [15, 16]. Currently, as Fig. 4.2 shows, a high energy resolution of less than 1 meV has been obtained using SR [17]. With this energy resolution, a superconducting gap in Pb is clearly visible (Fig. 4.3) [18].

If the angular resolution is held constant at $\Delta\theta = 0.3°$, the resolution of the wave vector (k) of the photoelectron is dependent on E_K as $\Delta k \sim 0.0027\sqrt{E_K}$ Å^{-1}. As the excitation photon energy decreases, one can obtain

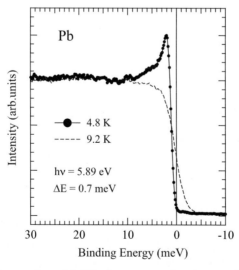

Fig. 4.3. Direct observation with SR of a superconducting gap formation of Pb [18]

a smaller Δk: at $E_K = 5\,\text{eV}$, Δk becomes $6 \times 10^{-3}\,\text{Å}^{-1}$ [19]. Note that Δk is directly related to the resolution of crystal momentum, $\Delta p = \hbar \Delta k$.

4.3 High-Resolution Photoemission Spectroscopy in the VUV and SX Regions

Rapid developments in energy and angular resolutions enable us to examine the fine electronic structures near E_F. In this section, we shall examine some examples of SR high-resolution photoemission spectra of Ni and Ce metals, and rare-earth compounds in the VUV and SX regions.

4.3.1 Quantitative Lineshape Analyses of Ni Photoemission Spectra

So far, there have been many ARPES studies of Ni, and the importance of electron correlation in the spin-polarized Ni $3d$ bands has been clearly recognized from several unusual spectral features, such as the narrowing of the $3d$ bands by $\sim 25\%$ [20–24], the reduction of the exchange splittings by $\sim 50\%$ [20–24] compared with band-structure calculations carried out with the local spin-density approximation (LSDA) [25, 26], and the existence of a spin-polarized 6 eV satellite [27–31].

Since most of the majority-spin $3d$ states of Ni are occupied and less than one hole exists in the minority-spin $3d$ states, Kanamori considered the effects of electron correlation on the ferromagnetism of Ni, taking into account the multiple scattering of two particles [32]. Penn calculated a single-particle spectral function $A^\sigma(k,\omega)$ based on this picture and interpreted the 6-eV satellite as a "two-hole-bound state" [33]. The narrowing and reduced exchange splittings of the Ni $3d$ bands are explained in terms of the electron correlation effect [34–37]. However, the energy-band and the spin-dependent lifetime of the quasi-particles near E_F, in particular at low temperatures, have not been clarified yet.

In this section, on the basis of a high-resolution low-temperature ARPES study of Ni(110), we discuss the self-energy of the valence bands ($\Sigma_{2\downarrow}$, $\Sigma_{1\uparrow}$ and $\Sigma_{1\downarrow}$) forming the Fermi surface along the high-symmetry line ΓKX (Fig. 4.4) [38–40]. In order to detect these bands, we carried out ARPES measurements by tuning the incident photon energy and by rotating the polar axis of the sample, parallel to the [001] direction for the (110) surface, as illustrated in Fig. 4.4 [40]. The inner potential was assumed to be 10.7 eV [23].

Figures 4.5(a) and 4.5(c) show energy distribution curves (EDCs) and intensity plot for the $\Sigma_{2\downarrow}$ bands. In order to determine energy-band points, we used a Lorentzian on a linear background to fit the momentum distribution curves (MDCs) which are intensity distribution curves as a function of momentum for a given binding energy, as shown in Fig. 4.5(b). The evaluated peak positions are indicated by open circles in the intensity plots (Fig. 4.5(c)).

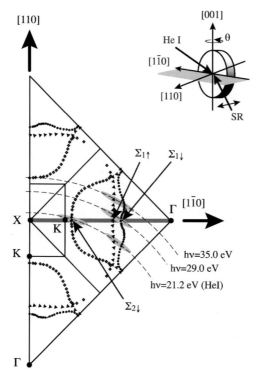

Fig. 4.4. Schematic view of the geometry of the present experiment, and the Brillouin zone of Ni in the extended zone scheme. The $\Sigma_{2\downarrow}$, $\Sigma_{1\uparrow}$, and $\Sigma_{1\downarrow}$ bands crossing the ΓKX line were examined. Small symbols (*filled triangles* and *circles*) indicate the Fermi surface obtained from a de Haas–van Alphen measurement [25, 39]

Figures 4.6(a)–4.6(c) exhibit the intensity plot for the $\Sigma_{1\sigma}(\sigma=\uparrow,\downarrow)$ band and the bands on the different points of the Fermi surface (Fig. 4.4).

A kink structure is apparent in the $\Sigma_{1\downarrow}$ and $\Sigma_{2\downarrow}$ bands, but is much less clear in the $\Sigma_{1\uparrow}$ band. In order to elucidate the origin of the kink structure, we will evaluate the self-energy (Σ), based on the ARPES results. The major spectral features are given by $A^{\sigma}(\boldsymbol{k},\omega)$, which is related to the imaginary part of the single-particle Green's function:

$$A^{\sigma}(\boldsymbol{k},\omega) = -\frac{1}{\pi}\mathrm{Im}G^{\sigma}(\boldsymbol{k},\omega) = -\frac{1}{\pi}\mathrm{Im}\frac{1}{\omega - \epsilon_{\boldsymbol{k}}^{0} - \Sigma^{\sigma}(\boldsymbol{k},\omega)}, \quad (4.1)$$

where $\epsilon_{\boldsymbol{k}}^{0}$ represents the energy of the non-interacting band [3, 9]. The imaginary part ($\mathrm{Im}\Sigma^{\sigma}$) and the real part ($\mathrm{Re}\Sigma^{\sigma}$) of the self-energy can be evaluated from the spectral width (δE) and the energy shift from the non-interacting band, respectively [3, 9]. In the present analyses, in stead of using the EDC widths, we used the MDC widths, δk's, to estimate the imaginary part, on the basis of the relation: $|2\mathrm{Im}\Sigma| = \delta E = (dE/dk)\delta k$ [12].

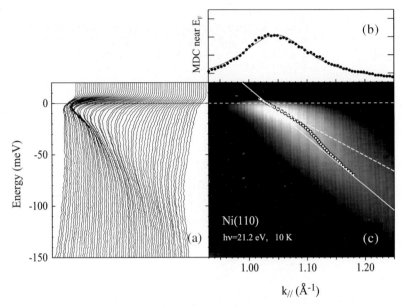

Fig. 4.5. The ARPES results for Ni(110) at 10 K at $h\nu = 21.2$ eV. (**a**) shows EDCs; (**b**) MDC at E_F, and (**c**) intensity plot for the $\Sigma_{2\downarrow}$ band. The *circles* and *solid line* in (c) respectively indicate evaluated peak positions, and linear dispersion (ϵ_k^0) without the kink structure. The *dashed line* shows the gradient at E_F [39]

In order to evaluate ReΣ, we have assumed that the band dispersion is linear when the kink structure is absent, as in the $\Sigma_{1\uparrow}$ band, and the linear band dispersion represents ϵ_k^0 (solid lines in Figs. 4.5(c) and 4.6(a)–4.6(c)). The real part of the self-energy is deduced from the energy shifts from ϵ_k^0.

Figure 4.7(a) provides the evaluated $|2\text{Im}\Sigma^\sigma|$ of the $\Sigma_{2\downarrow}$ and $\Sigma_{1\sigma}$ bands. We should note that $|2\text{Im}\Sigma^\downarrow|$ decreases for the energy of $\omega > -40$ meV, a result which implies that the kink structures originate from the many-body interaction and not from the energy dispersion. Since the energy scale of the kink structures coincides well with the Debye temperature, $\Theta_D = 450$ K ($k_B \Theta_D = 39$ meV) [41], it is reasonable to assume that the structure is derived from the electron–phonon interaction.

When the electron scattering processes due to the electron–phonon, electron–electron, and electron–impurity interactions are independent, the lifetime broadening ($\Gamma^\sigma = |2\text{Im}\Sigma^\sigma|$) of the quasi-particle can be expressed by the sum of each contribution: $\Gamma^\sigma = \Gamma^\sigma_{el-ph} + \Gamma^\sigma_{el-el} + \Gamma^\sigma_0$, where Γ^σ_{el-ph} and Γ^σ_{el-el} are the lifetime broadenings caused by electron–phonon and electron–electron interactions, respectively [9, 42]. The variable Γ^σ_0 represents an energy-independent term which should include the lifetime broadening derived from the electron–impurity scattering (Γ^σ_{el-imp}). The broadening due to the electron–impurity scattering is much smaller for high purity single crystal sample than the other terms. In the case of photoemission from a three

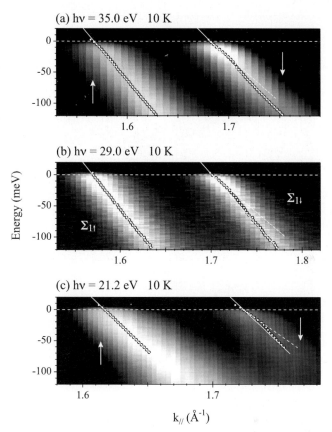

Fig. 4.6. The ARPES intensity of Ni(110) at 10 K taken at $h\nu =$ (a) 35.0 eV, (b) 29.0 eV and (c) 21.2 eV. (b) shows the $\Sigma_{1\uparrow}$, and $\Sigma_{1\downarrow}$ bands. The *circles* and *solid lines* indicate the evaluated peak positions, and linear dispersions (ϵ_k^0) without the kink structure, respectively. The *dashed lines* show the gradient of the down-spin bands at E_F [39]

dimensional system such as Ni, a substantial final-state broadening (Γ^σ_{final}) exists [11, 43, 44]. If we restrict our attention to a narrow energy region, we can neglect the energy dependence of Γ^σ_{final}. Therefore Γ^σ_{final} can be regarded as the dominant term in Γ^σ_0, namely, $\Gamma^\sigma_0 \sim \Gamma^\sigma_{final}$ [39].

The energy dependence of $|2\text{Im}\Sigma^\sigma|$ should derive, therefore, from Γ^σ_{el-ph} and Γ^σ_{el-el} terms. The lifetime broadening due to the electron–phonon interaction is given by,

$$\Gamma^\sigma_{el-ph} \cong 2\pi \int_0^\infty \alpha^2_{k\sigma} F(\nu)[2n(\nu, T) + f(\nu + \omega, T) + f(\nu - \omega, T)]d\nu , \quad (4.2)$$

where $n(\nu, T)$ and $f(\nu, T)$ represent the Bose–Einstein and Fermi–Dirac distribution functions, respectively, and $\alpha^2_{k\sigma} F(\omega)$ is the Eliashberg function [42].

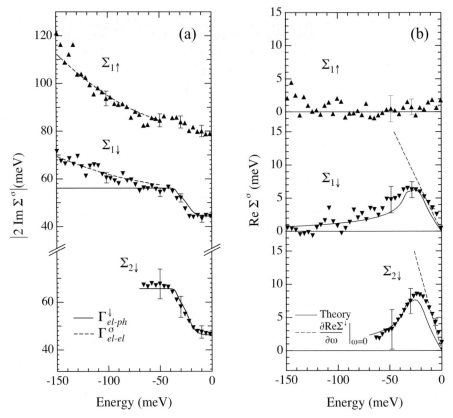

Fig. 4.7. Experimentally obtained imaginary (**a**) and real (**b**) parts of the self-energy of the $\Sigma_{2\downarrow}$, $\Sigma_{1\uparrow}$, and $\Sigma_{1\downarrow}$ bands of Ni. The symbols represent the observed imaginary/real parts of the self-energy. The *solid* and *dashed* lines in (**a**) exhibit the theoretical $\Gamma^{\downarrow}_{el-ph}$ and Γ^{σ}_{el-el}, respectively. The theoretical $\Gamma^{\downarrow}_{el-ph}$ and $\Gamma^{\uparrow}_{el-el}$ in (**a**) are shifted by the offset $\Gamma^{\downarrow}_{0}$ and $\Gamma^{\uparrow}_{0}$. The $\Gamma^{\downarrow}_{el-el}$ is shifted by the maximum value of the $\Gamma^{\downarrow}_{0} + \Gamma^{\downarrow}_{el-ph}$. The *solid* and *dashed* lines in (**b**) indicate the theoretical $\mathrm{Re}\Sigma^{\downarrow}$, and the gradient of experimental $\mathrm{Re}\Sigma^{\downarrow}$ near E_{F}, respectively [39]

For the simplicity, $\alpha^2_{k\sigma}$ is treated as an energy-independent constant and optimized to reproduce the observed self-energy. A calculated phonon density-of-states (DOS) of Ni [41] is used for $F(\omega)$.

The solid lines in Fig. 4.7(a) represent the calculated $\Gamma^{\downarrow}_{el-ph}$ for the $\Sigma_{2\downarrow}$ and $\Sigma_{1\downarrow}$ bands, which explain well the observed $|2\mathrm{Im}\Sigma^{\downarrow}|$ especially the drop for $\omega > -40\,\mathrm{meV}$. The $\mathrm{Re}\Sigma^{\downarrow}$'s are calculated to satisfy the Kramers–Kronig relation with $\mathrm{Im}\Sigma^{\downarrow}$, and account well for the observed $\mathrm{Re}\Sigma^{\downarrow}$'s (Fig. 4.7(b)). These results confirm that the kink structure is produced by the electron–phonon interactions.

The electron–phonon coupling constant (λ) can be evaluated from the expression $\lambda = |\partial \text{Re}\Sigma^\sigma/\partial\omega|_{\omega=0}$ using experimental $\text{Re}\Sigma^\downarrow$ values (the dashed lines in Fig. 4.7(b)). The coupling constants $\lambda = 0.57 \pm 0.06$ and 0.33 ± 0.05 are obtained for the $\Sigma_{2\downarrow}$ and $\Sigma_{1\downarrow}$ bands, respectively (Table 4.1). The electron mass (m) is enhanced due to the electron–phonon interaction [$m^* = (1+\lambda)m$]. On the other hand, the electron–phonon interaction is much weaker in the $\Sigma_{1\uparrow}$ band ($\lambda \sim 0$).

The electron–phonon coupling constant has been evaluated theoretically using the phase shifts of the wave function scattered by the core potential [45, 46]. In the case of Ni, the $d-f$ scattering contributes most strongly to the coupling constant [46]. Since the d electrons are located near the Ni ions, they are more easily influenced by the motion of the ions than are the sp electrons. Assuming that the d weight given by the LSDA calculation is a measure of the coupling constant, the $\Sigma_{2\downarrow}$ band is purely d-like, while the d weight at E_F for the $\Sigma_{1\downarrow}$ band, ~90%, is smaller due to $s-d$ hybridization. Although the $\Sigma_{1\uparrow}$ and $\Sigma_{1\downarrow}$ bands have the same symmetry, the d component in the $\Sigma_{1\uparrow}$ band at E_F, ~80%, is slightly smaller than that in the $\Sigma_{1\downarrow}$ band, due to exchange splitting. It seems that this explanation works well in a qualitative way. The difference of the d weight evaluated by the LSDA calculation, ~10%, however, is rather small to account for the large spin-dependence of the λ in the $\Sigma_{1\sigma}$ bands. One should take electron correlation into account when considering the spin-dependent spectral-weight distribution [34, 36, 37].

In order to see the electron correlation effects, we have evaluated the group velocity, v_F^{ARPES}, using non-interacting band $\epsilon_{\mathbf{k}}^0$, namely, $v_F^{\text{ARPES}} = 1/\hbar (d\epsilon_{\mathbf{k}}^0/dk)_{k=k_F}$. Obtained v_F^{ARPES} values are smaller than those given by the LSDA calculation, v_F^{LSDA}, by a factor of $v_F^{\text{ARPES}}/v_F^{\text{LSDA}} \sim 47\text{--}69\%$, an amount which is consistent with the narrowing of the experimental Ni $3d$ band width [20–24]. Note that the Fermi wave numbers given by the LSDA calculation (k_F^{LSDA}) [25], de Haas–van Alphen measurements (k_F^{dHvA}) [48], and ARPES measurements (k_F^{ARPES}) [12, 40] do not differ so much along ΓKX line; $k_F^{\text{dHvA}}/k_F^{\text{LSDA}} = 0.9-1.1$ [25, 48], and $k_F^{\text{ARPES}}/k_F^{\text{LSDA}} = 0.8-1.0$ [12, 25, 40]. In this experiment, the $k_F^{\text{ARPES}}/k_F^{\text{LSDA}}$ ratios are estimated to be 1.0 ± 0.1 and 0.9 ± 0.1 for the Σ_1 and Σ_2 bands, respectively. The significant deviation from the LSDA calculation is, therefore, in the Fermi velocity. The mass-enhancement factor due to electron correlation (η) can be evaluated by

Table 4.1. The electron–phonon coupling constant (λ), mass-enhancement factor due to electron correlation (η), and effective mass (m^*) of Ni compared with the mass given by the band-structure calculation (m_b) [39]

Bands	λ	η	$m^*/m_b = (1+\lambda)\eta$
$\Sigma_{2\downarrow}$	0.57±0.06	1.8±0.2	2.8±0.2
$\Sigma_{1\uparrow}$	~0	2.2±0.1	2.2±0.1
$\Sigma_{1\downarrow}$	0.33±0.05	1.4±0.1	1.9±0.1

$\eta = m/m_b = (k_F^{\rm ARPES}/k_F^{\rm LSDA})(v_F^{\rm LSDA}/v_F^{\rm ARPS})$, where m_b represents an electron mass given by the LSDA calculation.

Table 4.1 exhibits the effective mass enhancement $m^*/m_b = (1+\lambda)\eta$ due to both the electron–phonon interaction and electron correlation. The m^*/m_b ratio for the d-like $\Sigma_{2\downarrow}$ band is larger than those for the Σ_1 bands. The present results for the m^*/m_b ratio (1.9–2.8) agree well with those from the de Haas–van Alphen measurement (1.8–2.3) [48]. Note that the m^* enhancement is mainly derived from electron correlation η in the $\Sigma_{1\uparrow}$ band ($\eta \sim 2.2$, $1+\lambda \sim 1$), while both the electron–phonon interaction ($1+\lambda \sim 1.3$) and electron correlation ($\eta \sim 1.4$) contribute to the m^* enhancement in the $\Sigma_{1\downarrow}$ band.

The lifetime broadening due to the electron–electron interaction can be expressed as follows: $\Gamma_{el-el}^{\sigma} \sim 2\beta^{\sigma}[(\pi k_{\rm B}T)^2 + \omega^2]$ [42, 49]. The $2\beta^{\sigma}$ value gives a measure of the electron–electron interaction. By the fit of the lower energy side of $|2{\rm Im}\Sigma^{\sigma}|$, one can evaluate $2\beta^{\uparrow} \sim (1.4 \pm 0.3)\,{\rm eV}^{-1}$ and $2\beta^{\downarrow} \sim (0.6 \pm 0.2)\,{\rm eV}^{-1}$ for the $\Sigma_{1\uparrow}$ and $\Sigma_{1\downarrow}$ bands, respectively. These results indicate that the quasi-particles with an up-spin are strongly scattered compared with those with a down-spin. The relation $\beta^{\downarrow} < \beta^{\uparrow}$ is also consistent with the theoretical ${\rm Im}\Sigma^{\sigma}$ [36, 37].

By means of high-resolution ARPES, we have clarified that many-body interactions such as the electron–phonon and electron–electron interactions act on the quasi-particles in different ways depending on the identity of the energy band and the spin direction.

4.3.2 Ce Metal: the Ce $4f$ Spectrum

The fcc metal cerium is one of the simplest $4f$ electron systems. Its isostructural $\alpha - \gamma$ phase transition, where the unit cell volume decreases by $\sim 15\%$ on cooling, has attracted much interest [8, 50–52]. It has been claimed that this transition is closely related to the temperature dependence of the hybridization between the Ce $4f$ and the conduction electrons ($c-f$ hybridization) [8, 50–52]. To date, many high-resolution photoemission studies have been reported for Ce metal in the α- and γ-phases [8, 53–59]. Since the photoemission spectral features of Ce form the basics for understanding those of the rare-earth compounds in general, we shall describe the Ce $4f$ spectral feature in this section [60, 61].

Figure 4.8 shows the photoemission spectra of α-Ce taken with $h\nu = 21.218\,{\rm eV}$ (He I$_\alpha$, $\Delta E = 7\,{\rm meV}$), $40.814\,{\rm eV}$ (He II$_\alpha$, $\Delta E = 7\,{\rm meV}$), and $h\nu = 123\,{\rm eV}$ ($\Delta E = 28\,{\rm meV}$) photons. Based on the photoionization cross-section dependence [62], as the excitation energy decreases, the spectral intensity of the Ce $5d$ state is increased with respect to the Ce $4f$ state. The He I$_\alpha$ spectrum can be regarded as a Ce $5d$-derived spectrum since the photoionization cross section of Ce $5d$ is expected to be two order of magnitude larger than that of Ce $4f$ [62]. The $h\nu = 123\,{\rm eV}$ spectrum is on-resonance in the Ce $4d$–$4f$ resonance regime, and therefore, the Ce $4f$ state dominates the spectral intensity. One can clearly see the difference between the spectral

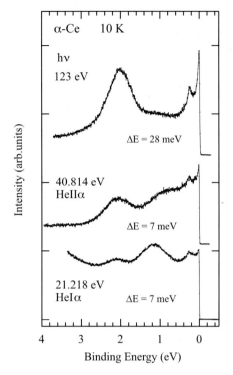

Fig. 4.8. Photoemission spectra of α-Ce taken at $h\nu = 21.218\,\text{eV}$ (He I$_\alpha$), $40.814\,\text{eV}$ (He II$_\alpha$) and $123\,\text{eV}$ at a temperature of $10\,\text{K}$ [60]

features of these states in Fig. 4.8. The spectral feature observed at a binding energy of ∼1.2 eV in the He I$_\alpha$ spectrum is mostly derived from the Ce $5d$ states.

Figures 4.9(a) and 4.9(b) display the $h\nu = 123\,\text{eV}$ spectra in the α- and γ-phases [60]. The spectral features at ∼2 eV and E_F are interpreted as the Ce $4f^0$ (Ce $4f^1 \rightarrow$ Ce $4f^0$) and $4f^1$ (Ce $4f^1 \rightarrow$ Ce $4f^1\underline{c}$) final states, respectively. Here $\underline{c}$ denotes holes in the wide conduction bands. The $4f^1$ final state is split into two peaks separated by Δ_{SO} due to the spin-orbit interaction (Fig. 4.9(b)). If one assumes a localized $4f$ state, the peak at E_F (ground state) corresponds to the Ce $4f^1_{5/2}$ state, and the peak at ∼300 meV to the Ce $4f^1_{7/2}$ state. The Ce $4f^1_{5/2}$ state can be further split due to crystal field, which is not clearly observed in the Ce metal case.

Although Ce metal is a simple system, its spectral feature deviates significantly from the DOS given by the band-structure calculation [57, 63]. In particular, the Ce $4f^0$ spectral feature cannot be explained, which is attributed to strong electron correlation in narrow $4f$ bands. Assuming that the dispersion of the Ce $4f$-derived band is small, and regarding the $4f$ state as an impurity level, the single-impurity Anderson model (SIAM) [64–67] has been

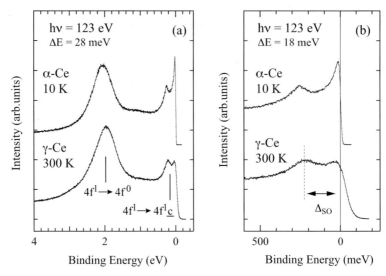

Fig. 4.9. (a) $4f$ spectra in the α- and γ-phases of Ce; (b) Ce $4f^1$ spectra near E_F [60]

applied to explain the observed spectral features. The SIAM is believed to give a reasonable spectral function to account for the unusual spectral features including that from the Ce $4f^0$. The SIAM predicts a sharp peak just above E_F, which is called a Kondo resonance (KR). The energy of this peak, measured from E_F, should scale to the Kondo temperature (T_K), which is an important parameter for the description of the physics of Kondo systems [64, 68]. The observed temperature dependence of the KR of the metallic Ce compounds, which have $T_K < \sim 100$ K, seem to be explained well by the SIAM [69].

If one compares the Ce $4f$ spectra in the α- and γ-phases in Figs. 4.9(a) and 4.9(b), the spectral intensity for the $4f^1$ with respect to that for the $4f^0$ decreases from the α-phase to the γ-phase, which seems to indicate that T_K is lowered in the γ-phase. However, in order to relate the Ce $4f$ spectrum with the bulk physical properties, it is necessary to separate the bulk and surface components [55, 57, 59].

The $4d$–$4f$ and $3d$–$4f$ resonance photoemission spectroscopies have been widely used to evaluate the "surface" and "bulk" components of the Ce $4f$ spectra [70]. On the basis of the on-resonance spectra, we will try to separate surface and bulk components of the Ce metal spectra [60, 61]. Here, we assume that variables d and l are the distance from the surface, and the length of the IMFP, respectively. The integrated spectral weight for $0 < d < l$ is given by $I(d/l) = 1 - e^{-d/l}$. We further assume a characteristic length d_0 which effectively separates the electronic states into two regions: specifically, the electronic states in the region of $0 < d < d_0$ and $d_0 < d$ reflect the surface and bulk electronic states, respectively. Introducing a parameter $\mu = d_0/l$, the integrated spectral weight of the surface component is given by $I_s(\mu) = 1 - e^{-\mu}$ and that of the bulk component by $I_b(\mu) = 1 - I_s(\mu) = e^{-\mu}$. In the case of Ce metal and Ce compounds, the l values for the 4d–4f and 3d–4f resonance regimes are usually assumed to 3–5 Å and 12–15 Å, respectively [57]. In the case of $d_0 \sim 3$–6 Å, which is of the order of the mean Ce-Ce distance or the size of the unit cell, we can obtain $\mu >\sim 1$ (surface sensitive) for $E_K \sim 100$ eV, and $\mu < 1$ (bulk sensitive) for $E_K \sim 900$ eV.

Figures 4.10(a) and 4.10(b) exhibit on-resonance spectra of Ce metal in the α- and γ-phases in the 4d–4f and 3d–4f resonance regimes [60, 61]. A significant change in the spectral weight of the Ce $4f^0$ and Ce $4f^1$ final states is observed between the 4d–4f and 3d–4f resonance photoemissions. This change occurs because the μ value is larger or smaller than unity between these excitation photon energies, which is favorable for the precise evaluation of the surface and bulk components. We have assumed that the 4d–4f on-resonance spectra $[I_{4d-4f}(\epsilon)]$ and 3d–4f on-resonance spectra $[I_{3d-4f}(\epsilon)]$ can be expressed by the linear combination of the surface component $[S(\epsilon)]$ and bulk component $[B(\epsilon)]$:

$$I'_{4d-4f}(\epsilon) = I_b(\mu)B(\epsilon) + I_s(\mu)S(\epsilon) = \sqrt{\frac{c}{\pi}} \int e^{-c\eta^2} I_{4d-4f}(\epsilon - \eta) d\eta , \quad (4.3)$$

$$c = 4\ln 2/[(\Delta E_{3d-4f})^2 - (\Delta E_{4d-4f})^2] \quad (4.4)$$

$$I_{3d-4f}(\epsilon) = I_b(\mu')B(\epsilon) + I_s(\mu')S(\epsilon) . \quad (4.5)$$

In this expression, the 4f–4d resonance spectrum $[I'_{4d-4f}(\epsilon)]$ has been broadened taking into account the difference of energy resolution between the 4d–4f

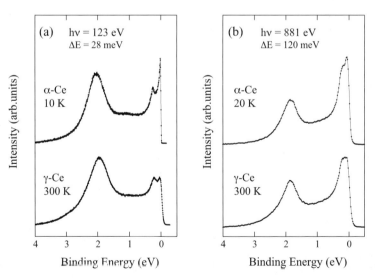

Fig. 4.10. High-resolution 4d–4f (**a**) and 3d–4f (**b**) resonance photoemission spectra of Ce metal in the α- and γ- phases [60, 61]

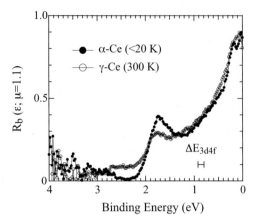

Fig. 4.11. Bulk weight $R_b(\epsilon; \mu = 1.1)$ as a function of binding energy evaluated for the 4d–4f resonance photoemission spectra of α- and γ-Ce. ΔE_{3d-4f} indicates the energy resolution of the 3d–4f resonance photoemission spectroscopy

($\Delta E_{4d-4f} = 28$ meV) and 3d–4f ($\Delta E_{3d-4f} = 120$ meV) resonance photoemission spectra. The intensities of I_{4d-4f} and I_{3d-4f} are normalized to the area, and $\mu = 1.1$ and $\mu' = 0.3$ have been assumed. Using these equations, one can evaluate $B(\epsilon)$ and $S(\epsilon)$ whose energy resolution was determined by the 3d–4f resonance spectra.

Next we introduce a function $R_b(\epsilon; \mu) = I_b(\mu)B(\omega)/[I_b(\mu)B(\epsilon)+I_s(\mu)S(\epsilon)]$, which represents the bulk weight as a function of the binding energy ϵ and the μ value. Figure 4.11 shows the evaluated $R_b(\epsilon; \mu = 1.1)$ for the 4d–4f on-resonance spectra in the α- and γ-phases. Since the observed R_b function does not oscillate rapidly, one may assume that the R_b function does not significantly depend on energy resolution between ΔE_{4d-4f} and ΔE_{3d-4f}. Then we can extract the bulk component by multiplying R_b directly to the $h\nu = 123$ eV spectra, namely, $B'(\epsilon) = R_b(\epsilon; \mu = 1.1)I_{4d-4f}(\epsilon)$. The nominal energy resolution of $B'(\epsilon)$ is ΔE_{4d-4f}, which allows us access closer to the fine spectral features near E_F.

Figure 4.12 exhibits the Ce 4f bulk components [$B'(\epsilon)$] obtained in this manner for the α- and γ-Ce together with the surface components ($S'(\epsilon) = [1 - R_b(\epsilon; \mu = 1.1)]I_{4d-4f}(\epsilon)$). One can see more clearly the differences in the Ce 4f-derived spectral features in these phases; namely, the Ce $4f^1$ weight is enhanced with respect to the Ce $4f^0$ weight on going from the γ to the α phase. The KR peak is significantly enhanced in the α phase consistent with a drastic increase of T_K.

There have been ARPES experiments on the Ce compounds [71–74], and some of them indicated a finite dispersion in the Ce 4f-derived states [72]. There has been also an attempt to measure ARPES of γ-Ce(100) single crystal [54], but the 4f-derived energy band dispersion in α-Ce has not yet been addressed due to the complicated crystal structure at the surface [55, 56].

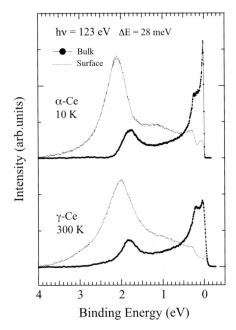

Fig. 4.12. Evaluated bulk and surface components for α- and γ-Ce

Since the Kondo temperature of α-Ce is thought to be order of 1000 K, it is not a trivial question to enquire to what extent the SIAM approach is effectively applicable. An alternative approach is to start from the itinerant model and take into account strong electron correlation. Recent dynamical mean-field theory (DMFT) based on the band-structure calculation [75, 76] was able to reproduce the Ce $4f$ photoemission spectral feature even for the γ-phase, to which the conventional itinerant model is hard to apply due to the localized nature of the $4f$ electron. However, more improvement may be needed to compare theoretical spectral functions with the photoemission spectra in a quantitative way. The electronic states of Ce metal, being so simple but yet very important, should be examined in more detail.

4.3.3 Ce-Based Kondo Semiconductors

There are a few systems in which the itinerant nature of the $4f$ state manifests itself as an appearance of the $c-f$ hybridization gap on cooling without a magnetic order; such systems are called the Kondo semiconductor [77–79]. In this section, we will describe a high-resolution photoemission study with SR of the Kondo semiconductor CeRhAs, the semimetal CeRhSb, and the metal CePtSn, which all have the same orthorhombic ϵ-TiNiSi-type crystal structure [79,80]. The Kondo temperatures for CeRhAs, CeRhSb, and CePtSn are estimated to be $T_\mathrm{K} \sim$ 1500 K ($\sim$130 meV), $\sim$360 K ($\sim$30 meV), and $\sim$10 K

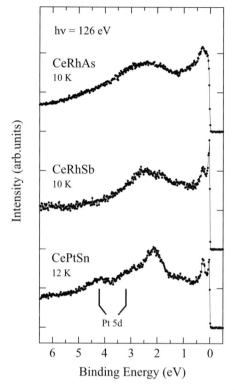

Fig. 4.13. High-resolution photoemission spectra of CeRhAs, CeRhSb and CePtSn taken at $h\nu = 126\,\text{eV}$ and at 10–12 K. The spectral intensities are normalized to the intensity of the Ce $4f^0$ states at ∼2–2.5 eV. Bars indicate the Pt 5d-derived spectral features in CePtSn [79]

(<1 meV) [81], respectively. The former two temperatures were inferred by assuming the relation $T_K \sim 3T_m$ [82], where T_m is the temperature where the magnetic susceptibility goes through a maximum [83,84]. It should be noted that the unit-cell volume increases on going from CeRhAs (239 Å^3) [85], to CeRhSb (269 Å^3) [85], and to CePtSn (276 Å^3) [81]. The volume expansion should weaken the $c - f$ hybridization.

Figure 4.13 shows the photoemission spectra of CeRhAs, CeRhSb, and CePtSn, taken at $h\nu = 126\,\text{eV}$ [79]. As shown below, the surface and bulk contributions respectively dominate the spectral weights of the Ce $4f^0$ and Ce $4f^1$ final states. Here, we will first examine fine Ce $4f^1$ spectral features near E_F.

Figure 4.14(a) displays the Ce $4f^1$-derived spectra whose intensities are normalized to the peak at ∼300 meV. The Ce $4f^1$ spectra of CeRhSb and CePtSn are split by the spin-orbit interaction into two peaks at ∼300 meV and ∼E_F. We cannot see, however, a peak at E_F for CeRhAs.

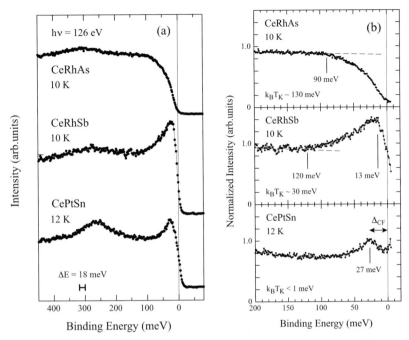

Fig. 4.14. (a) High-resolution photoemission spectra of CeRhAs, CeRhSb, and CePtSn near E_F. The spectral intensities are normalized to the intensity of the peak at ~300 meV; (b) Photoemission spectra of CeRhAs, CeRhSb, and CePtSn divided by a broadened FDD. These spectra are assumed to reflect the SDOS broadened with the instrumental resolution. The SDOS of CeRhAs decreases monotonically below ~90 meV forming a large energy gap. The SDOS of CeRhSb is enhanced below ~120 meV, but decreases below ~13 meV, forming a pseudogap. The SDOS of CePtSn has a weak KR, and a peak at ~27 meV corresponding to crystal field excitations [79]

In order to estimate the spectral DOS (SDOS), we divided the photoemission spectra by a Fermi–Dirac distribution (FDD) function, convoluted with a Gaussian which represents the instrumental resolution [69]. Figure 4.14(b) exhibits resulting SDOS spectra. There is no KR at E_F for CeRhAs. The spectral intensity decreases monotonically below ~90 meV, forming a gap structure.

In the case of CeRhSb, the spectral intensity shows an enhancement below the binding energy of $E_B \sim 120$ meV, a result which is similar to that of Kondo metals with high T_K [58]. However, below $E_B \sim 13$ meV, the spectral intensity decreases steeply, an important observation which differs from that seen in CePtSn and other Kondo metals [58]. The observed Ce $4f^1$ SDOS feature of CeRhSb agrees with the V-shaped pseudogap in the conduction bands as proposed in the analyses of the specific heat [87] and the value of $1/T_1$ found from NMR [88] measurements. It is also noted that the size of the pseudogap

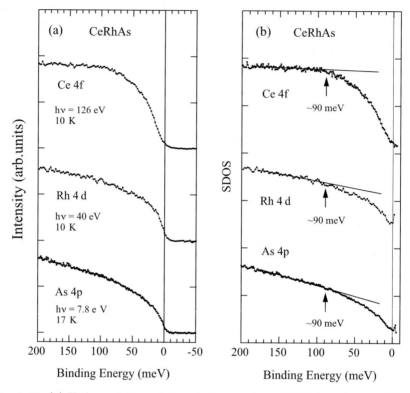

Fig. 4.15. (a) High-resolution photoemission spectra of CeRhAs taken at different incident photon energies. (b) SDOS's obtained from (a) [80]

$\sim$13 meV coincides well with the Δ_{p-p} values of 10–13.5 meV obtained by tunneling spectroscopy [89].

The spectral intensity of CePtSn exhibits no remarkable enhancement near E_F, except for a peak structure at $\sim$27 meV. The peak structure is in good agreement with the crystal field excitations (Δ_{CF}) observed in inelastic neutron scattering [90]. The weak KR is consistent with either the low $k_B T_K < 1$ meV, or a weak $c-f$ hybridization.

In order to better describe the CeRhAs and CeRhSb spectral features, the Periodic Anderson Model (PAM) may provide us with an insight into the $c-f$ hybridization gap. Since the dispersion of the Ce $4f$ states is assumed to be small in the PAM, that model leads to a sharp peak structure near E_F. The spectral feature in the Kondo semimetal CeRhSb seems to be explained very well by the spectral function in the PAM [91, 92]. However, with regards to the Kondo semiconductor CeRhAs, the situation seems to be quite different. Below we will evaluate the Ce $4f$ spectral feature of CeRhAs in the bulk and compare it with the DOS given by the band-structure calculation [95].

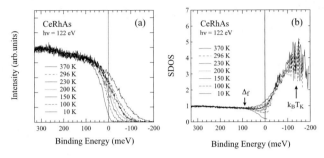

Fig. 4.16. (a) High-resolution temperature-dependent photoemission spectra of CeRhAs taken at $h\nu = 122\,\text{eV}$; (b) SDOSs evaluated for each temperature [80]

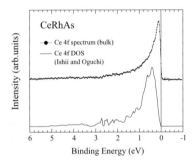

Fig. 4.17. Evaluated bulk component at 10 K compared with the Ce $4f$ DOS from a band-structure calculation [94]

Figure 4.15(a) shows photoemission spectra taken at $h\nu = 7.8\,\text{eV}$, $40\,\text{eV}$, and $122\,\text{eV}$ at 10–17 K [80]. Taking the photoionization cross-sections [62] into account, most of the spectral intensity at $h\nu = 7.8\,\text{eV}$ is derived from the As $4p$ state and that at $40\,\text{eV}$ from the Rh $4d$ state. The SDOSs found for the Ce $4f$, Rh $4d$ and As $4p$ states in Fig. 4.15(b) are reduced below $\sim 90\,\text{meV}$. Clearly, the magnitudes of the energy gap for the Ce $4f$ (Δ_f), Rh $4d$ (Δ_d), and As $4p$ (Δ_p) states are the same, $\Delta_f \sim \Delta_d \sim \Delta_p$. The present result suggests that the energy gap is formed by the f, d, and p hybridized bands near E_F.

In order to confirm a temperature-dependent energy gap formation, one should measure photoemission spectra as a function of temperature. As Figs. 4.16(a) and 4.16(b) indicate, the spectral intensity of the Ce $4f$ spectrum at E_F is gradually reduced on cooling and forms an energy gap ($\Delta_f \sim 90\,\text{meV}$). Similar temperature-dependent energy gap formation in the Rh $4d$ states was reported elsewhere [93]. At a temperature of 370 K, in the metallic phase of CeRhAs, the KR peak is observed at ~ 130–$140\,\text{meV}$ above E_F (Fig. 4.16(b)). We should note that this energy scale coincides well with T_K estimated from $3T_m$, which suggests the SIAM can be applied to the metallic phase.

Figure 4.17 shows the Ce $4f$ bulk component obtained from the $4d$-$4f$ and $3d$-$4f$ resonance photoemission spectroscopy [94]. We cannot clearly see a structure corresponding to the Ce $4f^0$ final state, totally different from the Ce metal case. If one compares the $4f$ spectrum with the $4f$ DOS given by the LSDA calculation [95], the overall spectral features are quite similar, which implies the itinerant nature of the Ce $4f$ state in this case. The peak structure in the LSDA is, however, located at slightly higher binding energy compared with the observed one. This difference is most likely derived from the renormalization effect.

4.4 High Energy Resolution Photoemission Spectroscopy with HX Combined with VUV and SX

In the case in which the thickness of the surface region (d_0) is larger than the IMFP (l) even for $E_K \sim 1000\,\text{eV}$ ($l \sim 15\,\text{Å}$), which is true for some Yb compounds as shown below, the $\mu = d_0/l$ value exceeds unity, $\mu > 1$ (surface sensitive). Here, the photoemission intensity from the surface region cannot be neglected.

Due to the long probing depth of more than $50\,\text{Å}$, hard x-ray photoemission (HXPES) spectra reflect the bulk-derived electronic states. The HXPES technique is a powerful tool for investigating the DOS of the valence band and the core levels, almost free from the electronic states in the surface region ($0 < d < d_0$). The recent rapid development of third-generation synchrotron radiation provides us with high flux, high-energy resolution photons which are a prerequisite for performing high-resolution HXPES spectroscopy. A total energy resolution (photon+electron energy analyzer) of $\Delta E \sim 70$–$90\,\text{meV}$, at photon energies up to 6–10 keV, has been achieved [7, 96].

Figure 4.18 exhibits the valence-band HX and VUV photoemission spectra of CeRhAs and CeRhSb, taken at $h\nu \sim 5.95\,\text{keV}$, and at $h\nu = 40\,\text{eV}$ [6]. A consideration of the photoionization cross-sections [62] shows that the Rh $4d$, As $4p$ and Sb $5p$ states are mainly represented at $h\nu \sim 5.95\,\text{keV}$ and that the Rh $4d$ dominates the $h\nu = 40$-eV spectra. While the $h\nu = 40$-eV spectra of the two compounds are quite similar, the $h\nu \sim 5.95$-keV spectra differ between these compounds in the energy ranges from E_F to $E_B \sim 1\,\text{eV}$ and from $E_B \sim 3\,\text{eV}$ to $5\,\text{eV}$, where the p-d-f hybridized states are located (Fig. 4.18). The variation in the spectral features is mainly derived from the different As $4p$ and Sb $5p$ partial DOS that exist in this energy region. Thus in the HXPES spectra, we could clarify the DOS of the anion p states located at the top of the valence band, which play an important role for the energy-gap formation.

The next example is the valence transition of YbInCu$_4$ at $T_V \sim 42\,\text{K}$, which has been studied extensively by various spectroscopic methods [97–101] (see also contribution by Takata in this volume). The transition is an isostructural,

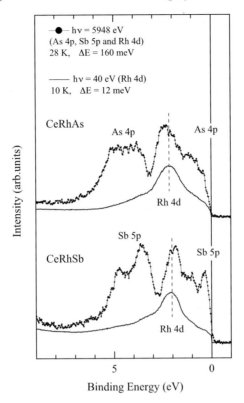

Fig. 4.18. Photoemission spectra of CeRhAs and CeRhSb taken at $h\nu = 5.95\,\text{keV}$ and 40 eV [6]

first-order phase transition similar to the $\alpha - \gamma$ transition in Ce metal. Figure 4.19 shows the temperature-dependent Yb $3d$ HXPES spectra [99]. The Yb^{2+} and Yb^{3+} signals are well separated, and the kinetic energy of the Yb $3d$ states, $E_K \sim 4.4\,\text{keV}$, is large enough ($l > \sim 50\,\text{Å}$) to probe the intrinsic bulk spectral intensity ratio of Yb^{2+}/Yb^{3+}. This ratio rises drastically between 55 K and 30 K, resulting in a change in the average valence from $Yb^{2.90+}$ down to $Yb^{2.74+}$. The temperature dependence of the valence transition of Yb measured in this manner was relatively sharp compared with those obtained using lower photon energies (inset of Fig. 4.19) [97, 98]. The present result is closer to the value that was obtained by another bulk-sensitive spectroscopic probe, resonant inelastic x-ray scattering [101]. It confirms that the HXPES probing depth is large to detect the bulk electronic properties.

Figure 4.20 provides the valence-band photoemission spectra. We should note that a drastic transition has been also observed above and below T_V, which was less significant for photoemission spectra with lower excitation energies. In order to evaluate the Yb valence using the valence-band HXPES

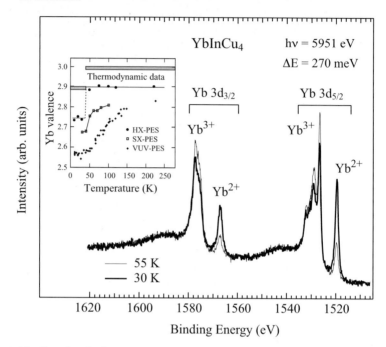

Fig. 4.19. Core-level photoemission spectra of YbInCu$_4$ above and below the valence transition ($T_V \sim 42$ K). The inset shows the valence transition estimated by PES [97–99]

spectra, however, a careful treatment may be needed taking into account enhanced photoionization cross-sections of p- or d-derived spectral features.

Finally we shall describe the electronic states of a typical Yb-based Kondo semiconductor, YbB$_{12}$. The metal-insulator crossover of YbB$_{12}$ takes place around temperature of ~ 80 K, on the basis of the magnetic and transport measurements [102–104]. Photoemission-spectroscopy studies in the VUV region indicate that an energy gap is formed as the temperature decreases [102–104].

Figure 4.21 presents Yb $3d_{5/2}$ spectra of this compound [103]. On cooling, the Yb^{2+} intensity slightly increases with respect to the Yb^{3+} intensity, but there is no sudden change as observed in YbInCu$_4$, which is consistent with the gradual cross-over.

Figure 4.22(a) exhibits the valence-band HXPES spectra of YbB$_{12}$ together with SX photoemission spectra taken at $h\nu = 100$ eV ($\Delta E \sim 15$ meV) [103, 104]. It is known that the spectral features derived from the surface region have a larger binding energy, by ~ 1 eV, compared to those from the bulk [103, 104]; thus, we can compare the Yb^{2+} $4f_{7/2}$ spectra at E_F in HXPES with those in the $h\nu = 100$-eV spectra. The intensity of the Yb^{2+} $4f_{7/2}$ peak in both HX and SX PES spectra increases on cooling in agreement with the Yb $3d$ spectra. Furthermore, in $h\nu = 100$-eV spectra, one can see a peak

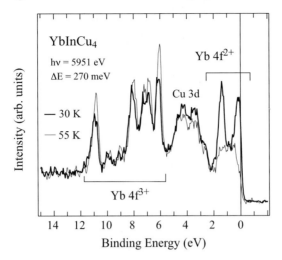

Fig. 4.20. Valence-band HXPES spectra of YbInCu$_4$ above and below the valence transition $T_V \sim 42$ K. The valence transition occurs in a narrow temperature range, with the Yb^{2+}-derived spectral feature appearing clearly below T_V [99]

appears at ~ 15 meV, which grew rapidly below ~ 60 K (Fig. 4.22(b)). The appearance of the peak can be discerned as a shoulder at 60 K (see the difference spectrum in Fig. 4.22(b)). The 15-meV peak intensity is strong at the L point of the Brillouin zone but becomes weaker away from the L point according to the ARPES measurement [104]. The appearance of this spectral feature is closely related to the formation of an energy-gap in the Yb $4f$-derived energy band near E_F [104]. In the future, we hope to observe similarly fine spectral features in the HXPES region.

4.5 Summary

The reported results of high-resolution photoemission spectroscopy with SR emphasize the merits of a tunable photon energy. In the case of ARPES, such tunability enables us to choose the initial state as well as the final state, and to observe fine electronic structures related to the low-energy excitations near E_F at a specific point of the Brillouin zone. By means of quantitative spectral lineshape analyses, one can evaluate the self-energy experimentally, and clarify the respective magnitudes of the many-body interactions such as the electron–phonon and electron–electron interactions. In the case of transition-metal compounds and rare-earth materials, the kinetic-energy dependence of the photoelectron of the IMFP can be utilized to decompose the surface and bulk electronic structures. It has been shown that the narrow $4f$ states are very sensitive to the strength of the $c - f$ hybridization, which is weaker near the surface. For some compounds, the thickness of the surface layer in

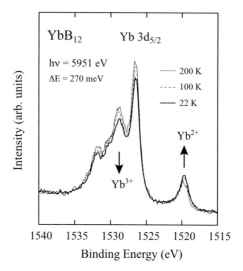

Fig. 4.21. Core-level photoemission spectra of YbB$_{12}$. The valence of Yb is estimated to be ∼2.9 [103]

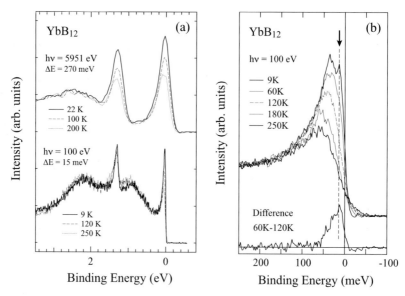

Fig. 4.22. (a) Valence-band HXPES and ultraviolet photoemission spectra of YbB$_{12}$ as function of temperature. The Yb^{2+} 4f peaks are enhanced on cooling; (b) valence-band ultraviolet photoemission spectra of YbB$_{12}$ near E_F. The additional spectral feature at ∼15 meV develops below ∼60 K [103, 104]

which the electronic states are different from those in the bulk is very large, and photoemission spectroscopy with hard x-ray is effective for deducing bulk-derived spectra. To date, as a result of the rapid improvements in HXPES, the linewidths of the core-level HXPES spectra are determined by lifetime broadening of several hundred meV [105]. However, higher energy resolution below < 100 meV is indispensable for the study of the valence band, especially near E_F. Based on developments in the energy resolution of photoemission spectroscopy, further improvements may be expected in the future as indicated by the dashed line in Fig. 4.1(b).

Acknowledgment

I would like to thank Prof. Masaki Taniguchi and Prof. Hirofumi Namatame for providing me an opportunity to perform synchrotron radiation high-resolution photoemission projects. I have profited by enlightening conversations with Prof. Friedrich Reinert and Prof. Stefan Hüfner. I would also like to thank Prof. Toshiro Takabatake, Prof. Fumitoshi Iga, Prof. Kenichi Kojima, Prof. Koichi Hiraoka, Prof. Hiroshi Negishi, Prof. Hitoshi Sato, Dr. Yukiharu Takeda, Dr. Yoshihiro Aiura, Mr. Masashi Arita, Mr. Mitsuharu Higashiguchi, Mr. Keisuke Nishiura, Mr. Yuichi Miura, Mr. Naohisa Tobita, Dr. Xiaoyu Cui for collaborative VUV and SX PES experiments with SR. In addition, I would like to thank all the staff members in Hiroshima Synchrotron Radiation Center, Hiroshima University. SX and HX PES experiments in the energy range $h\nu > 500$ eV have been done in collaboration with JASRI, RIKEN, and JAERI/SPring-8. I would like to acknowledge Prof. Keisuke Kobayashi, Dr. Eiji Ikenaga, Dr. Makina Yabashi, Dr. Yasutaka Takata, Prof. Shik Shin, Dr. Kenji Tamasaku, Dr. Yoshinori Nishino, Dr. Daigo Miwa, Prof. Tetsuya Ishikawa, Dr. Shin-ichi Fujimori, Dr. Yuji Saitoh, and Prof. Atsushi Fujimori for collaborative efforts for SX and HX PES experiments. I also would like to thank Dr. Stefan Schmidt for the compilation of this manuscript. This work is partly supported by a Grant-in-Aid for Scientific Research (No.17654060) and for COE Research (13CE2002) from the Ministry of Education, Culture, Sports, Science and Technology of Japan.

References

1. F. Reinert and S. Hüfner: New J. Phys. **7**, 97 (2005)
2. S. Hüfner et al: Nucl. Instrum. Meth. Phys. A **547**, 8 (2005)
3. A. Damascelli et al: Rev. Mod. Phys. **75**, 473 (2003), and references therein.
4. A. Sekiyama et al: Nature **403**, 396 (2000)
5. A. Yamasaki et al: Nucl. Instrum. Meth. Phys. Res. A **547**, 136 (2005), and references therein.
6. K. Shimada: Nucl. Instrum. Meth. Phys. A **547**, 169 (2005)

7. K. Kobayashi: Nucl. Instrum. Meth. Phys. A **547**, 98 (2005)
8. J. W. Allen: J. Phys. Soc. Jpn. **74**, 34 (2005), and references therein.
9. S. Hüfner: *Photoelectron Spectroscopy*, 3rd ed., (Springer-Verlag, Berlin, 2003).
10. E.-E. Koch et al: in *Handbook on Synchrotron Radiation Vol. 1A* (North-Holland Pub. Co., Amsterdam, 1983) Chap. 1.
11. T.-C. Chiang: Chem. Phys. **251**, 133 (2000), and references therein.
12. F. J. Himpsel et al: J. Magn. Magn. Mater. **200**, 456 (1999), and references therein.
13. C. R. Brundle: J. Vac. Sci. Technol. **11**, 212 (1974)
14. M. P. Seah and W. A, Dench: Surf. Interf. Anal. **1**, 2 (1979)
15. A. Chainani et al: Phys. Rev. Lett. **85**, 1966 (2000)
16. F. Reinert et al: Phys. Rev. Lett. **91**, 186406 (2003)
17. The data were taken at Helical undulator beamline BL-9A on a compact electron storage ring (HiSOR) at Hiroshima Synchrotron Radiation Center, Hiroshima University. For the beamline description, see M. Arita et al: Surf. Rev. Lett. **9**, 535 (2002)
18. M. Arita, unpublished data. The photoemission spectra were taken at HiSOR BL-9A using helical undulator radiation.
19. T. Yamasaki et al: cond-mat/0603006
20. E. Dietz et al: Phys. Rev. Lett. **40**, 892 (1978)
21. D. E. Eastman et al: Phys. Rev. Lett.**40**, 1514 (1978)
22. F. J. Himpsel et al: Phys. Rev. B **19**, 2919 (1979)
23. W. Eberhardt and E. W. Plummer: Phys. Rev. B **21**, 3245 (1980)
24. K. Ono, A. Kakizaki et al: Solid State Commun. **107**, 153 (1998)
25. C. S. Wang and J. Callaway: Phys. Rev. B **15** 298 (1977)
26. F. Weling and J. Callaway: Phys. Rev. B **26**, 710 (1982)
27. S. Hüfner and G. K. Wertheim: Phys. Lett. **51**, 299 (1975)
28. R. Clauberg et al: Phys. Rev. Lett. **47**, 1314 (1981)
29. Y. Sakisaka et al: Phys. Rev. Lett. **58**, 733 (1987)
30. T. Kinoshita et al: Phys. Rev. B **47**, 6787 (1993)
31. A. Kakizaki et al: Phys. Rev. Lett. **72**, 2781 (1994)
32. J. Kanamori: Prog. Theor. Phys. **30**, 275 (1963)
33. D. R. Penn: Phys. Rev. Lett. **42**, 921 (1979)
34. A. Liebsch: Phys. Rev. Lett. **43**, 1431 (1979); A. Liebsch: Phys. Rev. B **23**, 5203 (1981)
35. L. Kleinman: Phys. Rev. B **19**, 1295 (1979)
36. L. Kleinman and K. Mednick: Phys. Rev. B **24**, 6880 (1982)
37. G. Treglia et al: Phys. Rev. B **21**, 3729 (1980); G. Treglia, F. Ducastelle, D. Spanjaard: J. Phys. (Paris) **43**, 341 (1982)
38. M. Higashiguchi et al: J. Electron Spectrosc. Relat. Phenomen. **144-147**, 639-642 (2005)
39. M. Higashiguchi et al: Phys. Rev. B **72**, 214438 (2005)
40. P. Aebi et al: Phys. Rev. Lett. **76**, 1150 (1996)
41. R. J. Birgeneau et al: Phys. Rev. **136**, A1359 (1964)
42. T. Valla et al: Phys. Rev. Lett. **83**, 2085 (1999)
43. N. V. Smith et al: Phys. Rev. B **47**, 15476 (1993)
44. S. Sahrakorpi et al: Phys. Rev. B **66**, 235107 (2002)
45. G. D. Gaspari and B. L. Gyorffy: Phys. Rev. Lett. **28**, 801 (1972)
46. D. A. Papaconstantopoulos et al: Phys. Rev. B **15**, 4221 (1977)

47. F. Manghi, V. Bellini et al: Phys. Rev. B **59**, R10409 (1999)
48. E. I. Zornberg: Phys. Rev. B **1**, 244 (1970)
49. P. Nozieres and D. Pines: *The Theory of Quantum Liquids* (Perseus books, Cambridge, 1999) Chap. 1.
50. J. W. Allen and R. M. Martin: Phys. Rev. Lett. **49**, 1106 (1982), and references therein.
51. T. Naka et al: Physica B **121-126**, 205 (1995), and references therein.
52. K. Maezawa et al: J. Phys. Soc, Jpn. **65**, 151 (1996), and references therein.
53. F. Patthey et al: Phys. Rev. Lett. **55**, 1518 (1985), and references therein.
54. G. Rosina et al: Phys. Rev. **33**, 2364 (1986)
55. E. Weschke et al: Phys. Rev. B **44**, 8304 (1991)
56. E. Weschke et al: Phys. Rev. B **58**, 3682 (1998)
57. L. Z. Liu et al: Phys. Rev. B **45** 8934 (1992), and references therein.
58. M. Garnier et al: Phys. Rev. Lett. **78**, 4127 (1997)
59. Yu. Kucherenko et al: Phys. Rev. B **66**, 155116 (2002), and references therein.
60. M. Higashiguchi et al: Physica B **351**, 256-258 (2004)
61. The results of the $3d$-$4f$ resonance photoemission spectra taken at SPring-8 BL23SU are unpublished data by M. Higashiguchi et al.
62. J. J. Yeh and I. Lindau: At. Data Nucl. Data Tables **32**, 1 (1985); J. J. Yeh: *Atomic Calculation of Photoionization Cross-Sections and Asymmetry Parameters* (Gordon and Breach, New York, 1993)
63. W. E. Pickett et al: Phys. Rev. B **23**, 1266 (1981)
64. O. Gunnarsson and K. Schönhammer: in *Handbook on the Physics and Chemistry of Rare Earth*, ed. K. A. Gschneidner, L. Eyring, S. Hüfner (North-Holland, Amsterdam 1987) Vol. 10, p. 103
65. Y. Kuramoto: Z. Phys. B **53**, 37 (1983)
66. P. Coleman: Phys. Rev. B **29**, 3035 (1984)
67. N. E. Bickers et al: Phys. Rev. B **36**, 2036 (1987)
68. A. A. Abrikosov: *Fundamentals of the Theory of Metals*, (North-Holland, Amsterdam, 1988) Chap. 13.
69. F. Reinert et al: Phys. Rev. Lett. **87**, 106401 (2001), and references therein.
70. L. Duò: Surf. Sci. Rep. **32** 233 (1998), and references therein.
71. A. B. Andrews et al: Phys. Rev. **51**, 3277 (1995)
72. A. J. Arko et al: Phys. Rev. **56**, R7041 (1997)
73. J. D. Denlinger et al: J. Electron Spectrosc. Relat. Phenomen. **117-118**, 347 (2001)
74. D. Ehm et al: Phys. Rev. **64**, 235104 (2001)
75. M. B. Zölfl et al: Phys. Rev. Lett. **87**, 276403 (2001)
76. O. Sakai et al: J. Phys. Soc. Jpn. **74**, 2517 (2005), and references therein.
77. T. Takabatake et al: J. Magn. Magn. Mater. **177-181**, 277 (1998), and references therein.
78. Z. Fisk et al: Physica B **206-207**, 798 (1995), and references therein.
79. K. Shimada et al: Phys. Rev. B **66**, 155202 (2002) and references therein.
80. K. Shimada et al: J. Electron Spectrosc. Relat. Phenomen. **144-147**, 857 (2005)
81. T. Takabatake et al: Physica B **183**, 108 (1993)
82. N. E. Bickers et al: Phys. Rev. Lett. **54**, 230 (1985)
83. T. Susakawa et al: Phys. Rev B **66**, 041103(R) (2002)
84. T. Takabatake et al: Physica B **206-207**, 804 (1995); T. Takabatake et al: *ibid.* **223-224**, 413 (1996)

85. P. Salamakha et al: J. Alloy and Compounds **313**, L5 (2000)
86. H. Kumigashira et al: Phys. Rev. Lett. **87**, 067206 (2001)
87. S. Nishigori et al: J. Phys. Soc. Jpn. **65**, 2614 (1996)
88. K. Nakamura et al: J. Phys. Soc. Jpn. **63**, 433 (1994)
89. T. Takabatake et al: Physica B **206-207**, 804 (1995); T. Ekino et al: Physica B **223-224**, 444 (1996)
90. M. Kohgi et al: Physica B **186-188**, 409 (1993)
91. H. Ikeda and K. Miyake: J. Phys. Soc. Jpn. **65**, 1769 (1996)
92. J. Moreno and P. Coleman: Phys. Rev. Lett. **84**, 342 (2000)
93. H. Kumigashira et al: Phys. Rev. Lett. **82**, 1943 (1999)
94. K. Shimada et al, unpublished data.
95. F. Ishii and T. Oguchi: J. Phys. Soc. Jpn. **73**, 145 (2004)
96. Y. Takata et al: Nucl. Instrum. Meth. Phys. Res. A **547**, 50 (2005)
97. F. Reinert et al: Phys. Rev. B **58**, 12808 (1998)
98. H. Sato et al: Phys. Rev. B **69**, 165101 (2004)
99. H. Sato et al: Phys. Rev. Lett. **93**, 246404 (2004)
100. S. Schmidt et al: Phys. Rev. B **71**, 195110 (2005)
101. C. Dallera et al: Phys. Rev. Lett. **88**, 196403 (2002)
102. T. Susaki et al: Phys. Rev. Lett. **82**, 992 (1999)
103. Y. Takeda et al: Physica B **351**, 286 (2004)
104. Y. Takeda et al: Phys. Rev. B **73**, 033202 (2006)
105. J. C. Fuggle and S. F. Alvarado: Phys. Rev. A **22**, 1615 (1980)

Part II

Low-Dimensional Systems

5

Photoemission on Quasi-One-Dimensional Solids: Peierls, Luttinger & Co.

R. Claessen, J. Schäfer, and M. Sing

Experimentelle Physik 4, Universität Würzburg, D-97074 Würzburg, Germany
claessen@physik.uni-wuerzburg.de

Abstract. The article reviews the key issues associated with many-body effects in low-dimensional materials accessible by angle-resolved photoemission (ARPES). Peierls instabilities driven electronically by the Fermi surface topology lead to band-backfolding and opening of energy gaps. Precursor fluctuations and altered spectral weight resulting from superimposed potentials become detectable, and low-temperature gap spectra can be studied in detail. ARPES also allows to search for theoretically predicted spin-charge separation in quasi-one-dimensional systems which results from electron-electron interaction. Such data can be analyzed in terms of a 1D-Hubbard model. Some materials show a more complex behavior due to the simultaneous interaction between spin, charge, and lattice degrees of freedom. Examples of recent and ongoing research which elucidate these phenomena are discussed.

5.1 Introduction

Quantum phenomena resulting from electron–electron interaction or coupling of electrons to other degrees of freedom in the solid are central topics of modern condensed matter physics. Many-body effects in the electronic structure can often be absorbed into a phenomenological Fermi-liquid (FL) description in which the low-lying excitations of the conduction electron system are described as only weakly interacting quasi-particles of renormalized dynamical properties. However, many fascinating phenomena of topical interest – such as high-temperature superconductivity or quantum-critical behavior in heavy fermion systems – are thought to be related to a breakdown of the FL picture. On the theoretical side, a microscopic description of these phenomena represents a tremendous challenge due to the complexity of the problem. In this context, one-dimensional (1D) electron systems have become prototypical objects of study, as they are expected to display generic non-FL behavior and at the same time are theoretically easier to access.

If conduction electrons are confined to only one spatial dimension, conventional metallic behavior may break down due to a variety of mechanisms.

The best established one is the Peierls instability [1], in which nesting of the 1D Fermi surface in combination with electron–phonon coupling leads to the spontaneous formation of a charge-density wave (CDW) and often a concomitant transition into an insulating phase. While the very existence of Peierls transitions has for long been established, there remain open problems concerning, e.g., the role of fluctuations above the Peierls temperature. Another fascinating phenomenon is the separation of spin and charge degrees of freedom in a 1D electron system due to electron–electron interaction, originally described by Tomonaga [2] and Luttinger [3]. While theoretically well elaborated over the past 15 years, the search for direct experimental evidence of Tomonaga–Luttinger liquid (TLL) behavior is still going on.

In the theoretical treatment of these topics the single-particle excitation function $A(\mathbf{k}, \varepsilon)$ plays a central role. Since it is measured within the sudden approximation by angle-resolved photoelectron spectroscopy, this technique has become the key experimental method in the study of FL instabilities in 1D electron systems. A very comprehensive review on high-resolution photoemission studies of low-dimensional systems up to the year 2000 has been given by Grioni and Voit [4]. Here we will focus on more recent developments, in particular on ARPES studies of quasi-1D bulk materials. Photoemission experiments on 1D physics in artificial atomic chains on surfaces ("nanowires") will be discussed in a separate chapter in this book by Ortega and Himpsel.

This chapter is organized as follows. Section 5.2 offers an introduction to the physics of the Peierls instability and of spin-charge separation in 1D electron systems. Photoemission studies on CDW-related phenomena such as umklapp and fluctuation effects are presented in Sect. 5.3. Section 5.4 focusses on the search for TLL behavior and spin-charge separation. Finally, in Sect. 5.5 we summarize and address open questions left for future investigations.

5.2 Electronic Instabilities in One Dimension

5.2.1 Peierls Instability

As just mentioned, in an ideal 1D solid the coupling of the electrons to the lattice can lead to a phase transition where at low temperatures long-range periodic modulations of the charge density are observed. The possibility of such a CDW ground state has first been recognized theoretically by Peierls [1]. The CDW is intimately coupled to a periodic lattice distortion (PLD) as shown schematically in Fig. 5.1(a). Bragg scattering of conduction electrons by the new periodic structure leads to the occurrence of band gaps at the Fermi vectors and consequently to the destruction of the Fermi surface. As a result the spontaneous formation of CDWs is often accompanied by a metal-insulator transition.

Regarding the electronic energies, the simplest case is a 1D system with a single half-filled band [1]. An essential characteristic is that the CDW wave

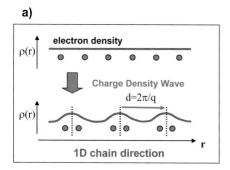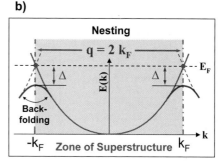

Fig. 5.1. Schematic view of a CDW. (a) The electronic charge density is modulated when atoms undergo e.g. a dimerization; (b) Electron energies for the 1D case with a single band. A CDW wave vector $q = 2k_F$ spans the two Fermi level crossings and new zone boundaries are formed at $\pm k_F$. This leads to band backfolding and the opening of energy gaps

vector q assumed by the 1D system is such that q connects the Fermi level crossings at $+k_F$ and $-k_F$. This mechanism, namely that the distortion vector q is spanned across the Fermi surface is referred to as "nesting". In reciprocal space this implies new zone boundaries imposed at $\pm k_F$, and the electron bands are thereby backfolded to continue symmetrically in the next Brillouin zone, as in Fig. 5.1(b). For the example of a half-filled band, such nesting implies a reduction of the Brillouin zone to half its size, which in real space corresponds to a period doubling. However, the mechanism is not restricted to the half-filled case, and many examples even of incommensurate superstructure vectors are known (see, e.g., NbSe$_3$ in Subsect. 5.3.1). As a consequence of the CDW-induced backfolding, energy gaps are opened at the Fermi vectors. The distortion thereby leads to a reduction of the total energy of the system. The electronic energy of the system is reduced, however, the lattice distortion is limited because of the cost in elastic energy. The optimum balance between electronic and lattice energies results in the equilibrium distortion amplitude.

The shape of the Fermi surface determines whether the system is susceptible to a CDW instability. Specifically, the density of states at the Fermi level which can be connected via the nesting vector $q = 2k_F$ must be sufficiently high to ensure an energy gain that results from the opening of an energy gap there. This requirement is met by large parallel sections of the Fermi surface. As can be seen from the free-electron Fermi surfaces sketched in Fig. 5.2(a), such sections are frequently found in quasi-1D systems. The ideal 1D Fermi surface (embedded in 3D) simply consists of two parallel sheets. The degree to which any system is inclined to undergo a transition is characterized by the generalized electronic susceptibility

$$\chi(q) - \sum_k \frac{f_k - f_{k+q}}{\varepsilon_{k+q} - \varepsilon_k}, \tag{5.1}$$

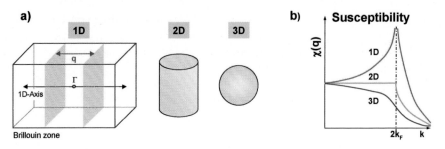

Fig. 5.2. Role of dimensionality for Fermi surface nesting in a free electron gas. (a) A Fermi surface in 1D provides the largest number of states for nesting; (b) electronic susceptibility as a function of wave vector q. In 1D the susceptibility assumes a singularity at $2k_F$ and can thus lead to a CDW instability

where f_k is the occupation function at wave vector $\boldsymbol{k}$ and ε_k the band energy. From this expression it becomes obvious that contributions result only from states at the Fermi level. A singularity in $\chi(\boldsymbol{q})$ corresponds to the condition of Fermi-surface nesting for the Peierls instability, and a 1D electron gas with such a singularity at $2k_F$ will thus be unstable against formation of a CDW state. There is a drastic dependence on dimensionality, as illustrated in Fig. 5.2(b), where in 1D $\chi(\boldsymbol{q})$ becomes singular at $\boldsymbol{q} = 2k_F$. For higher dimensions, the peak in the susceptibility at $\boldsymbol{q} = 2k_F$ becomes weaker. For the free electron gas in 2D, the maximum is already completely flattened. For specific Fermi-surface topologies that contain piecewise parallel sections, even in 3D a maximum in the susceptibility at $\boldsymbol{q} = 2k_F$ can be obtained. A prominent example is the *spin* density wave (SDW) that develops in bulk chromium [5] where the same topological argument applies to a singularity in the spin susceptibility, albeit based on electron–electron interaction [6]. In general, however, nesting and the associated instability is particularly likely to occur in low dimensions.

The mechanism described depends on the existence of electron–phonon coupling, as is reflected in two aspects. First, the intuitive picture of the electron gas does not consider the Coulomb repulsion, which tends to spread out the electrons evenly, and for a solid it has been shown by Chan and Heine [7] that the electron–phonon interaction must be included in order to stabilize a CDW, linking it to the PLD. Second, the effect of electron–phonon coupling is also reflected in the phonon dispersion. In the particular case of CDW nesting at k_F, the phonon dispersion will be modified with a soft phonon mode and exhibit a dip at $2k_F$, known as Kohn anomaly [8].

A key feature of a CDW system is the formation of an energy gap. Due to the backfolding, an idealized Peierls instability for a single band in 1D implies a metal-insulator transition. The magnitude of the energy gap 2Δ scales with the electron–phonon coupling constant λ as $2\Delta(T=0) \sim e^{-1/\lambda}$ [9]. At finite temperature, the low-temperature ground state is destabilized and the energy gap becomes smaller until eventually a phase transition occurs. In mean-field theory analogous to the BCS theory of superconductivity, a relation between the low-temperature energy gap and the transition temperature is given by $2\Delta(T=0) = 3.53\,k_B\,T_c$. However, a mean-field description neglects the effect of order-parameter fluctuations which are very strong in 1D systems. As a consequence the actual transition temperatures are often found to be much lower than expected from mean-field theory. It should be mentioned that a truly 1D system does not exhibit any phase transition at $T > 0$. CDW transitions at finite T are possible only because of some 3D coupling ("quasi 1D systems").

5.2.2 Spin-Charge Separation

As already discussed in the introduction, in conventional metals the effect of electron–electron interaction results in a mere renormalization of the single-particle excitations with respect to the independent particle picture. In FL theory we can think of the quasi-particle excitations as bare particles (electrons or holes) dressed by a cloud of virtual excitations induced by the interaction, i.e. in the case of electron–electron interaction by virtual electron–hole pair excitations. In terms of the spectral function $A(\mathbf{k},\varepsilon)$ the sharp peaks expected from band theory, $\delta(\varepsilon-\varepsilon_0(\mathbf{k}))$, will shift closer to the chemical potential due to the effective mass enhancement of the quasi-particles and broaden in energy as result of their finite lifetime. However, there is still a symmetry-conserving one-to-one correspondence between the quasi-particles of the real system and the Bloch states of the non-interacting reference system.

This picture no longer applies to 1D electron systems. In fact, here we encounter a generic failure of FL theory. It can be shown quite generally that in 1D the spin and charge degrees in the Hamiltonian decouple from each other, at least at low excitation energies (see [10] and references therein). As a consequence the low-lying excitations are found to be collective spin and charge modes which are completely distinct from quasi-particle excitations. While the theory of spin-charge separation has been worked out in much detail, it is difficult to give an intuitive picture of this phenomenon similar to that of a quasi-particle as a dressed particle. However, spin-charge separation can be best understood in the limit of strong local Coulomb interaction, i.e. within the framework of the 1D Hubbard model [11]:

$$\hat{H} = t \sum_{<ij>,\sigma} c^{\dagger}_{i\sigma} c_{j\sigma} + U \sum_{i} n_{i\uparrow} n_{i\downarrow} \,. \tag{5.2}$$

If the on-site Coulomb energy is larger than the hopping amplitude ($U/t \gg 1$), double occupancy of sites will be largely suppressed, and for a band close to half-filling there will be essentially one electron per site of the 1D chain. Note that due to virtual hopping processes there is a magnetic exchange energy $J \sim t^2/U$ which favors antiferromagnetic alignment of electron spins

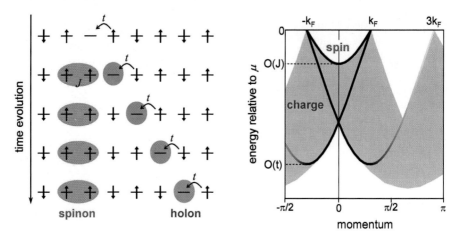

Fig. 5.3. *Left*: Cartoon showing the decay of a (photo-)hole into spinon and holon in a 1D Hubbard chain. *Right*: Schematic structure of the single-particle spectrum of the 1D Hubbard chain

on neighboring sites, leading to the situation depicted in Fig. 5.3. Now suppose that we remove one electron from the chain, e.g. by photoemission as in ARPES. The resulting hole state is not stationary, but will decay as shown in the figure. Initially, a neighboring electron will hop onto the hole site, thereby creating a ferromagnetic spin pair, which will cost an energy J. Further hopping processes will maintain antiferromagnetic correlations so that they do not require additional energy, until eventually a situation is reached where the chain displays two anomalies: (i) a ferromagnetic spin perturbation close to the site of the initial hole creation, and (ii) a missing charge moving rapidly away. These two objects are identified as *spinon* and *holon*, respectively. This phenomenon occurs only in one dimension. In higher-dimensional systems the initial (photo-)hole remains rather immobile, because it costs far too much magnetic energy to move it away from its original site. It is hence better described as a heavy quasi-particle, in accordance with FL theory.

The effect of spin-charge separation on the single-particle spectrum is shown in Fig. 5.3 as well. Because there is a manifold of ways to divide energy and momentum of the original hole between spinon and holon, the spectrum will consist of a continuum in $(\mathbf{k}, \varepsilon)$-space, indicated by the grey-shaded area. However, the huge phase space for decomposing the hole into a zero-energy spinon and a finite-energy holon or *vice versa* leads to pronounced singularities in the spectrum, denoted in Fig. 5.3 as "charge" and "spin" branch, respectively. Typically, the energy scale of the charge branch is given by the hopping integral t, whereas the spinon energy scale is determined by the magnetic exchange integral J. Thus, the occurrence of *two* dispersive spectral features with common Fermi vector (while without Coulomb interaction only a single conduction band would be expected) is a signature of spin-charge separation in a 1D metal.

It is important to point out that spin-charge separation is generic to all 1D electron systems at low excitation energies, even if the interaction is not short-ranged or strong. This has been demonstrated by Tomonaga [2] and Luttinger [3] by simply linearizing the 1D conduction band about the Fermi vector leading to an exactly solvable model. As has been worked out in detail, the FL paradigm is in 1D replaced by a new generic low-energy phenomenology called Tomonaga–Luttinger liquid (TLL) behavior [10, 12–14]. A TLL is characterized by non-universal power law singularities which are characterized by the so-called charge stiffness constant K_ρ [10]. For example, the onset of the $\boldsymbol{k}$-integrated electron removal part of the spectral function (i.e., of the photoemission spectrum) is not given by a finite density of states cut off by a Fermi–Dirac distribution function as in normal metals, but displays a power law behavior $|\varepsilon - \mu|^\alpha$ with exponent $\alpha = \frac{1}{4}(K_\rho + K_\rho^{-1} - 2)$. The resulting pseudogap at the chemical potential μ is thus another hallmark of TLL behavior and has been intensively searched for in the early photoemission studies of 1D metals [4]. However, it is very difficult to distinguish it experimentally from the pseudogap induced by CDW fluctuations discussed in Subsect. 5.3.3. Unfortunately most 1D metals display a Peierls instability at sufficiently low temperatures. Furthermore, because the Tomonaga–Luttinger model has by construction no intrinsic energy scale, it remains unclear how low the excitation energies have to be in order to see TLL physics in real materials.

As shown above, in the limit of strong short-range interactions we can expect to see consequences of spin-charge separation also at high energy scales of the order of the band width. Peierls physics occurs on the energy scale of the phonon spectrum and can thus not interfere there. There has recently been enormous progress in the treatment of the 1D Hubbard model [11] and especially in the calculation of its single-particle spectrum [15–17], thus providing detailed theoretical predictions. In Sect. 5.4 we will discuss several experimental tests of the theory.

5.3 Photoemission of Quasi-1D CDW Systems

5.3.1 Spectral Function and CDW Systems

The occurrence of CDW phase transitions has spurred an ongoing research activity with many unsettled issues, many of which should be accessible by ARPES. Among these is an account of the coupling strength between electrons and phonons. Within some reasonably justified approximations, ARPES gives direct access to the already mentioned spectral function $A(\boldsymbol{k}, \varepsilon)$ that describes the single-particle excitations of the electron system. Thereby ARPES provides e.g. the possibility to observe quasi-particle renormalizations resulting from electron–phonon coupling (see Subsect. 5.3.4). Furthermore, a conceptually challenging environment for CDWs is the presence of multiple electron bands which may lead to incomplete gap formation at the Fermi surface. Here one may also ask what happens to the quasi-particle state in the presence of an energy gap. Low-dimensional systems are also known to exhibit precursor fluctuations of the ground state already above the critical temperature, leading to a pseudogap [18] (see Subsect. 5.3.3). This is one of the most notable failures of a mean-field description for CDW systems.

Quite a few bulk materials are known that exhibit CDWs due to their intrinsic low-dimensional bonding nature. Among these are inorganic solids including e.g. transition metal dichalcogenides [19], transition metal bronzes [20], and platinum chain compounds [21]. Strong current relevance of these instabilities that occur preferably in reduced dimensions, stems from the heightened interest in nanoscale structures. Prominent examples in this respect are adatom structures on surfaces that can exhibit CDWs. A two-dimensional metallic reconstruction was reported to show indications of such instability [22], albeit that additional structural distortions seem to dominate the picture. Rather recently and somewhat more clear-cut, it has been discovered that quasi-1D *nanowires* can exhibit CDWs [23,24], as treated in the present book in a separate chapter (Ortega and Himpsel). In what follows we will restrict ourselves to bulk systems.

There is a number of quasi-1D systems that have been the object of ARPES studies in the past years. Among these are the molybdenum bronzes. The molybdenum blue bronze $K_{0.3}MoO_3$ is a 1D system that exhibits a Peierls transition at 180 K [25]. The Li purple bronze $Li_{0.9}Mo_6O_{17}$ is also of quasi 1D nature like its blue-bronze counter part, however, it remains metallic down to at least 24 K, and a CDW transition has not been clearly established [26] (see also Subsect. 5.4.1). Another compound where a CDW transition is not clearly seen is $BaVS_3$. This material exhibits a drastic loss of spectral weight near the Fermi level, and it remains a matter of debate whether it is due to TLL behavior or due to phonon satellites [27].

Yet another group of materials with 1D electronic properties and Peierls instabilities is formed by the transition-metal trichalcogenides. A rather well studied CDW phase transition is known from $NbSe_3$, which in fact exhibits two transitions at 145 K and 59 K [28,29]. As we shall see below in Subsect. 5.3.4 in more detail, the coexistence of two CDW superstructures becomes possible due to the multiband nature of this compound. Furthermore, $(TaSe_4)_2I$ is a 1D chain structure with a phase transition slightly below room temperature at 263 K [30–32].

5.3.2 CDW Umklapp and Spectral Weight Distribution

As discussed in Subsect. 5.2.1, the formation of the CDW states implies superimposed periodic potentials with corresponding zone boundaries. Therefore, one would expect an umklapp of the electron bands at these new boundaries at $q = 2k_F$, accompanied by a gap formation. However, the actual situation can be much more subtle, because the superstructure potential is weak compared to the underlying unperturbed lattice potential. Moreover, it may be incommensurate with it. This has consequences for the spectral weight seen in ARPES. A nice example that illustrates the influence of competing periodic potentials is $(TaSe_4)_2I$.

For CDW nesting in this compound, the bonding may be viewed as highly ionic with the consequence that each iodine atom accepts one electron. Assuming fully independent $TaSe_4$ chains, the conduction band of Ta d-electron character would be 1/4 filled. Yet the structure is distorted such that the length of the unit cell is four Ta spacings in the chain direction, $c = 4a$. This would in principle lead to a filled-band situation. However, this band filling is not found to occur due to finite interchain interaction and incomplete charge transfer to the iodine atoms. Effectively the Ta atoms contain slightly more charge and are thus electron-doped. This leads to a k_F that is marginally larger than $\pi/c = 1/4 \cdot \pi/a)$. As a result, the CDW boundary is slightly outside the bare Brillouin zone boundary [32].

Using this system as an example, one may ask what happens to the ARPES spectral weight in the presence of these two incommensurate potentials. Globally this implies a loss of translational invariance, and naively one might expect that conventional band structure cannot exist. Using such boundaries in a simple model system in 1D, it can be shown that the Brillouin zone looses its role as perfect scattering plane, and the spectral weight is essentially distributed along electron bands that correspond to a free-electron-like parabola. However, in the extended zone scheme the superimposed boundaries of crystal lattice and CDW, respectively, modulate these bands. They lead to umklapp scattering and the opening of energy gaps. This can be seen nicely in the ARPES data of $(TaSe_4)_2I$ [32], see Fig. 5.4.

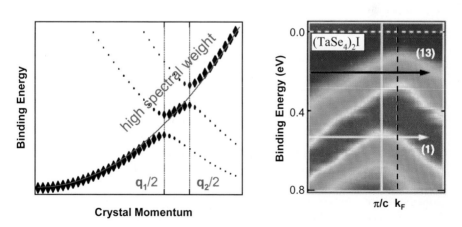

Fig. 5.4. Spectral weight distribution in the presence of two incommensurate potentials from crystal lattice and CDW [32]. (a) Calculated distribution of spectral weight for a 1D electron system with a single band; (b) ARPES data from $(TaSe_4)_2I$, showing umklapp occuring at two neighbouring scattering planes for the lower branch of the bands below the CDW gap. The intensity is interpreted as arising from two bands in $(TaSe_4)_2I$ that are energetically closely together

In other words, the *dispersion* obeys the scattering planes, while the *spectral weight* is peaked in the extended zone scheme [32]. The backfolded branches with weak spectral weight are usually referred to as "shadow bands". Thus, despite the incommensurability that arises from the presence of a CDW nested at $q = 2k_F$ and the subsequent loss of translational invariance, the weight distribution in the spectral function mimicks an "effective" band dispersion which reflects these periodicities. This allows the identification of CDW-related electron bands with their specific periodicity, as well as of the CDW energy gap.

5.3.3 CDW Fluctuation Effects in ARPES

As a further implication of CDWs on the electronic structure, precursor fluctuations can be present in the temperature regime above the phase transition. They modify the potential and periodicity on a short range scale. A very yielding model system in this respect is NbSe$_3$ with multiple metallic bands and two incommensurate CDW distortions [28]. NbSe$_3$ crystals take the form of fine fibers as a result of their directional bonding. Owing to the difficulty to prepare the minute samples, photoemission data have not been available until recently. Using a micro-focused synchrotron beam, ARPES can be performed and serves to determine the bands that supply the nesting conditions for the CDW wave vectors [29].

The CDW transitions of NbSe$_3$ at $T_1 = 145\,\text{K}$ and $T_2 = 59\,\text{K}$ both diminish the area of the Fermi surface, as known from resistivity measurements [33]. Merely the T_1 CDW is directed along the main axis, and it nicely illustrates the deviation of the system from mean-field behavior. Using an estimate for the low-temperature energy gap obtained from STM of $2\Delta = 170\,\text{meV}$ [34,35], one calculates a mean-field transition temperature of $T_{MF} \sim 580\,\text{K}$. This has to be contrasted with the fact that for an ideal 1D system thermal fluctuations suppress a charge ordered state, and no phase transition should occur above $T = 0$. For real-world systems of quasi-1D character, the observed transition temperature is thus considerably lower than T_{MF}, typically by a factor of two or more [36]. For NbSe$_3$ the large discrepancy between T_{MF} and T_1 by a factor of four is a strong indication how well the states at the Fermi level are confined to 1D.

Photoemission data provide direct information on the effect of fluctuations on the band structure. The Fermi surfaces of NbSe$_3$ consists of five sheets of Fermi surfaces which all have a weak curvature in the two directions perpendicular to the 1D axis. The CDW nesting vectors are well known from diffraction measurements, both are incommensurate and notably the low-temperature one is diagonal [28,35]. Interestingly, no nesting condition is found within a single Fermi surface sheet. Instead, the nesting for $\boldsymbol{q}_1$ and $\boldsymbol{q}_2$ occurs between pairs of Fermi surface sheets, as outlined for the T_1 CDW in Fig. 5.5(a). Both CDW distortions introduce new zone boundaries, which are marked in Fig. 5.5 for the example of $q_1 = 0.44\,\text{Å}^{-1}$. In ARPES data

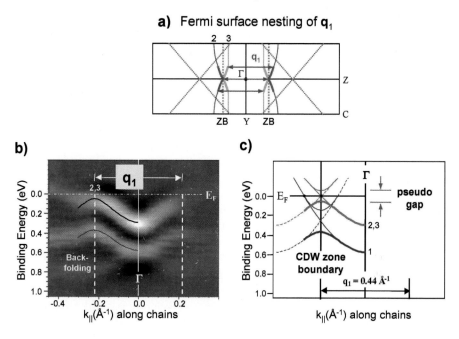

Fig. 5.5. Nesting of the q_1 CDW in NbSe$_3$. (a) The two pairs of Fermi surface sheets (plotted along chain direction and perpendicular momentum) nested with q_1 as obtained from band theory; (b) ARPES band map measured at room temperature. The nesting is identified, and a gapping and backfolding is found. While NbSe$_3$ is nominally metallic at 300 K, these observations result from precursor fluctuations of the CDW state; (c) band diagram, sketching the formation of energy gaps by backfolding. After [29]

along the chain direction taken at $T = 300$ K (above both phase transition temperatures) as in Fig. 5.5(b), two bands approach the Fermi level. The upper band is in fact formed by two degenerate states. Most importantly, it exhibits a close approach to the Fermi level at $k_F = 0.22$ Å^{-1}. This provides the nesting condition for q_1, consistent with neutron scattering and density functional calculations. The study in [29] thereby serves as a spectroscopic determination of the nesting condition of this CDW in the electron bands in NbSe$_3$.

Moreover, although close to E_F, the band does not exhibit a metallic crossing. Instead, a loss of spectral weight is observed. Both upper and lower band turn over at the same nested k-value and disperse downward beyond k_F. The behavior is illustrated in the schematic of Fig. 5.5(c). The q_1 zone boundary implies a band backfolding, which for the nested bands with $k_F = q_1/2$ implies a CDW pseudogap, while the lower band exhibits a hybridization gap. For increasing $k > k_F$, the intensity in the ARPES data gradually fades out. This is a characteristic of backfolded bands, as the superstructure potential is usually weak compared to the potential of the unperturbed lattice [32], for which reason these bands typically appear as shadow bands. A related backfolding behavior with diminishing intensity is found for the SDW in chromium [5]. The backfolding observation is somewhat exceptional because the data is recorded at room temperature and therefore far above the critical temperature, with $T = 2.1T_1$. This observation would not be possible without precursor fluctuations of the CDW state. These occur because the quasi-1D electronic structure is susceptible to thermal fluctuations above the critical temperature which lead to short-range order. Diffraction measurements on NbSe$_3$ have found diffuse intensity corresponding to the CDW vectors $\boldsymbol{q}_1$ and $\boldsymbol{q}_2$ at least up to room temperature [37, 38].

The fluctuating CDW is characterized by a coherence length as displayed in Fig. 5.6. As already mentioned, the actual transition temperature is strongly suppressed compared to the mean-field transition temperature T_{MF}. In addition, the ground state does not show an abrupt condensation, but has a finite coherence length above the critical temperature. The temperature-dependent coherence length ξ can be estimated from fluctuations of the electron energy on the scale of the thermal energy. The 1D coherence length ξ for $\boldsymbol{q}_1$ can be estimated as $\hbar v_F / \pi k_B T$ [9] using the Fermi velocity v_F and amounts to $\xi \sim 30$ Å at 300 K. Neutron scattering determines roughly similar values of 15–45 Å [37, 38]. In view of the electron states in reciprocal space, this implies Bragg scattering from the CDW fragments with a linewidth of the order of the inverse coherence length ~ 0.03 Å^{-1}. The electron bands become broadened, however, this is still much less than the nesting vector $q_1 = 0.44$ Å^{-1}. Under these conditions, the symmetry-broken band structure continues to exist at room temperature.

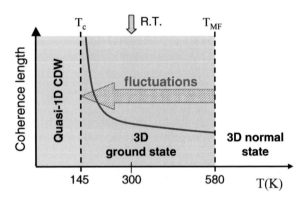

Fig. 5.6. Role of fluctuations in a dimensionally reduced system. A mean-field transition temperature can be stated for a 3D system. In a quasi-1D system, the ground state condenses at a much lower temperature, because it is hindered by fluctuations. However, even above T_c a finite coherence length remains (labeled temperatures correspond to NbSe$_3$)

Evidence for an incipient energy gap above the critical temperature, referred to as a pseudogap, has been independently obtained in tunneling spectroscopy of the T_1 CDW energy gap [34]. There a pseudogap of ∼100 meV was found to still exist at 250 K. In photoemission, a pseudogap has been observed near the CDW transition in inorganic and organic 1D compounds [25,39]. The formation of a pseudogap in the density of states can also be modeled theoretically based on fluctuations of the order parameter [18,40,41]. However, there are difficulties in determining the pseudogap. Experimental read-out of the gap magnitude in the fluctuation regime is not straightforward because a sharp edge is absent [9]. Moreover, in a 1D system a pseudogap may coexist with Tomonaga–Luttinger liquid behavior [3] which also removes the Fermi cutoff. Nonetheless, the ARPES data on NbSe$_3$ demonstrate that far above the critical temperature the symmetry of the electronic band structure is still broken, pointing towards a dominant role of CDW fluctuations.

5.3.4 Partial Gapping and Coupling to Bosonic Excitations

Quasi-1D materials will always have a small degree of coupling to the remaining dimensions, and therefore nesting conditions will not be perfect throughout the Fermi surface. Consequently, parts of the electron system can stay permanently metallic irrespective of CDW transitions. This is also the case for NbSe$_3$ where a finite metallic conductivity remains even at very low temperatures [33,42]. Such existence of residual metallic bands in the low temperature phase provides an excellent opportunity to study mass renormalization effects near the Fermi level.

Quasi-particle renormalization within FL theory arises from the interaction of conduction electrons with other elementary excitations such as phonons. At low binding energies, the electrons are dressed by these excitations, and their effective mass is increased. A rather well-known example is electron–phonon coupling [43]. One may visualize the ion lattice as distorted in the vicinity of the electron under consideration, leading to a "phonon cloud" surrounding the electron. As a result of the electrostatic distortion field, the effective mass of the electron is increased. For the electronic band structure this implies that the bare electron band is modified below the Fermi level in an energy window corresponding to the phonon spectrum. Here the slope of the band is reduced, reflecting the increased mass. Quasi-particle formation due to interaction with phonons has been observed on clean metal surfaces [44,45], as well as on adsorbate-covered surfaces with specific vibrational modes [46]. Recently, quasi-particles due to interactions with magnetic excitations have also been identified by ARPES [47]. In low-dimensional CDW systems electron–phonon interaction is expected to be strong, and hence one would expect to see such renormalization effects on ungapped parts of the Fermi surface, if they exist at all.

Recently, in a dedicated study [48], the Fermi surface of NbSe$_3$ was scanned at $T = 15$ K for the spectral shape of the low-temperature energy gaps as well

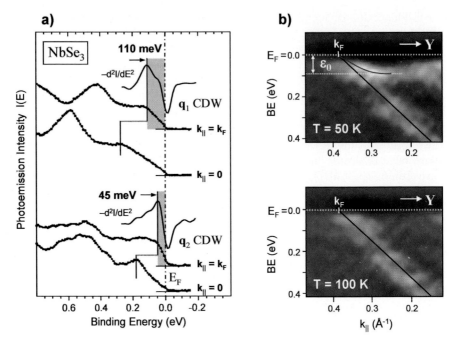

Fig. 5.7. Low-temperature spectra of the permanently metallic CDW compound NbSe$_3$. (a) CDW gap spectra measured at $T = 15$ K. For both CDWs, the spectra at k-space locations nested with q_1 and q_2 show a gap that is smeared out. Second derivatives can be used to estimate values for the energy gaps; (b) band maps recorded at the zone boundary where nesting does not occur. The nominally metallic band shows a renormalized branch that saturates at ~ 90 meV. It is seen only below $T_2 = 59$ K, suggestive of a relation to the q_2 CDW. After [48]

as renormalization effects. Energy-gap spectral functions have been recorded at the nested wave vectors k_F where backfolding occurs. As it turns out, the low-temperature spectra do not display well-defined energy gaps, but exhibit pseudogap behavior instead with the spectral weight continuously decreasing towards E_F, as displayed in Fig. 5.7(a). In the absence of other plausible influences one must conclude that the pseudogaps are *intrinsic* properties of the single-particle spectra in the CDW phase. A mechanism that generates intensity in the gap region is the occurrence of quantum fluctuations in the low-temperature ordered phase [49, 50]. Also, the imperfectly nested Fermi surface may cause additional interactions that contribute to the gap spectral function, such as to scatter intensity from the metallic regions into the gap region.

An effective gap (local in k-space) may still be estimated. Values of ~ 110 meV and ~ 45 meV as indicated in Fig. 5.7(a) are derived for the q_1 and q_2 CDW half-gap, respectively. Interestingly, their numerical ratio of ~ 0.4 is still equivalent to the temperature ratio T_2/T_1, as would strictly be expected only for a mean-field situation. The actual gap may be smaller than stated, because spectral weight can be shifted to higher binding energy by phonon satellites [51]. Other gap determinations were obtained from tunneling spectroscopy [34,35,52] with similar spectral shapes and slightly smaller gap values. This is conceivable because k-space averaged tunneling will lead to an underestimation of the maximum gap. Optical determinations of the energy gaps which also average over k are roughly consistent with the magnitude of the energy gaps obtained from ARPES [53].

Concerning possible quasi-particle formation, those parts of the Fermi surface that are not nested with a CDW indeed exhibit anomalies in the band dispersion, reminiscent of coupling to a bosonic excitation [48]. A nominally metallic character is expected for the non-nested bands, such as near the zone boundary. Yet one finds that here the intensity at the Fermi level is rather low, and a Fermi–Dirac edge is not observed. The bands dispersing towards E_F in Fig. 5.7(b) exhibit an additional low energy shoulder of much reduced dispersion. Its binding energy saturates at $\sim$90 meV. In temperature-dependent experiments one observes that the effect disappears above $T_2 = 59$ K when the system is still in the q_1 CDW state. The broadening of the observed peaks [48] is not limited by the experimental resolution, instead seems to be an intrinsic property reminiscent of the polaronic broadening concluded for other systems [51,54]. This effect may also be responsible for the lack of a Fermi edge.

The renormalized dispersion branch in Fig. 5.7(b) carries a strong analogy to self-energy effects like electron–phonon coupling. However, the energy scale observed here is much larger than that of the phonon modes in NbSe$_3$ with no more than 30 meV [55]. Therefore this effect must be attributed to coupling to a different bosonic excitation. The energy of $\sim$90 meV in fact coincides with the full excitation gap of the q_2 CDW. In addition, the effect is observed in a band partially nested with q_2, and the temperature dependence supports a connection to the q_2 CDW phase. This raises the question about the mechanism for dressing of the electrons. The energy of collective excitations of the CDW condensate is too small to be relevant [9], and excitonic states are not supported conclusively by the optical data [53]. Alternatively, self-energy effects might be caused by the reduced phase space available for electron scattering in the state with a gapped Fermi surface. Excitations of the CDW state that couple to phonons have been considered recently for a particular Fermi surface topology encountered in quasi-2D transition metal dichalcogenides [56]. They provide a good description of similar renormalization effects in TaSe$_2$ also exceeding phonon energies [57]. The ARPES data on NbSe$_3$ likewise point at a quasi-particle formation related to the CDW-gapped Fermi surface.

5.4 Electronic Correlation Effects in 1D

As discussed in Sect. 5.2.2 the decoupling of spin and charge degrees of freedom in 1D leads to the TLL phenomenology of the low-energy single-particle spectrum, signalled, e.g., by a power law onset at the chemical potential μ. Pioneering photoemission studies of 1D metals indeed observed a suppression of spectral low-energy weight [25, 58, 59], and it has since been confirmed for practically all quasi-1D metals, including carbon nanotubes [60]. However, such behavior can not easily be distinguished from the pseudogap behavior caused by CDW fluctuations (see Sect. 5.3.3). Hence, the search for anomalous non-FL behavior must either be directed at energy scales beyond that of the Peierls dynamics (which is of the order of the phonon energy scale), or to systems which do not display a Peierls instability. In the following we will first discuss the latter strategy. The possibility of signatures of spin-charge separation on higher energy scales in Hubbard model-like materials will be explored in the subsequent section.

5.4.1 Tomonaga–Luttinger-Liquid Lineshapes

By a systematic comparison of quasi-1D and quasi-2D metals it was observed that additional to the power-law onset also the ARPES *line shapes* often display anomalous behavior distinctly different from that of well-established FL model compounds [61,62]. This opens the possibility to identify TLL behavior by a detailed line shape analysis, as the theoretical electron removal spectra of the Tomonaga–Luttinger model can be calculated exactly. Again, for such approach it is highly desirable to exclude any possible interference with CDW effects. Unfortunately, however, almost all quasi-1D metals display a transition into a CDW state at sufficiently low temperatures.

$Li_{0.9}Mo_6O_{17}$, also known as Li purple bronze, is a rare exception to this rule. It displays T-linear resistivity and temperature-independent magnetic susceptibility for temperatures down to $T_X \approx 24\,\mathrm{K}$, where a phase transition of unknown origin is signalled by a very weak anomaly in the specific heat [63]. The properties of the transition are not consistent with CDW (or SDW) gap formation, and in any case, the small value of T_X allows lineshapes to be studied up to $10\,T_X$, where any putative CDW fluctuations should be absent. The quasi-1D nature of the electronic structure has been confirmed by measuring the band anisotropy and the Fermi surface topology [26]. Furthermore, it has been verified that the ARPES spectra are not affected by surface effects [64] or by the crystal-growth method [65], a crucial requirement for a meaningful lineshape analysis.

Figure 5.8 shows ARPES spectra measured along the easy transport axis of $Li_{0.9}Mo_6O_{17}$ in comparison to the spectral function calculated for a TLL [62]. As revealed by a detailed analysis, the experimental line shape does not follow the behavior expected for a FL [26,61,62]. In fact, non-FL-like spectra seem to be a hallmark of many low-dimensional Mo bronzes and oxides [61,62,66,67].

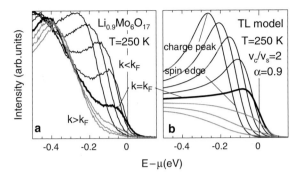

Fig. 5.8. *Left*: High-resolution ARPES data for the Li purple bronze taken along the easy transport axis. *Right*: Theoretical spectra calculated for the Tomonaga–Luttinger model. In the data, the peaks with energy $< -0.3\,\text{eV}$ arise from a non-μ-crossing band, which is excluded from the theoretical simulation. Taken from [62]

For the case of the Li purple bronze the comparison to TLL theory [68] in Fig. 5.8 indeed reveals striking similarities in shape and k-dependence. Best agreement is found for a TLL exponent $\alpha = 0.9$ and a ratio of charge and spin velocities of $v_c/v_s = 2$. Note that for these parameters ($\alpha > 1/2$) the spinon branch does not appear as a singular peak but degenerates into an edge singularity. In this case spin-charge separation cannot be directly inferred from a splitting of the non-interacting conduction band into two dispersive (charge and spin) branches (as in Fig. 5.3) but has to be identified by the line shape.

A power-law behavior in the spectral onset has independently been confirmed by scanning tunneling spectroscopy (STS) [69]. However, the exponent derived from the STS data is $\alpha = 0.62 \pm 0.17$, quite different from the value obtained by ARPES. From the temperature dependence of the tunneling conductance an even smaller exponent is extracted ($\alpha \approx 0.5$). The main difference between both experiments lies in the different temperature: While ARPES was performed in the range 250–300 K, the STS experiment was conducted between 5 and 55 K. The contrasting results for α may thus be reconciled by a possible T dependence of the exponent. However, it turns out that such a large T renormalization of α is inconsistent with the transport properties of $Li_{0.9}Mo_6O_{17}$ implying T-independent α within usual one-band TLL theory [70]. In a very recent combined ARPES and theoretical study [70] it was confirmed that a large T dependence of α does indeed occur. Moreover, this new effect could be traced back to the occurrence of interacting charge-neutral critical modes that emerge naturally from the two-band nature of the Li purple bronze. This not only establishes the role of this compound as a paradigm material for studying non-FL physics in 1D but also adds an unexpected twist to standard one-band TLL theory.

5.4.2 Spin-charge Separation in Strongly Correlated 1D Systems

As already discussed in Subsect. 5.2.2 the effects of spin-charge separation are no longer limited to low excitation energies but become observable over the entire conduction band width, if the Coulomb interaction is strong and short-ranged. In this situation the electronic structure can appropriately be described by the 1D Hubbard model, for which detailed theoretical results exist [10, 11]. In particular, the exponent α is found to be rather small ($\alpha \leq 1/8$) so that the spin and charge branches in the single-particle spectrum should be clearly observable as two separate dispersive singularities. There are two classes of quasi-1D materials which fall into this category, organic conductors and 3d transition metal oxides.

Solids composed of organic molecules are interesting candidates for strong electronic correlation effects, because their intra-molecular Coulomb energy U often by far exceeds their inter-molecular hopping integral t. Among these the so-called charge-transfer salts are particularly interesting. Here, the building blocks are molecules of planar geometry, so that covalent bonding via π-like molecular orbitals is possible only in one spatial direction, resulting in the formation of molecular stacks. Often the HOMO of the isolated molecule is fully occupied, and charge transfer to (or from) suitable acceptor (or donor) complexes is required to render the stacks metallic and thus induce a quasi-1D conductivity.

A case in point is the 1D conductor tetrathiafulvalene tetracyanoquinodimethane (TTF-TCNQ), whose structure is shown in Fig. 5.9. In this compound there are actually two types of molecules, TTF and TCNQ, which form separate and parallel stacks. Here TTF acts as donor and TCNQ as acceptor, with a charge transfer of 0.59 electrons per formula unit from TTF to TCNQ [71] leading to a slightly filled TCNQ-derived conduction band and a hole-like TTF-derived band. TTF-TCNQ is consequently a strongly anisotropic metal, with the conductivity along the stack direction (crystallographic b-axis) three orders of magnitude larger than in any other direction. At $T_P = 54$ K TTF-TCNQ undergoes a Peierls transition into a $2k_F$ CDW phase. Here we will focus on the metallic high-temperature phase which deviates from simple FL behavior in many respects [71]. For example, the magnetic susceptibility is strongly enhanced compared to simple Pauli paramagnetism. Strong electronic correlations have also been inferred from the observation of $4k_F$ fluctuations in diffuse scattering.

Optical spectroscopy [72] on the closely related Bechgaard salts has recently given some indirect evidence of possible TLL behavior. A more direct probe is ARPES. However, the high surface sensitivity of the method requires good control of quality and long-time stability of the surfaces on molecular length scales. This has proven to be problematic for the Bechgaard salts due to the polarity of their surfaces and the rapid radiation damage in the exciting light field [73]. In contrast, the surfaces of TTF-TCNQ crystals are non-polar and hence unreconstructed (though not necessarily unrelaxed, see below), and radiation damage remains slow, as long as the photon energy stays below ~30 eV [73,74]. Hence, the resulting high quality and stable long-range order of TTF-TCNQ surfaces allows the observation of intrinsic k-dispersive ARPES spectra, as first reported by Zwick et al. [39].

Figure 5.9 shows a false-color map of the ARPES signal measured along the 1D transport axis, covering the central region of the Brillouin zone where according to band theory the TCNQ-derived band is occupied [74, 75]. The theoretical TCNQ band is actually doubled, because the unit cell contains two TCNQ stacks whose degeneracy is slightly lifted by a small interstack coupling. Clearly, the comparison of ARPES and band theory reveals a number of qualitative and quantitative discrepancies. Instead of the TCNQ band doublet the experiment yields *three* dispersive features, labeled **a**, **b**, and **d** in the figure (feature **c** originates in the TTF band and is discussed further below). Neither the shape nor the width of the experimental dispersions agree with band theory. Concerning the band width one could speculate that due to surface relaxation the molecules in the topmost layers are differently tilted than those in the bulk (cf. Fig. 5.9), leading to a larger hopping integral t between neighboring molecules within a stack and hence an enhanced band width.

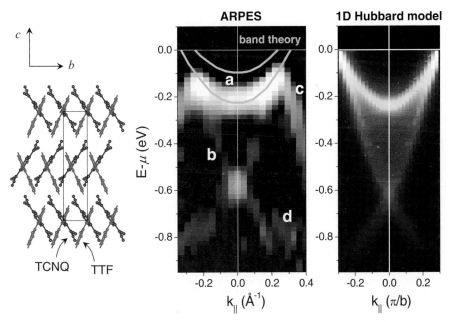

Fig. 5.9. *Left*: Crystal structure of TTF-TCNQ. The molecular stacks are oriented along the crystallographic b axis. *Center*: ARPES data measured along the b axis, represented as a false-color map of the second energy derivative, $-d^2I/d\varepsilon^2$, for enhanced contrast. Also shown is the TCNQ-derived conduction band doublet obtained by band theory. *Right*: Electron removal spectrum of the 1D Hubbard model at $T=0$ calculated by DDMRG [17]. For model parameters see text

However, surface effects can not explain the curious X-shaped dispersion. Rather, we find that the experimental dispersion is remarkably similar to that of the 1D Hubbard model discussed in Subsect. 5.2.2. In fact, this model even yields a quantitative explanation of the data, as was first shown by a Bethe ansatz calculation of the spectral dispersions [15, 74, 76]. The right panel of Fig. 5.9 shows the electron removal spectrum of the 1D Hubbard model calculated within the dynamical density-matrix renormalization group (DDMRG) [17], assuming a band filling of $n = 0.6$, close to the actual filling 0.59 of the TCNQ band. Best agreement with experiment is found for $U/t = 4.9$ and $U = 1.96\,\text{eV}$. The ratio U/t corresponds to intermediate coupling strength, and the parameter U is very close to the value $2.4\,\text{eV}$ for the screened Coulomb interaction on a TCNQ molecule obtained by *ab initio* calculations [77]. In contrast, the hopping integral t resulting from the quantitative comparison of ARPES and the Hubbard model is about twice as large as that expected from band theory. This may be a consequence of a possible molecular surface relaxation as discussed above. Alternatively, very recent theoretical studies [77, 78] have suggested that the seemingly too large value of t is an artefact of neglecting long-range Coulomb interaction. They found that the inclusion of the nearest-neighbor Coulomb interaction V in an extended Hubbard model results for a given t in a strongly enhanced width of the electron removal spectrum without the need to invoke surface effects. First attempts to calculate the relevant V-value for TTF-TCNQ indeed indicated that it may be non-negligible relative to the on-site Coulomb repulsion. On the other hand, the characteristic X-shaped dispersion of the spinon and holon singularities remains conserved in the extended 1D Hubbard model. Therefore, its observation in the ARPES spectra of TTF-TCNQ is a direct experimental manifestation of spin-charge separation in this 1D conductor.

With these results obtained for the TCNQ-related conduction band one would expect to see similar 1D correlation effects also in the TTF-derived band. Surprisingly, the ARPES data on this band (shown in Fig. 5.10) display only a single dispersive feature, with the same qualitative shape as in band theory except again for a renormalized band width [74, 75]. This behavior was initially thought to reflect weak coupling ($U \ll t$) in the TTF band [79], which however would be in conflict with other experimental evidence for strong electronic correlations in the TTF chains, e.g., the observation of $4k_\text{F}$ CDW fluctuations [71]. These conflicting observations have finally been resolved by the 1D Hubbard model [80]. The right panel of Fig. 5.10 shows DDMRG results for similar strong coupling parameters ($U/t \sim 5$) as for the TCNQ case. The essential difference to the TCNQ band is the complimentary higher band filling of $n = 1.41$, implying that the spectral weight near the chemical potential corresponds to the *upper* Hubbard band (the lower Hubbard band is offset by approximately $U \sim 2\,\text{eV}$ to higher binding energies where in the real material it becomes obscured by lower lying bands). Although a Bethe ansatz analysis of the Hubbard model spectra confirms the existence of separate spinon, holon, and secondary branches, the DDMRG finds the spectral weight

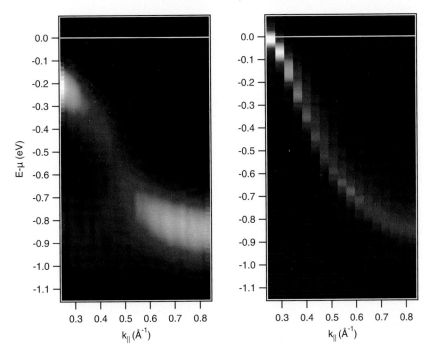

Fig. 5.10. *Left:* ARPES data on the TTF-derived conduction band of TTF-TCNQ ($-d^2I/d\varepsilon^2$). *Right:* Corresponding Hubbard model spectrum at $T=0$ calculated by DDMRG [80]

concentrated almost exclusively at the lower dispersive edge of the excitation continuum, giving rise to the single-band-like appearance of the spectrum.

While thus the ARPES spectra of TTF-TCNQ show signatures of spin-charge separation at high excitation energies of the order of the conduction band width, the experimental data on the low-energy behavior are not consistent with simple TLL theory. For the Hubbard model parameters used in the above DDMRG calculation the spectral onset is expected to follow a power-law with $\alpha = 0.038$ [17,81] which would be undistinguishable from a conventional Fermi edge. In contrast, the experimental spectra taken at the Fermi vector k_F (shown in Fig. 5.11) display an almost linear onset ($\alpha \sim 1$), i.e. a pseudo-gap behavior which prevents the experimental "bands" in Figs. 5.9 and 5.10 to reach the chemical potential. This behavior could be taken as a hint that the Coulomb interaction is not purely local, as the exponent α becomes indeed larger in the extended t-U-V Hubbard model [82]. Other explanations for a large exponent involve the effect of defects, localizing the electrons on 1D strands of finite length leading to the concept of a *bounded* TLL [83], or the possible role of additional electron–phonon coupling [74].

Figure 5.11 also demonstrates the unusual temperature dependence of the ARPES spectra. With increasing temperature from 60 K to nearly room

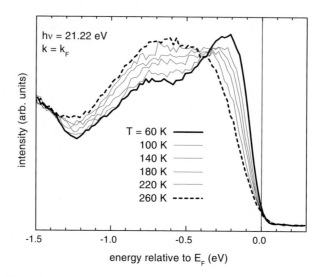

Fig. 5.11. ARPES spectra of TTF-TCNQ measured at the Fermi vector k_F as function of temperature. Note the linear suppression of intensity towards the chemical potential and the substantial spectral weight transfer as function of temperature. From [74]

temperature the data reveal a dramatic transfer of spectral weight from low to high binding energies over a scale of approx. 1 eV. The effect is fully reversible and conserves the integrated weight within experimental accuracy. An intensity transfer over an energy so much larger than the thermal energy is difficult to account for within a conventional electron–phonon coupling scenario or by Peierls fluctuations. Rather, it may be another signature of 1D correlation physics. Calculations on the temperature dependence of the 1D Hubbard model confirm that transfer of spectral weight takes place over a wide energy range above a characteristic temperature of the order of the magnetic exchange integral J but not below [78,84]. However, the Hubbard model parameters used for TTF-TCNQ yield $J = 110$ meV [74], way above the temperature range of Fig. 5.11. Thus, an only local 1D Hubbard model is not sufficient for explaining the unusual T dependence. It has been suggested that the long-range part of the Coulomb repulsion can also resolve this problem [78], as the effective magnetic energy scale (and hence the characteristic temperature for T-dependent spectral weight transfer) is found to decrease in the 1D extended Hubbard model. These calculations give theoretical support to the notion that the anomalous T dependence in the photoemission spectrum of TTF-TCNQ is indeed due to strong 1D correlation effects.

Spin-charge separation is not restricted to 1D metals but may also be observed in the spectral function of 1D Mott insulators. Interesting candidates are the transition metal oxides Sr_2CuO_3 and $SrCuO_2$ whose crystal structures consist of single and double Cu-O chains, respectively. Both compounds have

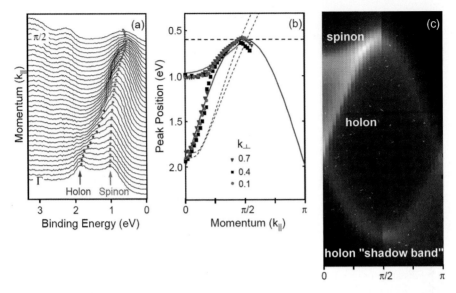

Fig. 5.12. (a) ARPES spectra measured along the 1D axis of SrCuO$_2$ (from [85]); (b) dispersion of the peak maxima (*symbols*). Also shown are the band dispersion resulting from band theory (*dashed curves*), where SrCuO$_2$ is a metal, and the spinon and holon dispersions expected for the 1D t–J model (*solid red and blue curves*, respectively); (c) single-particle spectrum of the Hubbard model at half-filling and for $U/t = 14$ [86]

been widely studied by ARPES [87–92], with spectra that could be interpreted in terms of a spinon-holon excitation continuum. However, direct observation of separate spinon and holon edge singularities has only recently been reported for SrCuO$_2$ [85]. The energy distribution curves measured along the Cu-O chain direction and shown in Fig. 5.12(a) display two separate peaks near the Γ-point of the Brillouin zone which merge at $k = \pi/2$, exactly as expected for the 1D Hubbard model at half-filling (Fig. 5.12(c)) [86]. However, the lower part of the holon branch (denominated as holon "shadow band" in Fig. 5.12(c) and which caused the X-shape of the experimental and Hubbard model spectra in Fig. 5.9) is not seen here. Whether it is obscured by other bands or if its apparent absence points to a shortcoming of the 1D Hubbard model, has not yet been clarified.

5.4.3 Photoemission on a 1D Spin-Peierls System: TiOCl

Another interesting material falling in the class of strongly correlated low-dimensional systems is the spin 1/2 quantum magnet TiOCl. Early in the nineties it was discussed as a possible candidate for a material where doping a resonating valence bond state might result in exotic superconductivity as had been proposed at this time for the cuprate high-T_c superconductors [93].

The current interest in TiOCl stems from the strong magneto-elastic coupling combined with magnetic frustration both of which seemingly dominate the unusual phase diagram and are eventually reconciled in a commensurate spin-Peierls type ground state.

TiOCl crystallizes in an orthorhombic quasi-two-dimensional structure of the FeOCl type, where buckled bilayers of Ti-O are separated by Cl ions [94]. The bilayers, which are stacked along the crystallographic c-axis, only weakly interact through van der Waals forces (cf. Fig. 5.13(a)). Magnetically, spin 1/2 Heisenberg chains form along the crystallographic b axis, mediated by direct exchange ($J = 660$ K) of the Ti $3d_{xy}$ orbitals as evidenced by the observation of a Bonner-Fisher-type magnetic susceptibility at high temperatures and inferred from LDA $+U$ calculations of the electronic structure, respectively [95]. Nonetheless, if one projects the Ti sites of one bilayer onto the (a,b) plane one recognizes the triangular arrangement of Ti ions between neighboring chains (cf. Fig. 5.13(b)) with the inherent possibility of geometrical frustration of the magnetic interactions *across* the chains.

Electronically, TiOCl is a Mott insulator. Its electronic properties are determined by the octahedral coordination of the Ti ion in a $3d^1$ configuration. The strongly distorted octahedra are formed by four O and two Cl ions (s. Fig. 5.13(a)), and share corners along the a-axis and edges along the b-axis. Thus, the low-lying local charge excitations occur within the Ti $3d$ t_{2g} triplet while the on-site Coulomb repulsion causes a charge gap of about 2 eV for electrons hopping from site to site. This basic picture is essentially confirmed by LDA $+U$ calculations [95,96], which identify the d_{xy} derived band as slightly split off from the bands with d_{xz} and d_{yz} character [97].

As already mentioned, the ground state of TiOCl has been identified as a spin-Peierls phase: the magnetic susceptibility shows a sudden drop at a temperature $T_{c1} = 67$ K [95] concomitant with a dimerization of the Ti atoms

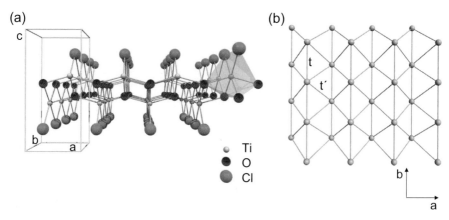

Fig. 5.13. Central projection view of (**a**) the crystal structure of TiOCl along the b- and (**b**) its bare Ti-network along the c-axis.

along the crystallographic b direction [98, 99]. The unusual scenario of this spin-Peierls transition is signalled by both the hysteresis in the susceptibility [86], which identifies the transition as being of first instead of second order, and the existence of another transition of second order at $T_{c2} = 91$ K, which shows up in the susceptibility as a kink anomaly. This non-canonical behavior was correlated with the results from magnetic resonance [100, 101], Raman [102] and infrared spectroscopy [103], and specific-heat measurements [104] which point to the importance of strong spin- and/or phonon-induced orbital fluctuations in the high-temperature phase up to 130 K. In partial contradiction, there was evidence from cluster calculations in connection with polarization-dependent optical data [105] and more recently from electron–spin resonance [106] that the orbital degrees of freedom are actually quenched. Based on Ginzburg-Landau arguments the transition at T_{c_2} instead was ascribed to the onset of incommensurate order which arises due to the structural frustration of the magnetic interchain interactions and eventually commensurately locks in below T_{c_1} [105]. This in a way brings back through the back door the idea of unconventional superconductivity induced by bond dimer fluctuations if one is able to drive the system metallic. We note that indeed it has been shown recently by means of optical spectroscopy that an insulator-to-metal transition can be induced applying pressures around 12 GPa [107].

Thus, under the perspective of doping charge carriers into the system and, more general, in order to achieve a complete understanding of the relevant competing interactions a thorough investigation of the electronic properties in the normal state is imperative. In the following we will focus on two aspects, on polarization-dependent measurements revealing the orbital character and symmetry of the Ti $3d$ states [86] and on their electronic dispersion.

The unusual spin-Peierls scenario involving two phase transitions in TiOCl was linked to the effects of strong phonon-induced orbital fluctuations [96]. Motivated by the observation of an anomalous broadening of certain Raman active phonon modes [102] corresponding frozen-phonon calculations for several slightly distorted structures were performed which show that the energetic order of the t_{2g} orbitals indeed can change although significantly probably only at temperatures much higher than 300 K [96, 108]. However, it was argued that aspects of a dynamical Jahn-Teller effect involving orbital fluctuations may still play an important role for the non-canonical spin-lattice coupling in TiOCl. As we will show, polarization-dependent photoemission spectroscopy is capable of answering this question within experimental limits.

The idea is to exploit selection rules realized for special experimental geometries. If one takes the direction of the incident linearly polarized light and the emission direction of the photoelectrons to lie within the same crystal mirror plane, then with the polarization vector within and perpendicular to the mirror plane the ejected electrons can stem only from states with even or odd parity with respect to this plane, respectively [109]. Using the experimental geometry as sketched in Fig. 5.14, the d_{xy}-derived band states are even with respect to the (b,c) mirror plane while the $d_{xz,yz}$ states are odd. Thus

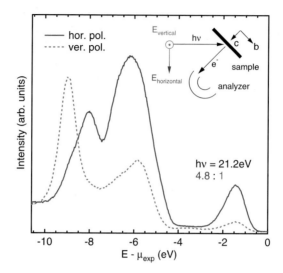

Fig. 5.14. Photoemission spectra measured at the Γ-point with horizontal and vertical light polarization ($T = 300$ K). The experimental setup is sketched in the upper right corner

with the (b,c)-plane lying horizontally, photoemission from the d_{xy} states is dipole-allowed only for horizontal light polarization, whereas $d_{xz,yz}$-emission is observed for vertical polarization only.

Looking at the corresponding polarization-dependent spectra at the Γ point one indeed notices strong effects in the entire valence band. Focusing on the Ti 3d-derived spectral weight above about -3 eV a quantitative analysis gives a 4.8:1 ratio for the spectral weight measured with horizontal and vertical polarization, respectively. Patently, there is no sizable admixture of d_{xz} and d_{yz} derived states to the ground state at room temperature. The fact that not all the spectral weight is suppressed for vertical polarization can be explained by a possible small sample misalignment, mirror-symmetry breaking phonons, and the finite polarization degree of the laboratory He lamp ($\approx 85\%$). Thus, in agreement with earlier reports we arrive at the conclusion that phonon-induced orbital fluctuations are not effective in TiOCl at least up to room temperature.

In Fig. 5.15(a) we show ARPES spectra in the near-E_F region taken along the crystallographic b axis. At the Γ-point an intense single peak starts moving upwards to the experimental chemical potential [110]. Its dispersion maximally approaches μ_{exp} at about half-way of ΓY. Further towards the Brillouin zone edge the peak shifts back again away from the chemical potential while rapidly losing intensity. The dispersion is clearly asymmetric reaching further down to higher binding energies at the Y-point. Concomitantly with the bend-over of the dispersion around $\frac{1}{2}$ΓY a new feature appears at ≈ -2.5 eV which gives rise to an overall broad spectral shape at the Y-point. These distinct spectral

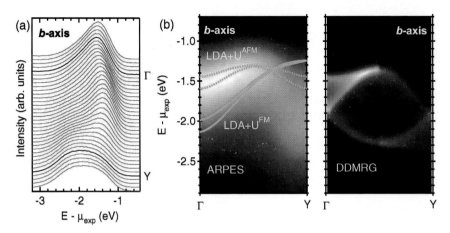

Fig. 5.15. (a) ARPES spectra of TiOCl along the crystallographic b axis; (b) *left:* Intensity plot $I(\mathbf{k},\varepsilon)$ of the same data as in (a) with overlaid LDA+$U^{\text{FM/AFM}}$ bands for the high-temperature structure ($U = 3.3\,\text{eV}$, $J_0 = 1\,\text{eV}$). *Right:* Single-particle spectral function of the Hubbard model calculated by the DDMRG method ($U = 3.3\,\text{eV}$, $t = 0.23\,\text{eV}$)

changes with momentum along the b axis on the one hand and their absence along a (not shown) on the other clearly indicate the one-dimensional nature of the electronic structure in TiOCl.

The same data is plotted in the left panel of Fig. 5.15(b) as an intensity plot $I(\mathbf{k},\varepsilon)$, together with LDA+$U$ bands assuming a ferromagnetic (solid line) and an antiferromagnetic (dotted line) spin alignment, respectively. Comparing the experimental dispersions with the theoretical LDA+U results based on a ferromagnetic spin configuration the discrepancy is obvious. Neither the energy position of the band starting out at Γ is reproduced correctly nor is the overall energy-vs.-momentum relation along the ΓY direction. Turning over to the results assuming antiferromagnetic spins along the b axis the agreement with experiment is satisfactory for the first half of ΓY. Indeed, in a spin-Peierls system an antiferromagnetic spin arrangement seems more appropriate. However, these calculations clearly fail to reproduce the asymmetry of the dispersion in the second half from Γ towards the zone boundary. Led by considerations that the ARPES spectra could be dominated by strong spin-Peierls fluctuations we also compared to LDA+U^{AFM} calculations for the low-temperature dimerized phase (not shown). However, except for band doubling the effects of dimerization in the calculations are rather small and cannot explain the ARPES data.

To better account for electron correlation effects beyond the scope of LDA+U calculations we compare in the right panel of Fig. 5.15(b) to the momentum-resolved spectral weight distribution of the one-dimensional single band Hubbard model. The calculations [17, 86] were performed for half-filling using DDMRG with the same values for U and t as in the LDA+U

calculations. Here the entire spectral weight is incoherent but not at all structureless, and reflects the probability of how to distribute the energy and momentum of a hole created in the photoemission process to the separate collective spinon and excitations (cf. Subsect. 5.2.2). Actually, the main qualitative features of the experimental data are well reproduced in these calculations: The initial upward dispersion as well as the intensity distribution due to the spinon and holon branches, the dispersion maximum at $\frac{1}{2}\Gamma Y$, the asymmetry towards the Y-point, and the overall correct energetics. On the other hand, the experimental data does not show several important details [10, 17] of the calculations such as the pronounced spin-charge splitting at Γ or the so-called "shadow band" at higher binding energies. In conclusion, the overall fair and – compared to the LDA $+\,U$ calculations –much better agreement of these model calculations with experiment manifests the importance of electronic correlations beyond the LDA$+\,U$ approach in TiOCl. We cannot expect a full description of the data since the lattice and orbital degrees of freedom identified as being crucial in this material are completely ignored. One would expect that an extension of the Hubbard model as to account for multiband effects, next nearest neighbor interactions, and phonons will describe the involved correlation dominated physics in this compound more satisfactorily.

5.5 Conclusions and Open Questions

The preceding sections have highlighted the key issues associated with quasi-1D materials, as are the inherent instability of such compounds against the Peierls state, as well as the potential breakdown of the Fermi-liquid picture. To date, with the examples prominently underlined by ARPES data, these major phenomena can be considered as reasonably established. However, this does not mean that important open questions could deceive unnoticed.

Concerning the CDW phase of 1D compounds we are still far from a detailed understanding of the spectral line shape near the chemical potential. For example, the value for the Peierls gap inferred from ARPES data often largely exceeds that obtained by other methods. It has been suggested that this seeming discrepancy can be resolved by assuming strong polaronic contributions to the spectra. However, a coherent microscopic calculation of the single-particle spectrum of a 1D metal with strong polaronic coupling is still lacking. The theoretical understanding is much better developed concerning the effect of thermal and quantum CDW fluctuations on the spectra (albeit in the absence of phonons), because here detailed predictions already exist. As discussed in this chapter, ARPES qualitatively confirms the fluctuation-induced pseudogap behavior in a wide temperature window above the actual Peierls transition, but a thorough study of the pseudogap shape and its temperature dependence has yet to be done. The most surprising ARPES result, however, is the curious temperature-dependence observed in $NbSe_3$ which strongly suggests the existence of anomalous renormalization effects beyond simple electron–phonon coupling. This urges further theoretical treatment, and very likely implies more complex scattering scenarios where e.g. the k-dependence of the coupling parameter or the scattering phase space have to be taken into account.

With respect to Tomonaga–Luttinger physics and the description of electron correlations in 1D, detailed comparison between calculations for correlated systems such as the LDA + U, DDMRG and others on one hand, and experimental ARPES data on the other hand still unveil considerable discrepancies. Therefore, the various low-dimensional compounds can be used as test case for such models. Moreover, while such theories aim at describing correlation effects properly, there is still a gap to bridge to the accurate description of the actual spectral line shape. As an example, the high power law exponent ($\alpha \sim 1$) in 1D metals is still lacking a satisfactory explanation. Moreover, an accurate treatment of correlated systems in low dimensions must not limit itself to electronic interactions, but has to also include phonons. An open issue here is whether strong electron–phonon interaction may modify or even inhibit the formation of a TLL. This intimately relates to the question which role finite temperature will play.

Finally, thus far both experimentalists and theorists have attempted to treat CDW physics and the occurrence of a TLL rather separately. This artificial distinction, however, does not keep up to the real world situation, where almost all 1D systems at low temperature are at least close to a CDW instability. Therefore, the ultimate description would tackle the existence of a TLL in the presence of CDW precursor fluctuations. Fortunately, a number of model compounds has been identified in the last years that are promising candidates to yield more data, especially employing high-resolution ARPES. In parallel, the vast growth of computing power has enabled advanced numerical treatments of unprecedented predictive power. Therefore, one can be cautiously optimistic that the uncertainties especially associated with the oftentimes elusive Tomonaga–Luttinger liquid will soon be lifted.

References

1. R. E. Peierls: *Quantum Theory of Solids*, Clarendon, Oxford, 1964.
2. S. Tomonaga: Prog. Theor. Phys. **5**, 544 (1950)
3. J. M. Luttinger: J. Math. Phys. **4**, 1154 (1963)
4. M. Grioni and J. Voit: in *Electron Spectroscopies Applied to Low-Dimensional Materials*, edited by H. Starnberg and H. Hughes, Kluwer, Dordrecht, 2000.
5. J. Schäfer et al: Phys. Rev. Lett. **83**, 2069 (1999)
6. A. W. Overhauser: Phys. Rev. **128**, 1437 (1962)
7. S.-K. Chan and V. Heine: J. Phys. F.: Metal. Phys. **3**, 795 (1973)
8. W. Kohn: Phys. Rev. Lett. **2**, 393 (1959)
9. G. Grüner: *Density Waves in Solids*, Addison-Wesley Publishing, Reading, 1994
10. J. Voit: Rep. Prog. Phys. **58**, 977 (1995)

11. F. Gebhard: *The Mott Metal-Insulator Transition*, Springer, Berlin, 1997
12. F. D. M. Haldane: J. Phys. C **14**, 2585 (1981)
13. V. Meden and K. Schönhammer: Phys. Rev. B **46**, 15753 (1992)
14. K. Schönhammer and V. Meden: Phys. Rev. B **47**, 16205 (1993)
15. J. M. P. Carmelo et al: Europhys. Lett. **67**, 233 (2004)
16. E. Jeckelmann: Phys. Rev. B **66**, 045114 (2002)
17. H. Benthien et al: Phys. Rev. Lett. **92**, 256401 (2004)
18. H. Monien: Phys. Rev. Lett. **87**, 126402 (2001)
19. H. P. Geserich et al: Physica B+C **143**, 174 (1986)
20. G. Travaglini et al: Solid State Commun. **37**, 599 (1981)
21. P. Bresch et al: Phys. Rev. B **12**, 219 (1975)
22. J. M. Carpinelli et al: Nature **381**, 398 (1996)
23. H. W. Yeom et al: Phys. Rev. Lett. **82**, 4898 (1999)
24. S. J. Park et al: Phys. Rev. Lett. **93**, 106402 (2004)
25. B. Dardel et al: Phys. Rev. Lett. **67**, 3144 (1991)
26. J. D. Denlinger et al: Phys. Rev. Lett. **82**, 2540 (1999)
27. M. Nakamura et al: Phys. Rev. B **49**, 16191 (1994)
28. J. A. Wilson: Phys. Rev. B **19**, 6456 (1979)
29. J. Schäfer et al: Phys. Rev. Lett. **87**, 196403 (2001)
30. H. P. Geserich et al: Physica B+C **143**, 198 (1986)
31. R. Claessen et al: Phys. Rev. B **56**, 12643 (1997)
32. J. Voit et al: Science **290**, 501 (2000)
33. N. P. Ong and P. Monceau: Phys. Rev. B **16**, 3443 (1997)
34. H. Haifeng and Z. Dianlin: Phys. Rev. Lett. **82**, 811 (1999)
35. A. Fournel et al: Phys. Rev. Lett. **57**, 2199 (1986)
36. P. Monceau et al: in *Physics and Chemistry of Low-Dimensional Inorganic Conductors*, edited by C. Schlenker et al., volume 254 of *Proc. of NATO-ASI*, 1996.
37. J. P. Pouget et al: J. Phys. **44**, 1729 (1983)
38. A. H. Moudden et al: Phys. Rev. Lett. **65**, 223 (1992)
39. F. Zwick et al: Phys. Rev. Lett. **81**, 2974 (1998)
40. A. J. Millis and H. Monien: Phys. Rev. B **61**, 12496 (2000)
41. A. Bartosch and P. Kopietz: Phys. Rev. B **62**, 16223 (2000)
42. E. Canadell et al: Inorg. Chem. **29**, 1401 (1990)
43. G. Grimvall: *The electron–phonon interaction in metals*, volume XVI of *Selected topics in solid state physics*, North-Holland, Amsterdam, 1981.
44. M. Hengsberger et al: Phys. Rev. Lett. **83**, 592 (1999)
45. T. Valla et al: Phys. Rev. Lett. **83**, 2085 (1999)
46. E. Rotenberg et al: Phys. Rev. Lett. **84**, 2925 (2000)
47. J. Schäfer et al: Phys. Rev. Lett. **92**, 097205 (2004)
48. J. Schäfer et al: Phys. Rev. Lett. **91**, 066401 (2003)
49. S. I. Matveenko and S. A. Brazovskii: Phys. Rev. B **65**, 245108 (2002)
50. A. Weiße and H. Fehske: Phys. Rev. B **58**, 13526 (1998)
51. L. Perfetti et al: Phys. Rev. Lett. **87**, 216404 (2001)
52. Z. Dai et al: Phys. Rev. B **45**, 9469 (1992)
53. A. Perucchi et al: Phys. Rev. B **69**, 195114 (2004)
54. D. S. Dessau et al: Phys. Rev. Lett. **81**, 192 (1998)
55. T. J. Wieting: *Quasi One-Dimensional Conductors*, volume 95 of *Lecture Notes in Physics*, Springer, 1979

56. A. H. C. Neto: Phys. Rev. Lett. **86**, 4382 (2001)
57. T. Valla et al: Phys. Rev. Lett. **85**, 4759 (2000)
58. B. Dardel et al: Europhys. Lett. **24**, 687 (1993)
59. R. Claessen et al: J. Electron Spectrosc. Rel. Phen. **76**, 121 (1995)
60. H. Ishii et al: Nature **426**, 540 (2003)
61. G.-H. Gweon et al: J. Electron Spectrosc. Rel. Phenom. **117-118**, 481 (2001)
62. G.-H. Gweon et al: Phys. Rev. B **68**, 195117 (2003)
63. C. Schlenker et al: Physica B **135**, 511 (1985)
64. F. Wang et al: Phys. Rev. B **74**, 113107 (2006)
65. G.-H. Gweon et al: Phys. Rev. B **70**, 153103 (2004)
66. G.-H. Gweon et al: J. Phys.: Condens. Matter **8**, 9923 (1996)
67. G.-H. Gweon et al: Phys. Rev. B **72**, 035126 (2005)
68. D. Orgad et al: Phys. Rev. Lett. **86**, 4362 (2001)
69. J. Hager et al: Phys. Rev. Lett. **95**, 186402 (2005)
70. F. Wang et al: Phys. Rev. Lett. **96**, 196403 (2006)
71. S. Kagoshima et al: *One-dimensional conductors*, Springer, Berlin, 1987, and references therein.
72. A. Schwartz et al: Phys. Rev. B **58**, 1261 (1998)
73. M. Sing et al: Phys. Rev. B **67**, 125402 (2003)
74. M. Sing et al: Phys. Rev. B **68**, 125111 (2003)
75. R. Claessen et al: Phys. Rev. Lett. **88**, 096402 (2002)
76. J. M. P. Carmelo et al: J. Phys.: Condens. Matter **18**, 5191 (2006)
77. L. Cano-Cortes et al: cond-mat/0609416
78. N. Bulut et al: Phys. Rev. B **74**, 113106 (2006)
79. R. Claessen et al: J. Phys. IV France **114**, 51 (2004)
80. H. Benthien: private communication.
81. H. J. Schulz: Phys. Rev. Lett. **64**, 2831 (1990)
82. A. K. Zhuravlev and M. I. Katsnelson: Phys. Rev. B **64**, 033102 (2001)
83. J. Voit et al: Phys. Rev. B **61**, 7930 (2000)
84. A. Abendschein and F. F. Assaad: Phys. Rev. B **73**, 165119 (2006)
85. B. J. Kim et al: Nature Physics **2**, 397 (2006)
86. M. Hoinkis et al: Phys. Rev. B **72**, 125127 (2005)
87. C. Kim et al: Phys. Rev. Lett. **77**, 4054 (1996)
88. C. Kim et al: Phys. Rev. B **56**, 15589 (1997)
89. H. Fujisawa et al: Phys. Rev. B **59**, 7538 (1999)
90. T. Valla et al: cond-mat/0403486.
91. S. Suga et al: Phys. Rev. B **70**, 155106 (2004)
92. A. Koitsch et al: Phys. Rev. B **73**, 201101 (2006)
93. R. J. Beynon and J. A. Wilson: J. Phys.: Condens. Matter **5**, 1983 (1993)
94. H. Schäfer et al: Z. Anorg. Allg. Chemie **295**, 268 (1958)
95. A. Seidel et al: Phys. Rev. B **67**, 020405(R) (2003)
96. T. Saha-Dasgupta et al: Europhys. Lett. **67**, 63 (2004)
97. For the orientation of the orbitals we use the notation of Ref. [96]
98. M. Shaz et al: Phys. Rev. B **71**, 100405(R) (2005)
99. A. Krimmel et al: Phys. Rev. B **73**, 172413 (2006)
100. T. Imai and F. C. Chou: cond-mat/0301425 v1
101. V. Kataev et al: Phys. Rev. B **68**, 140405(R) (2003)
102. P. Lemmens et al: Phys. Rev. B **70**, 134429 (2004)
103. G. Caimi et al: Phys. Rev. B **69**, 125108 (2004)

104. J. Hemberger et al: Phys. Rev. B **72**, 012420 (2005)
105. R. Rückamp et al: Phys. Rev. Lett. **95**, 097203 (2005)
106. D. V. Zakharov et al: Phys. Rev. B **73**, 094452 (2006)
107. C. A. Kuntscher et al: Phys. Rev. B **74**, 184402 (2006)
108. L. Pisani and R. Valentí: Phys. Rev. B **71**, 180409(R) (2005)
109. A. Damascelli et al: Rev. Mod. Phys. **75**, 473 (2003)
110. For a strongly insulating system the intrinsic chemical potential μ is difficult to define by experiment. Hence we rather use the notion *experimental* chemical potential μ_{exp}

6

Atomic Chains at Surfaces

J. E. Ortega[1,2] and F. J. Himpsel[3]

[1] Departamento de Física Aplicada I, Universidad del País Vasco, Plaza de Oñate 2, E-20018 San Sebastian, Spain
enrique.ortega@ehu.es
[2] DIPC and Centro Mixto CSIC/UPV, Paseo Manuel Lardizabal 4, E-20018 San Sebastian, Spain
[3] Department of Physics, University of Wisconsin Madison, 1150 University Ave., Madison, Wisconsin 53706, USA
fhimpsel@facstaff.wisc.edu

Abstract. It has become possible to assemble one-dimensional atom chains at stepped surfaces with atomic precision. These form a new class of materials for exploring electrons in one dimension. Theory predicts a radically different behavior compared to higher dimensions. The single-electron picture has to be abandoned, because electrons cannot avoid each other when moving along a line. This article gives an overview of the phenomena that have been observed for electrons in one-dimensional chain structures, many of them quite unexpected, such as a fractional electron number per chain atom, a doublet of nearly half-filled bands instead of a single filled band, and spin-polarized bands in non-magnetic materials. First, the basic methods for analyzing electrons in atomic wire structures are outlined. Metal surfaces with free-electron-like surface states serve as model cases for explaining the quantization phenomena induced by steps and terraces. These self-assemble into lateral superlattices at vicinal surfaces. The periodicity can be tuned by the miscut angle. One can distinguish two regimes, i.e., quantum-well states confined within each terrace and superlattice states extending over the whole step array. Then, we move on to semiconductor surfaces, where metal atom chains and broken bond chains can be combined into more complex structures. The chain atoms are locked rigidly to the substrate, but the electrons near the Fermi level completely decouple from the substrate, because they lie in the band gap of the semiconductor. The dimensionality can be controlled by adjusting the step spacing with intra- and inter-chain coupling ratios ranging from 10 : 1 to > 70 : 1.

6.1 Introduction to One-Dimensional Systems

6.1.1 Physics in One Dimension

One-dimensional (1D) physics is particularly elegant and simple. Many problems can be solved analytically in one dimension, but not in higher dimensions. Some problems can only be solved in one dimension. This is the lowest dimension, where electrons are able to propagate. In that sense, 1D is the lowest non-trivial dimension. For these reasons, quite a few books have been written about physics in one dimension [1, 2].

It has been predicted for some time that electrons exhibit fundamentally different properties when confined to move along a single dimension. One-electron excitations are replaced by collective excitations since the electrons cannot avoid each other when moving along a single line. Their wave functions overlap completely and become highly correlated. This strong interaction has startling consequences on the physics of one-dimensional systems leading to a variety of unusual phases at low temperatures [3–7]. In a one-dimensional metal even the identity of electrons is lost, and becomes replaced by separate spin and charge excitations, the spinons and holons [3–5]. The Chap. 4 by R. Claessen, J. Schäfer, and M. Sing discusses these phenomena in detail. While the predictions give glimpses of exotic physics, finding direct evidence for spin-charge separation has proven elusive. It has been very difficult to systematically tailor natural crystals with 1D character, such that the electronic properties can be optimized for reaching the appropriate part of the 1D phase diagram and avoiding other low-dimensional instabilities, such as a Peierls transition to an insulator, charge density waves, spin density waves, singlet and triplet superconductivity, etc. [6, 7].

6.1.2 Creating 1D Structures

While theorists have it easier in 1D, experimentalists have to work harder. In theory, the ideal 1D system is a string of atoms freely suspended in space. Experimentally, that is only possible for strings of a few atoms [8,9], and these atoms vibrate enormously. Nevertheless, there have been encouraging developments in surface physics, where stepped surfaces or otherwise anisotropic surfaces have been used as templates for creating 1D structures.

Particularly interesting are metallic chain structures on semiconductor surfaces [10]. The atoms are locked rigidly to the substrate by covalent bonds, but the electrons near the Fermi level completely de-couple from the substrate. They cannot hybridize with any of the three-dimensional bulk states, as long as they lie in the band gap of the substrate. Only the electrons in back-bond states are able to hybridize, but these lie well below the Fermi level (typically 5–10 eV). The dimensionality can be controlled by adjusting the step spacing. 1D/2D coupling ratios ranging from 10:1 to well above 70:1 have been achieved [11, 12]. The surface band structure exhibits unexpected features, such as a fractional electron count per chain atom, two half-filled bands instead of one completely filled band, and nanoscale phase separation into metallic and insulating chain sections [13]. Spin chains can be created by using transition-metal and rare-earth atoms [14, 15], and the orbital and spin moment changes dramatically in these structures.

While semiconductor surfaces provide a variety of exotic phenomena, they are more difficult to understand because of the complicated rearrangement of their broken bonds, which may involve more than hundred atoms, such as in the famous Si(111)7×7 surface, the most stable surface of silicon. Therefore, it is advisable to begin with metal surfaces, where the bonding is isotropic and extensive rearrangements are rare. The electronic states are delocalized and free-electron-like, in contrast to the strongly localized broken bond orbitals of semiconductors. Stepped metal surfaces will serve us as textbook examples for explaining the type of low-dimensional states that can be induced by steps.

6.1.3 Mapping 1D Electrons

Angle-resolved photoemission has the distinction of being capable to determine the complete set of quantum numbers for electrons in a solid and at a surface. These are energy, momentum, spin, and the point group symmetry. In a 2D system, both in-plane momentum components are conserved in the photoemission process up to a reciprocal lattice vector. Therefore, measuring the energy and the in-plane momentum of the emitted photoelectron provides directly the energy and momentum of the electron inside the solid using energy and momentum conservation. The resulting data set consists of the photoemission intensity as a function of three variables, the "intensity curve" $I(E, k_x, k_y)$. Typically, two-dimensional slices of this cube are plotted on a gray scale or in color code. (A popular scheme is that of a topographic map, where the deep blue sea represents lowest intensity and white mountains the highest.) The $I(E, k_x)$ and $I(E, k_y)$ slices represent the familiar $E(k)$ band dispersion, while the $I(k_x, k_y)$ slices are constant energy "surfaces". The most important of these is the Fermi surface, where the electronic states reside that are responsible for most of the interesting electronic phenomena, such as charge density waves, magnetism, superconductivity, and transport in general. An example is shown in Fig. 6.1, where the two types of slices are combined for a nearly one-dimensional chain structure on silicon. Comparing Fermi surfaces of two- and one-dimensional structures in Fig. 6.2 illustrates the dramatic effect of dimensionality on the topology of the Fermi surface. In 3D, the Fermi surface of a free electron is a sphere with radius k_F, in 2D a circle with radius k_F, and in 1D it degenerates into two points at $\pm k_F$. By measuring a 1D system embedded into a 2D surface, the two k_F points become spread out into lines perpendicular to the 1D atom chain. This embedding process is analogous to the mapping of a 2D surface diffraction pattern in 3D space, which leads to a spreading of the 2D Bragg spots into "rods" perpendicular to the surface. In the example in Fig. 2, the lines are not completely straight, which indicates a residual 2D coupling between 1D chains.

While high energy resolution is always a benefit for resolving sharp states at the Fermi level, even more important for 2D and 1D systems at surfaces is k-resolution. The unit cells on semiconductors are rather large in real space, such as the 7 × 7 cell. De-coupling the wires in a surface arrays requires a large wire spacing, which leads to a small Brillouin zone perpendicular to the wires.

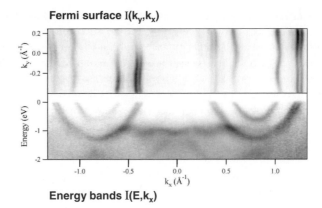

Fig. 6.1. Two ways of displaying an angle-resolved photoemission data set, demonstrated for the one-dimensional chain structure of Si(553)-Au (from [11]). *Bottom:* Photoemission intensity versus energy E and wave vector k_x (with k_x in the chain direction), representing a band dispersion. The metallic bands crossing the Fermi level E_F are free-electron-like, i.e., nearly parabolic. *Top:* Photoemission intensity at E_F versus k_y and k_x, representing the Fermi surface. High photoemission intensity is shown dark. A horizontal cut through the bottom part corresponds to a MDC, a vertical cut to an EDC. (The discussion of EDCs versus MDCs is also given in Chap. 1 by Reinert and Hüfner in this volume.)

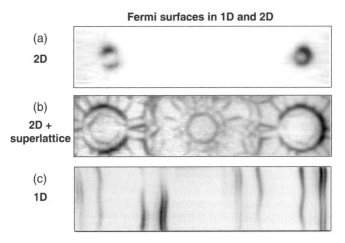

Fig. 6.2. Comparison between Fermi surfaces in 2D an 1D (from [16]). The topology of the Fermi surface changes completely from circles to lines. A superlattice creates replicas of the the Fermi circles that are shifted reciprocal superlattice vectors. **(a)** Si(111)$\sqrt{3} \times \sqrt{3}$-Ag; **(b)** Si(111)$\sqrt{21} \times \sqrt{21}$-(Ag+Au); **(c)** Si(553)-Au. High photoemission intensity is shown dark. EDCs and MDCs across the large fermi circle in (b) are given in Fig. 5.18

6.2 One-Dimensional Quantum Wells at Metal Surfaces

Recent developments in high energy and angular resolution in photoemission have an enormous impact on the study of electronic states of low-dimensional nanostructures on metal surfaces, which are readily accessible by this technique. Noble metal surfaces, in particular, offer a relatively simple electronic structure, which has been investigated in great detail by photoemission and scanning tunneling spectroscopy. As a consequence, there is a wealth of data and a detailed knowledge of the electronic properties for a variety of metallic surfaces. Knowing the electronic states of a flat surface is crucial for disentangling the complexity that arises in photoemission experiments from lateral nanostructures grown on top the surface. Low-dimensional structures on well-known metal surfaces are very valuable as model systems to investigate electronic states with angle-resolved photoemission. Additionally, one can envision potential applications for magnetic quantum stripes and atom chains [14], or for nanostructured metallic templates, which can be used for selective chemical adsorption in molecule/metal systems [17, 18], such as those searched for in catalysis, solar cell, or light-emitting diode device technologies.

6.2.1 The Vicinal Noble Metal Surface as a Model 1D System

One of the simplest nanostructured metallic systems is a vicinal surface with a regular array of monatomic steps, which give the surface a one-dimensional (1D) character. Vicinal surfaces are miscut by a few degrees from a high symmetry direction and can be readily prepared in situ, with the standard procedure of ion sputtering and mild annealing. The quality of the resulting step superlattice depends on a variety of factors, such as surface energy versus step and kink energy. Homogeneous arrays of steps with lattice constant d are frequently obtained over micron-size patches of the surface. Nonetheless, such well-defined areas are always characterized by a finite terrace size distribution $\sigma = \Delta d/d$, which in turn varies upon the strength of the interaction between steps [19]. The size distribution σ in lateral nanostructures and arrays self-assembled on solid surfaces is a key parameter that is necessary to control, since it introduces an intrinsic spectral broadening in averaging techniques, such as angle-resolved photoemission.

Vicinal surfaces are particularly useful as templates for the self-assembly of 1D metallic nanostructures. The general strategy consists in depositing submonolayer amounts of distinct materials from the gas phase under the right conditions (temperature and flux) to promote rapid diffusion on the surface and sticking to the steps. Ideally, one-dimensional arrays of linear structures of varying thickness are produced in the row-by-row, or "step-flow" growth regime. Thickness control is not an issue, since fine tuning is readily achievable by smoothly varying the coverage across a macroscopic sample. The ultimate limit is the atomic chain, like the one shown in Fig. 6.3(a). These are 1D Co atomic chains deposited on the stepped Pt(997) surface [14]. The

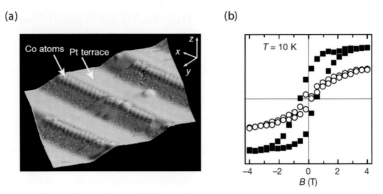

Fig. 6.3. (a) STM topography showing 1D atomic chains of Co grown by step decoration on Pt(997). (b) The magnetization of such chains is measured along the easy (*filled symbols*) and hard (*open symbols*) directions. The magnetic response at $T = 10$ K reveals non-zero remanent magnetization and hence long-range ferromagnetic order (from [14])

reduced atomic coordination of the monatomic chains causes a remarkable increase of the magnitude of both the orbital and the spin magnetic moments compared to bulk Co. The orbital magnetic moment increases from $0.15 \pm 0.01\mu_B$/atom in the bulk-like Co film to $0.68 \pm 0.05\mu_B$/atom in the monatomic chain. Moreover, as demonstrated by the hystheresis cycle in Fig. 6.3(b), such Co chains sustain long-range ferromagnetic order. That is possible thanks to the strong magnetic-anisotropy-energy barriers, which effectively block the relaxation of the magnetization at sufficiently low temperature.

The ideal step-flow regime is rather exceptional. Frequently molecules and adatoms form zero-dimensional aggregates attached to steps, and in other cases the adsorption induces chemical as well as structural changes in the vicinal substrate, such as faceting. In the latter the surface plane contains two separated 1D phases that are subject to mutual elastic repulsion, leading to a periodic 1D, hill-and-valley pattern. Such periodically faceted structures are also attractive as 1D metallic systems. They are stiffer than bare stepped surfaces and have periods that reach a few hundred nanometers. Additionally, they display an enhanced chemical contrast between phases, which is particularly useful to achieve 1D functional stripes (molecules, ferromagnets) by selective adsorption [17,18]. In Fig. 6.4(a) we show a characteristic example of adsorption-induced faceting, namely the Ag/Cu system. The faceted structure is formed by depositing submonolayer amounts of Ag on vicinal Cu(335) at 300 K, and then post-annealed at 450 K [20]. By varying the amount of Ag, the width of the stripes and the periodicity of the system can be changed. Despite the complexity of such self-assembly process, it does not show appreciable kinetic constraints, since the size distribution in Fig. 6.4(b) is analogous to that of the Cu(335) substrate. Such a sharp size distribution makes angle-resolved photoemission studies of surface states meaningful. It is indeed observed that

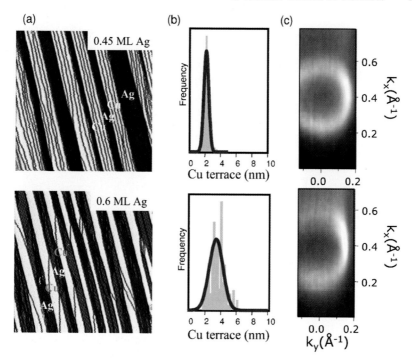

Fig. 6.4. One-dimensional periodic faceting in a vicinal Cu(111) surface induced by Ag adsorption. (**a**) STM image showing Ag covered facets that alternate with clean stepped Cu bands at different Ag thickness. The size distribution of (111)-oriented Cu terraces is analyzed in (**b**). The standard deviation σ is found to be analogous to that of the bare Cu substrate. The 1D character of surface states in narrow Cu nanostripes is shown in their Fermi surface in (**c**), which display strong asymmetry with large k_x broadening perpendicular to the stripes (adapted from [22, 25])

distinct Ag and Cu states characterize the two phases of the system [21, 22]. The Cu stripe thickness determines its surface-state dimensionality, which is observed to vary from 2D in relatively wide stepped stripes to 1D in narrow stripes. The 1D character in the latter is reflected in the large asymmetric broadening of the Fermi surface shown in Fig. 6.4(c).

6.2.2 Complex Scattering at Steps: From 2D to 1D Surface States

Vicinal metal surfaces and 1D striped nanostructures grown by step decoration or by periodic faceting are becoming benchmark systems to test 1D electronic states. In particular, noble-metal surfaces vicinal to the (111) plane, since these possess a free-electron-like surface state easily identified in scanning tunnelling spectroscopy and well characterized in photoemission. This surface state scatters strongly at step edges, giving rise to 1D confinement and superlattice effects (for review works see [24, 25] and references therein).

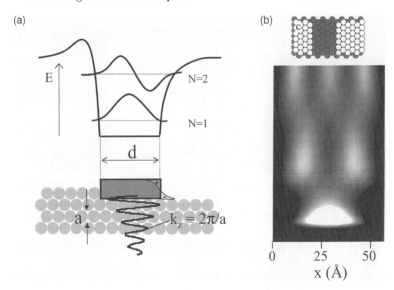

Fig. 6.5. (a) Schematic description of $N = 1$ and $N = 2$ wave functions and electron potentials for 1D electrons confined in a metallic nanostripe attached to a surface step. Asymmetric barriers, dipole-like step edge potentials and evanescent wave functions in the z-direction are characteristic features of the real system (b) 1D wave functions ($N = 1$ to $N = 3$) in a Ag(111) nanostripe (shown on top) mapped in real space by Scanning Tunneling Spectroscopy (from [26])

The fundamental parameter is the characteristic size d perpendicular to the step array, which corresponds to the terrace width in bare stepped surfaces or to the stripe width in nanostripes decorating steps, as shown in Fig. 6.5(a). There we represent the schematic side view of the resulting 1D quantum well in the direction perpendicular to the steps x, featured with the additional ingredients of the real system: asymmetric uphill and downhill electron potentials, dipole-like barriers at step edges and evanescent, bulk-like wave functions in the perpendicular direction z. All are important to understand angle-resolved photoemission data, as we shall see later. The presence of such 1D quantum well-levels in metallic nanostripes has been very well documented by STS experiments, such as those reproduced in Fig. 6.5(b) [26]. The STM image above shows a $d = 56$ Å wide trench limited by monatomic step edges on a Ag(111) surface. These behave as hard-wall potentials for surface-state electrons, leading to a succession of 1D QW levels that are spatially probed in the STS maps below.

In a similar way as in STS, 1D metallic QW levels and wave functions can be probed and mapped by angle-resolved photoemission in noble metal vicinal surfaces, as shown in Fig. 6.6 [27]. Surface-state bands in Au(23 23 21), i.e., a Au(111) vicinal surface with 56 Å wide terraces, break up into three QW levels below the Fermi energy. The data in Fig. 6.6(a) correspond

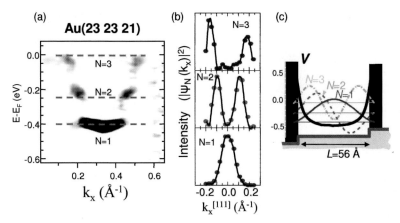

Fig. 6.6. Quantum wells in the Au(23 23 21) observed by angular photoemission. (**a**) Second derivative of the photoemission spectra shown as an intensity plot. The surface band breaks up in non-dispersing levels that fit to an infinite 1D QW of size $d = 56$ Å; (**b**) photoemission intensity map $|\Psi_N(k_x)|^2$ and (**c**) real space wave functions $\Psi_N(x)$ and effective electron potential $V(x)$ determined from an iterative phase recovery process using the data in (**b**). Note that k_x is defined with respect to the average surface in (a), but with respect to the (111) terrace in (b) (from [27])

to the second derivative of the photoemission spectra displayed in an intensity plot. The derivation enhances peaks while eliminating the background. The photon energy $h\nu = 60$ eV is chosen to minimize the intensity from nearby umklapp features produced by photoelectron interference with the step lattice. It has been shown that the $N = 1, 2,$ and 3 QW levels of the figure follow the $E_N = E_0 + \hbar^2\pi^2/2m^*d^2$ series of the infinite 1D QW of size $d = 56$ Å [27], proving that also in the step array of the vicinal surface uphill and downhill steps may behave as hard wall potentials that confine surface electrons.

The photoemission intensity from the three QW levels of Au(23 23 21) is strongly modulated along the k_x direction, as shown in Fig. 6.6(b). Under certain conditions, which are met in this case, the photoemission intensity maps in Fig. 6.6(b) represent the Fourier transform of the respective wave functions $|\Psi_N(k_x)|^2$ [27]. Moreover, the confined nature of the wave function in Au(23 23 21) terraces makes it possible to retrieve the real-space wave function $\Psi_N(x)$ from the experimental $|\Psi_N(k_x)|^2$ curve using an iterative formalism, called oversampling, borrowed from x-ray diffraction [28]. Iterative procedures overcome the well-known problem of the phase in momentum space, which is not measured in spectroscopic techniques, such as x-ray diffraction or photoemission. In iterative oversampling, the phase is obtained by repeatedly diminishing the amplitude of the wave function outside the confinement region, which in turn is estimated from the self-convolution of the $|\Psi_N(k_x)|^2$ curve. The three QW wave functions shown in Fig. 6.6(c) are obtained after applying the oversampling method to the data in Fig. 6.6(b). On the other hand, the

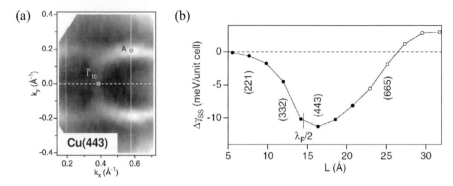

Fig. 6.7. (a) Fermi surface gap in the Cu(445) vicinal surface, where the terrace width $d = 16$ Å makes it possible Fermi surface nesting ($\pi/d = k_F$) in the direction perpendicular to the step array; (b) the surface tension change calculated for Cu(111) vicinals with superlatice bands shows a minimum around $d = 16$ Å, indicating that Fermi surface nesting conferes an extra stability to the Cu(443) plane (from [31])

effective one-electron potential (thick line) is obtained by simply dividing the Schrödinger equation by the wave function.

In the past few years, a large number of studies have been devoted to the study of the electronic structure in vicinal noble metal surfaces using high-resolution angle-resolved photoemission [24, 25]. All show the inherent complexity of surface states even in the simplest 1D step array. The clear-cut case of the 1D QW in the Au(23 23 21) surface is an exception, since surface states display a smoothly changing dimensionality from 1D to 2D within a wide range of d values. Also exotic structure/electronic interferences arise due to Fermi surface "nesting" that occurs when half of the superlattice vector π/d matches the Fermi wave vector k_F [29–31], as shown in Fig. 6.7. Additionally, data display strong photon-energy dependent intensity variations, and fine features, such as gaps, are obscured by size-distribution broadening. Yet photoemission spectra in vicinal surfaces always exhibit uniaxial anisotropy, with surface band changes in the direction perpendicular to the steps and flat surface behaviour in the parallel direction. In Fig. 6.8 we present a characteristic example of high resolution angle-resolved photoemission study in the sharply defined ($\sigma = 0.1$) step array of the Au(887) surface. The STM image indicates the presence of large (few micron size) homogeneous monatomic stepped areas, with well defined terrace size ($d = 39 \pm 4$ Å). Across the step array, one can observe the umklapped ($2\pi/d$) $N = 1$ and $N = 2$ mini-bands. These have been fitted with a periodic 1D Kronig–Penney model [23], which gives the strength of the repulsive barrier at the step edge $U_0 b = 2.2$ eV × Å. Along the step array, high resolution permits to observe the free-electron-like dispersion ($m^* = 0.25 m_0$) of the $N = 1$ band and its spin-orbit splitting, like in flat Au(111) [32–34].

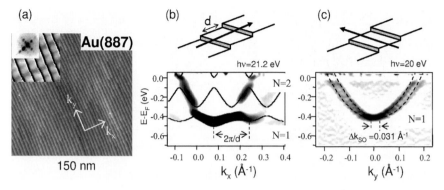

Fig. 6.8. High resolution photoemission experiments from the model, vicinal Au(887) surface. (**a**) STM topography showing the 1D step array. The periodic reconstruction pattern along the (111) terraces (inset) actually defines a square mesh, reflected in the fine electron diffraction pattern around the (0,0) spot shown in the left corner; (**b**) $N = 1$ and $N = 2$ surface bands measured along k_x and fitted with a 1D Kronig Penney model (*lines*); (**c**) Free-electron like dispersion along k_y, displaying the characteristic spin-orbit splitting of Au(111) surfaces [32–34] (adapted from [23, 24])

The complex nature and the repulsive character of the electron potential at steps is straightforwardly visualized by a direct comparison of high resolution photoemission spectra for flat and vicinal surfaces. The peaks in Fig. 6.9(a) correspond to the bottom of the surface band for Cu(10 10 11) ($d = 43\,\text{Å}$) and Cu(111) surfaces, prepared and measured under the same experimental conditions [23]. The upwards energy shift $\Delta E = E - E_0 = 30$ meV and the extra broadening $\Delta W = W - W_0 = 35$ meV (FWHM) with respect to Cu(111) are the signatures of the scattering at steps in the Cu(10 10 11) superlattice. Both peak shift and broadening increase as the density of steps (or miscut angle) increases, reflecting the repulsive nature of the step barrier and the local, step-edge absorption (or leaking) into bulk states [35]. Moreover, the effective strength of the barrier potential ($U_0 \times b$) sketched in Fig. 6.9(b) can be readily derived from ΔE using the simple 1D Kronig–Penney model [24,25]. The result of the fit in a variety of Cu(111) and Au(111) vicinals is shown in Fig. 6.9(c). The barrier strength is reduced by an order of magnitude from a surface with relatively wide terraces (Au(23 23 21), $d = 56\,\text{Å}$) to surfaces with smaller step lattice constant $d \sim 20\,\text{Å}$, thereby making surface states smoothly evolve from quasi-1D QW-s in wide terraces to 2D surface bands in narrow terraces. This phenomenon, although not totally understood yet, appears to be connected with the evolution of the bulk projected band gap that supports the surface state, which progressively shrinks for surfaces with smaller d values and vanishes around $d \sim 20\,\text{Å}$ [24, 25].

The terrace size distribution ($\sigma = \langle d \rangle/d$) in vicinal surfaces gives rise to both energy-level as well as wave-vector broadening. The latter is related to

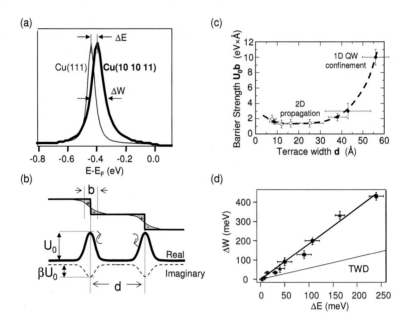

Fig. 6.9. (a) Surface state spectra of Cu(111) and Cu(10 10 11) measured at band minima. The peak shifts (ΔE) and broadens (ΔW) in the vicinal surface, as expected for a step potential with real (repulsive) and imaginary (absorptive) parts, like the one sketched in (b); (c) the barrier strength $U_0 \times b$ as determined from a number of stepped systems strongly varies as a function of the lattice constant d, such that surface states evolve from 1D QW-s at large d values to 2D superlattice states with small d; (d) ΔE versus ΔW measured in a variety of stepped noble metal systems. The linear relationship proves a local step edge scattering scenario, supporting the complex barrier potential description in panel (b) (adapted from [23, 35])

the fact that surface bands are located at the surface Brillouin zone edge (π/d) in the average surface plane. All in all, surface-state peaks become broader in a vicinal surface due to both size distribution and lifetime effects. In Fig. 6.9(d) we reproduce the recent peak width analysis of surface states collected from a variety of 1D stepped nanostructures. Independently of the system, a linear relationship between ΔE and ΔW is found [35]. The contribution from size distribution effects is minor (thin line), and hence the linear relationship in Fig. 6.9 (d) indicates that lifetime broadening of surface states in vicinal surfaces depends on the step density d and the barrier strength $U_0 \times b$ in the same way as ΔE. Thus local absorption at step edges appears to be the dominant source of inelastic scattering, supporting the model of complex step barrier potential proposed in Fig. 6.9(b).

The uphill-downhill asymmetry that characterizes vicinal surfaces, as well as nanostructures grown on top, has been shown to affect surface-state scattering through both the real and imaginary parts of the barrier potential [35–37].

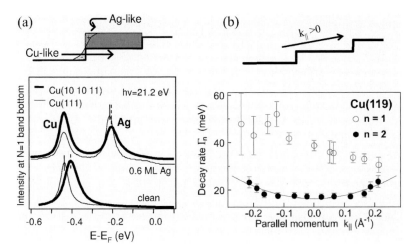

Fig. 6.10. Asymmetric scattering in stepped systems, proved in (**a**) surface states of Ag/Cu nanostructures, where peak shift and broadening with respect to flat systems affect Ag states (downhill step scattering) but not Cu states (uphill step scattering), and (**b**) image states in Cu(119), where lifetime is longer for $N=1$ states running uphill (adapted from [35, 36])

This is the case of the Ag/Cu nanostripe array shown in Fig. 6.10(a) [35]. At 0.6 ML coverage, one monolayer thick Ag stripes alternatively fill up terraces in the Cu(10 10 11) substrate, as schematized in the top, left panel. Such particular mode of growth is actually a consequence of the tendency to form large islands with triangular misfit dislocations on Cu(111), attached to surface steps [38]. The spectra on the bottom correspond to the surface state band minima compared to the flat system. One can observe step scattering, i.e., surface state shift and broadening, affecting electrons from Ag stripes (downhill scattering), but not electrons from Cu stripes (uphill scattering). The scattering asymmetry at steps is also found in pump-probe experiments for image states on Ag/Pt(997) nanostripes [37] and, as shown in Fig. 6.10(b), in Cu(119) [36]. It is observed that the $N=1$ image state lifetime, which is the inverse of the decay rate plotted in Fig. 6.10(b), is longer for electrons running uphill, in agreement with the Ag/Cu scattering asymmetry of Fig. 6.10(a).

6.2.3 Exploring the 3D Fourier Space: The Modulation Plane

Surface state properties, such as energies or wave functions, are mostly determined by the bulk projected band gap supporting the surface state. For example, the size of the gap determines the effective penetration inside the bulk. Relatively large gaps lead to a few atomic-layer damping, whereas surface states in narrow gaps and surface-bulk resonances tail deep inside the crystal. In the vicinal surface the formal projection picture breaks, since due to umklapp with the step superlattice, projected band gaps would not exist.

In reality, a "first-zone" projection scheme appears to hold [24, 25]. The major changes observed in surface states of vicinal surfaces seem to be connected to the shrinking size of the first-zone projected band gap, i.e., to the direct overlap between surface states and bulk states and the subsequent formation of surface resonances [24, 25, 39]. In particular, the change in the effective strength of the barrier potential shown in Fig. 6.9 [24, 25]. Pure surface states have a high probability within the surface layer and hence feel a relatively strong step edge potential, by contrast to surface-bulk resonances that are located away from the surface plane, where the electrostatic dipole potential is smoothen out [40]. On the other hand, surface-bulk mixing via step lattice umklapps is always possible in a vicinal surface, but this would rather influence the surface state lifetime [36].

The changes in both the effective penetration of the surface state and the step-barrier potential are reflected in a characteristic tilt of the wave-function modulation plane, i.e., the direction along which its damping tail decays inside the bulk. That is determined by probing the Fourier spectrum in the direction perpendicular to the bulk (k_z), as shown in Fig. 6.11. Such analysis, which requires photon-energy tuning and hence synchrotron radiation, is analogous to the $|\Psi_N(k_x)|^2$ mapping in the k_x direction shown in Fig. 6.6. However, the quantitative $|\Psi_N(k_z)|^2$ mapping is limited by the short (2–3 layer) photoelectron escape depth that causes a large k_z broadening, and hence only a qualitative analysis can be made [24, 25]. In Fig. 6.11(a) we show the Fermi surface for Cu(335) measured at increasing photon energies. In Fig. 6.11(b) we plot the corresponding (k_x, k_z) values calculated at surface band minima, i.e., at the Fermi ring center. The analogous plot for the $N = 1$ QW of Au(23 23 21) is shown in Fig. 6.11 (c). In both cases, the continuous spectral distribution of the surface state is being discretely probed in the $k_x - k_z$ plane, allowing one to determine the modulation plane. One can observe a remarkable qualitative difference between 2D bands and 1D QW-s, i.e., data points line up along the average surface normal in Cu(335), by contrast to the [111] direction in Au(23 23 21). This straightforwardly proves that the modulation plane switches from the average (optical) surface in 2D bands of Cu(335) to the (111) terrace in Au(23 23 21).

The effective proximity of the wave function to the outermost surface plane affects the modulation plane, as nicely proved for image states on the vicinal Cu(775) surface [40]. Figure 6.12 shows the band dispersion for the $N = 1$ and the $N = 2$ image states. The $N = 2$ parabola displays the characteristic symmetric dispersion around $\bar{\Gamma}$ for a noble-metal surface, by contrast to the $N = 1$ band, which is asymmetric and its minimum is shifted away towards the [111] direction. Such distinct behavior is explained by the different distance of the respective wave functions to the surface plane, as shown on the right panel. The $N = 2$ image state is located ~ 12 Å away from the surface, and hence it is being affected by a relatively smooth surface potential. By contrast, the $N = 1$ state lies ~ 3 Å above the surface plane, making it more sensitive to the step corrugation.

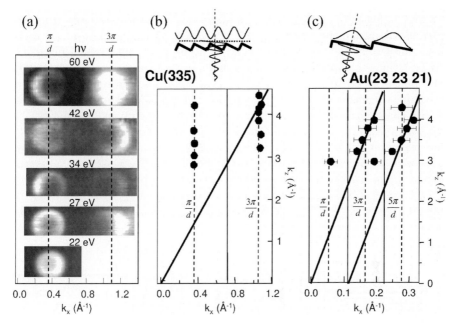

Fig. 6.11. Probing the modulation plane of surface states in vicinal surfaces by photon-energy-dependent photoemission. (**a**) Fermi surface rings for Cu(335) at increasing photon energies; (**b**) (k_x, k_z) plot, where data points correspond to the surface band minimum (ring center) in (a). The two umklapped sets of points line up perpendicular to the average surface plane; (**c**) analogous (k_x, k_z) plot for the $N = 1$ QW in Au(23 23 21) [25]. In this case, data points are aligned along the [111] direction, indicating a tilt in the wave-function modulation plane with respect to Cu(335), as sketched on top (from [25])

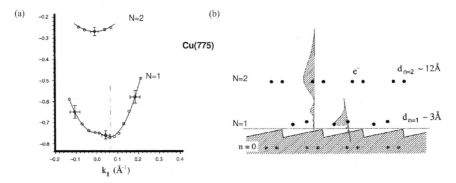

Fig. 6.12. Distinct modulation planes of image states in vicinal Cu(775), proved by the shift of the band with respect to the surface normal (*left*). $N = 1$ and $N = 2$ states lie close and away from the surface plane, respectively, defining the (111) terraces or the average surface as the modulation plane in each case (from [40])

The analysis of surface states in step arrays and lateral nanostructures carried out during the past few years in a variety of systems has provided the basic ideas and the analytical framework necessary to understand electronic states in more complex systems using angular photoemission. It is clear that, beyond the spectrum of energy levels, one also needs a thorough exploration of the three-dimensional Fourier space in order to probe electron wave functions in real space. Such measurements will require high energy and angular resolution, as well as photon-energy tuning. In Fig. 6.13 is shown a good example, i.e., the extensive analysis of electronic states in the Au(11 9 9) surface [25]. This is a periodically faceted surface made of wide $d_A = 42$ Å terraces and narrow $d_B = 14$ Å step bunches. By selecting an appropriate photon energy (17 eV) one is able of separating a sharp, non-dispersing $N=1$ peak from the broad, dispersing $N=2$ bands, as shown in Fig. 6.13(a). In order to unveil the

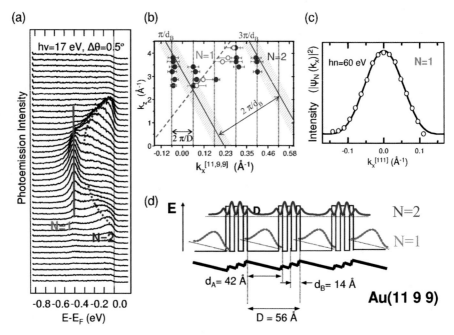

Fig. 6.13. Surface states in the faceted Au(11 9 9) surface, composed of terraces (d_A) and periodic step bunches (d_B), as shown in panel (d) [25]. (**a**) High resolution spectra showing distinct 1D $N = 1$ and 2D $N = 2$ features; (**b**) ($k_x - k_z$) plot for the 2D (*filled dots*) and the 1D (*open dots*) features observed in (a). The data points for the 2D bands group forming five sets of $2\pi/D$ ($D = d_A + 2d_B$) umklapps that line up perpendicular to the average surface, whereas the data for the non-dispersing feature follow the $4\pi/D$ umklapp line of the (111) direction [25]; (**c**) photoemission intensity as a function of k_x ($|\Psi(k_x)|^2$) for the 1D state in (a). The *line* represents $|\Psi(k_x)|^2$ calculated for the $N = 1$ state in the infinite QW of size $d_A = 42$ Å; (**d**) wave-function model deduced from (**b**) and (**c**) (from [25])

physical nature of both states, we examine the $(k_x - k_z)$ plot in Fig. 6.13(b), and analyze the k_x-dependent photoemission intensity in Fig. 6.13(c). The non-dispersing peak leads to a single set of data points along the [111] direction in the $(k_x - k_z)$ plot, and to a probability density $|\Psi(k_x)|^2$ that fits to that of the $N = 1$ state of the infinite QW of size d_A. Such behavior is indeed expected for a 1D QW mostly located inside terraces in Fig. 6.13(d). On the other hand, the photon-energy analysis of the dispersing $N = 2$ bands leads to five sets of vertical $2\pi/D$ umklapps ($D = d_A + d_B$) in the $(k_x - k_z)$ plot of Fig. 6.13(b). For such dispersing states, the photoemission intensity peaks at umklapps separated by $\sim 2\pi/d_B$ superlattice vectors [32], i.e., within the shaded blue stripes of Fig. 6.13(b). These are indeed the features for a 2D surface state strongly modulated within the step bunch, but propagating on the average surface plane along the k_x direction, as shown in Fig. 6.13(d).

6.3 Atomic Chains on Semiconductor Surfaces: The Ultimate Nanowires

With the analysis tools for low-dimensional surface states in hand, we now proceed to the atomic chain structures that can be assembled at semiconductor surfaces. Their electronic structure is more complex than that of stepped metal surfaces, due to the localized broken bond orbitals of covalently-bonded semiconductors. On the other hand, the electronic states at a semiconductor surface de-couple completely from the bulk, as long as they reside in the bulk band gap. At these energies, surface states cannot hybridize with three-dimensional states. A variety of metallic semiconductor surfaces have been found in recent years, both two-dimensional [41, 42] and one-dimensional [12, 43, 45]. These combine the best of both worlds: The surface electrons are de-coupled from the substrate, but the surface atoms are locked into place by highly-directional, covalent back-bonds. Those are formed by three-dimensional orbitals with energies well below the Fermi level. The interesting states are those at the Fermi level, not the low-lying back-bond. States within a thermal energy of the Fermi level ($k_\mathrm{B}T = 25\,\mathrm{meV}$ at room temperature) are relevant to 2D and 1D transport, superconductivity, magnetism, charge density waves, and other more exotic phases predicted for 1D systems. The surface states can be tuned systematically from 2D to 1D by varying the spacing between atomic chains via the step spacing. Moving the chains farther apart reduces the coupling exponentially with a decay constant of atomic dimensions. As a result, the 1D/2D coupling ratio can be varied from 12:1 to >70:1. At present, the detection limit of 70:1 is imposed by the finite angular resolution of photoelectron spectrometers.

6.3.1 Self-Assembly of Atom Chains on Stepped Si(111)7×7

The construction of atomic chain structures on vicinal Si(111) surfaces [15,47–49] is illustrated in Figs. 6.14 and 6.15. The clean surface exhibits a regular array of facets, which are stabilized by the deep 7 × 7 reconstruction and multiple steps. The data are from the Si(557)-Au structure, which consists of one gold chain every five atomic rows. It is among the best-studied 1D surfaces, and will serve as prototype for demonstrating the phenomena encountered in atomic chains on semiconductors. Another well-studied surface is Si(111)-4×1-In [43,44], where four indium chains per unit cell form two closely-spaced zigzag rows. The four chains produce three metallic bands that interact with each other. This surface serves as an example for the richness of the phenomena occurring in atomic chains. Chain structures can be obtained with a large variety of metal atoms, such as alkalis, alkaline earths, noble metals (Ag, Au), transition metals (Pt), and rare earths. Rare-earth chains bring f-electrons with large magnetic moments into the picture and thus become prototypes for

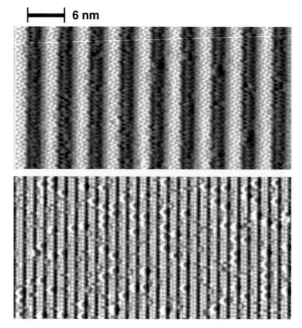

Fig. 6.14. Fabrication of atomic chains on a stepped Si(111) surface. The clean Si(557) surface forms regular facets that contain one 7 × 7 unit cell and a triple step (*top*). Deposition of 1/5 of a monolayer of Au removes the facets and creates atomic chains. The two chains seen by STM are Si atoms with broken bonds, not gold atoms. This and all following STM images display the derivative of the height in the direction perpendicular to the steps. This gives the appearance of a surface illuminated at grazing incidence, with steps casting dark shadows. From [46], [76], [73]

Fig. 6.15. One-dimensional growth pattern of the Si(557)-Au chain structure. Long strips of clean Si(111)7×7 remain, when the gold coverage is slightly below the optimum (by 2/100 of a monolayer). These are a single 7 × 7 cell wide. From [10]

spin chains. All rare-earth chain structures studied so far on Si(111) exhibit a similar 5 × 2 structure, consisting of alternating chains of filled and empty orbitals [15]. That provides the opportunity of systematically varying the spin by changing the f-count while keeping the structure the same.

Not all chain structures are metallic, but most of the Au-induced chains are. Essentially, gold forms a metallic chain structure on every vicinal Si(111) surface with odd Miller indices that has been studied. They consist of two classes with steps going uphill and downhill, as shown in Fig. 6.14. With a tilt of the surface normal towards $[\bar{1}\bar{1}2]$ there are two broken bonds at the step edge of the truncated bulk structure (such as (335), (337), (557)). With the opposite tilt towards $[11\bar{2}]$ there is only one broken bond at the step edge (such as (110), (553), (775), (995), (13 13 7) [12]). Even on the flat Si(111) surface, the three-fold symmetry is broken by gold and other metals. Three domains of a chain structure are formed. A single domain with chains parallel to the steps can be selected by a shallow miscut towards $[\bar{1}\bar{1}2]$. The most prominent example is Si(111)5×2-Au ([50–52] and references therein).

The one-dimensional character of these surfaces can be seen already during growth, as shown in Fig. 6.15. At a slight under-coverage (only 2/100 of a monolayer below the optimum of 1/5 monolayer), one finds long strips of clean Si(111)7×7 that have not been converted yet. These are exactly one 7 × 7 unit cell wide. Two 7×7 strips on the left side are in the process of being consumed by Au chains. This explains why an accurate Au coverage is the single most important criterion for preparing high-quality chain structures. The chains extend for long distances if the azimuthal orientation of the vicinal surface is correct (compare [53, 54] for the optimum preparation condition of stepped Si(111) surfaces). If the orientation is slightly off, the resulting kinks can be swept together into bunches by electromigration with a DC heating current parallel to the steps. Since this effect is uni-directional, it is advisable to try both directions in order to sweep out both left- and right-handed kinks. Kink-free chains have been achieved over distances comparable to the scanning range of a STM (~1 µm), with up to 20000 edge atoms between kinks. The least-critical parameters are the substrate temperature during Au deposition (600–700°C) and the subsequent post-anneal (800–900°C for a few seconds).

6.3.2 Atomic Structure

The atomic structure of these surfaces has been remarkably difficult to pinpoint. The best-studied surfaces are Si(557)-Au and Si(111)5×2-Au, which have been investigated by surface x-ray diffraction, electron diffraction, and local density calculations with total-energy minimization and modeling of the STM topography [51,55]. An increasing number of low-energy structures is being discovered in thorough theoretical searches [56]. The task is made difficult by the extra degrees of freedom introduced at a step edge, where additional silicon atoms can attach themselves rather easily. In fact, such silicon adatoms frequently serve as dopants that optimize the band filling of the chains [51].

With a certain caveat, one can isolate two features that are common to many 1D structures on Si(111), including those induced by gold, alkali metals, and alkaline earths. These are illustrated with the results of total-energy calculations for a few Au-induced chain structures in Fig. 6.16.

(1) A common structural element is the honeycomb chain [48], which consists of a graphitic strip of Si hexagons. π-bonding is unexpected for silicon, but many silicon surfaces exhibit π-bonding in one form or another, such as dimers on Si(100) and polyacetylene-like chains on cleaved Si(111). Broken bonds go through unusual contortions in order to regain some bonding energy. The honeycomb chains are highly one-dimensional. They can be extremely long (hundreds of nanometers), but they are less than two hexagons wide. The lattice match must be nearly perfect along the graphitic strip (in the [1$\bar{1}$0] direction), but very poor perpendicular to it (along [1$\bar{1}$2]). From these observations it is reasonable to conjecture that the honeycomb chains are the driving force that stabilizes 1D surface structures over potential 2D competitors, which are typically $\sqrt{3} \times \sqrt{3}$ structures. 2D structures do not appear until full monolayer coverage is reached.

(2) The second structural element is the metal chain. Judging from semiconductor homoepitaxy, one might expect the metal atoms to go to the step edge, as in step-flow growth. After all, there are extra bonding possibilities to the upper step edge. The gold chain structures defy such simple ideas and place themselves right in the middle of a terrace (Fig. 6.16). Instead of sitting on top, the Au chain is firmly incorporated into the Si(111) surface by substituting for a Si surface atom. For Si(557)-Au, this geometry is confirmed by both x-ray diffraction [57] and total-energy calculations [12]. STM is not able to see the Au chain, because each Au atom combines its s, p-electron with a

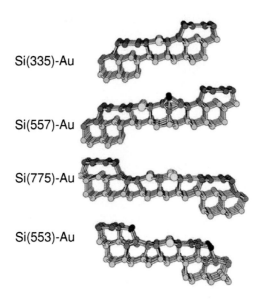

Fig. 6.16. A few of the chain structures induced by Au on vicinal Si(111), obtained from total-energy minimization. They have two structural elements in common: 1) A honeycomb chain of graphitic silicon (red), which drives the surface one-dimensional. 2) A chain of gold atoms at the center of the terrace (yellow), contrary to a simple model of step flow growth. The terrace width can be varied in increments of two row spacings, which allows tailoring of the inter-chain coupling. From [12]

Si broken bond and forms a filled band about 1–2 eV below E_F [55, 58]. Such a bound state has too little electron density outside the surface to be seen by STM.

The two atomic chains observed by STM on Si(557)-Au in Figs. 6.14, 6.15 actually originate from Si atoms with broken bonds, not from gold – another counter-intuitive feature of these 1D structures. One of them (the better-resolved chains on the right) can be assigned to the Si adatoms in the surface structure (dark blue in Fig. 6.16), the other to the step edge at the border of the honeycomb chain. This observation nay be generalized to chain structures formed by other metals, such as alkali metals and alkaline earths: The metal seems to play the role of a catalyst which facilitates the formation of the silicon honeycomb chain by bridging the gap between the honeycomb chain and the rest of the surface.

6.3.3 Metallic Surface States in 2D

The focus of this section will be on metallic surfaces, which have sharp states at the Fermi level E_F that are de-coupled from the 3D bulk. Before discussing one-dimensional chain structures, it is useful to consider 2D surface states.

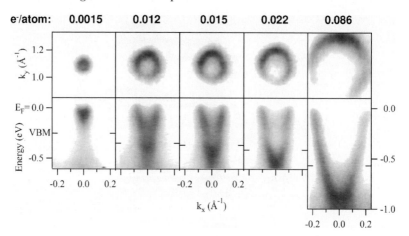

Fig. 6.17. Fermi surfaces and band dispersions for the two-dimensional structures on Si(111) shown in Fig. 6.17. A surface state band can be doped continuously by excess Ag or Au atoms on top of the stoichiometric Si(111)$\sqrt{3}\times\sqrt{3}$-Ag structure (from [42]). The analogous plot for one-dimensional structures is given in Fig. 6.1. One-dimensional chains are doped by excess Si atoms

They show already the importance of surface dopants in low-dimensional structures, which becomes even more dramatic in 1D. Natural examples are the $\sqrt{3}\times\sqrt{3}$ structures of Au and Ag on Si(111), which displace chain structures when the coverage comes close to a monolayer [41, 42].

An exactly stoichiometric $\sqrt{3}\times\sqrt{3}$ structure with 1 monolayer coverage is actually semiconducting, with a gap significantly smaller than the bulk band gap of Si [59–61]. However, extra metal atoms deposited on this surface beyond a monolayer dope the surface conduction band and create a metallic surface [16, 41, 42]. The metallicity of the surface becomes visible in high-resolution photoemission after adding as little as 1.5/1000 of a monolayer (Fig. 6.17). With increasing metal coverage the radius of the Fermi circle increases, enclosing the extra electrons from the dopants. The surface conduction band becomes visible as it fills up. However, it does not follow a simple rigid-band model. Instead of remaining fixed with respect to the bulk valence band maximum (VBM, tickmarks on the side), the bottom of the surface band drops in energy with increasing doping. It starts out above the VBM, crosses the VBM at about 1/100 of a monolayer and drops several tenths of an eV below the VBM at the highest doping levels. The maximum achievable doping is reached with a $\sqrt{21}\times\sqrt{21}$ superstructure on top of the $\sqrt{3}\times\sqrt{3}$ superlattice, with about 3 electrons per $\sqrt{21}\times\sqrt{21}$ unit cell [16].

The Fermi surface of the $\sqrt{21}\times\sqrt{21}$ structure is shown at the center of Fig. 6.2, together with the lightly-doped $\sqrt{3}\times\sqrt{3}$ structure at the top [16]. Both exhibit Fermi circles centered at the smallest reciprocal lattice vectors of the $\sqrt{3}\times\sqrt{3}$ lattice. Surprisingly, the equivalent circle at the center of the

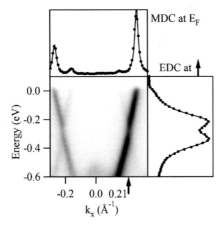

Fig. 6.18. Band dispersion $E(k_x)$ across the large Fermi circle of the Si(111) $\sqrt{21} \times \sqrt{21}$ superlattice formed by Au on top of the $\sqrt{3} \times \sqrt{3}$ structure formed by Ag (compare Fig. 6.2(b)). MDC and EDC cuts are shown. The double-peak in the EDC reveals a mini-gap induced by the $\sqrt{21} \times \sqrt{21}$ superlattice. From [42]

images ($\boldsymbol{k} = 0$) is absent for the $\sqrt{3} \times \sqrt{3}$ structure and very faint for the $\sqrt{21} \times \sqrt{21}$ structure. This could be due to strong polarization selection rules or to the proximity of bulk Si states at the VBM near $\boldsymbol{k} = 0$. An additional feature of the $\sqrt{21} \times \sqrt{21}$ structure is the intricate network of weaker Fermi circles. These can be explained rather simply as replicas of the dominant circle, shifted by the reciprocal lattice vectors of the $\sqrt{21} \times \sqrt{21}$ structure (see [16]). Furthermore, the corresponding energy bands exhibit mini-gaps at their intersections, whose magnitude is twice the superlattice potential acting on the surface electrons. These are shown in more detail in Fig. 6.18. The resulting potential is 55 meV in this case [16], a good example of a small energy scale at a surface requiring high resolution (both in energy and angle).

A quite different, but very prominent metallic semiconductor surface is Si(111)7×7 itself, the most stable silicon surface. The large unit cell leads to a very small Brillouin zone, which pushes the angular resolution capabilities of current spectrometers. Nevertheless, it has been possible to resolve its Fermi surface [62]. The electronic structure within a few meV of E_F is still largely unknown, but there have been intriguing hints about a possible small Hubbard gap with a Kondo peak at the center from local density calculations augmented by Coulomb interaction terms [63,64]. A Hubbard gap has been inferred as well from electron energy loss spectroscopy [65] and surface NMR measurements of the carrier density [66]. However the magnitude of the proposed Hubbard gap varies by an order of magnitude. Currently, the 7 × 7 surface is proving itself as ideal testing ground for electron–phonon interaction at semiconductor surfaces. The spectral function near E_F can be described remarkably well by a combination of a bare and a phonon-dressed bands with an electron–phonon coupling parameter $\lambda = 1.1$ [67]. There is a well-defined phonon mode at

70 meV from a vibration of a Si adatom against the atom underneath. It interacts with the band at E_F, which is derived from states localized on the adatom as well.

6.3.4 Electronic States from 2D to 1D

The topology of the Fermi surface changes dramatically between 2D and 1D, as demonstrated in Sect. 6.1.3 and Fig. 6.2. Two-dimensional Fermi surfaces are characterized by closed loops, such as the Fermi circles observed for the Si(111)$\sqrt{3} \times \sqrt{3}$-Ag and Si(111)$\sqrt{21} \times \sqrt{21}$-(Ag+Au) structures. A truly one-dimensional Fermi surface consists of two points at $\pm k_F$, but these become straight lines when plotted in two dimensions. The energy is independent of the momentum perpendicular to the chains (along the y-axis in Fig. 6.2). An actual chain structure, such as Si(553)-Au in Fig. 6.20, displays undulating lines. The amplitude of the undulations is a measure of the residual two-dimensional coupling. The complete data set can be reproduced by a simple tight binding calculation involving coupling energies to three sets of neighbor atoms, t_1 and t_3 for first and second neighbor along the chain and t_2 one between chains (see Fig. 6.19 left and [11]). The dimensionality ratio is given by t_1/t_2. For this particular structure one observes a doublet of closely-spaced Fermi lines with $t_1/t_2 = 39, 46$ and a single line with $t_1/t_2 = 12$. The Si(553)-Au surface has a fairly close chain spacing of 1.48 nm. By going to the Si(557)-Au surface with a somewhat larger chain spacing of 1.92 nm, the 1D/2D coupling ratio becomes so small that it cannot be measured any more (>70:1). The decay constant for the 2D coupling is so short, that the 1D limit can be approached rather rapidly. The absolute value of the dominant 1D coupling t_1 is about 0.7 eV for the doublet band that appears in both structures [12]. The effective mass is typically 1/2 in these chain structures, and the Fermi velocity about $1 \cdot 10^6$ m/s [11], which makes them rather free-electron-like along the chains.

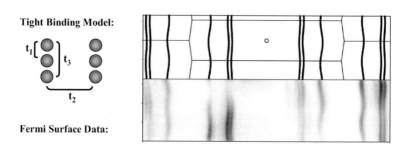

Fig. 6.19. Obtaining the 1D versus 2D coupling ratio from a tight binding fit to the Fermi surface of atomic chain structures, demonstrated for Si(553)-Au. The undulations in the Fermi lines are due to 2D coupling between the chains (from [11]). The 1D/2D coupling ratio can be varied from 10:1 to >70:1 in Au-induced chain structures belonging to the family shown in Fig. 6.16

In order to follow the transition from 2D to 1D, it is not even necessary to go from one structure to another with different chain spacing. The same surface can have bands with different dimensionality crossing the Fermi level, as demonstrated with the Si(553)-Au surface and with Si(111)4×1-In in Subsect. 6.3.8. And the Si(111)5×2-Au surface even contains a single band that changes its dimensionality from 1D at the top to 2D at the bottom [76]. Experiment and theory agree quantitatively on the change in the 1D/2D coupling ratio [51].

The game of gradually reducing dimensionality can be pursued further from 1D to 0D by breaking up the chains into sections with adsorbates. A zero-dimensional surface state has been found at the end of a finite chain on Si(553)-Au by scanning tunneling spectroscopy [68]. The atoms at the end of a finite chain appears either bright or dark, depending on the bias voltage used for the spectroscopic imaging. This phenomenon can be modeled by a simple calculation, thus making it a textbook example of quantum mechanics in very low dimensions. The Si(553)-Au surface is particularly useful for this purpose because it is the most perfect of the Au-induced chain structures studied so far. The length of its chains can be as long as 100 atoms and chains with finite length can be selected over a wide range [69, 71].

6.3.5 Fractional Band Filling

In addition to the 1D/2D coupling ratio, the band filling is an important quantity determining the position of these electronic states in universal diagrams of 1D phases [6]. The filling of a band can be determined from the occupied area between Fermi surfaces and normalizing it to the Brillouin zone [12]. The latter is indicated in Fig. 6.19, with the circle near the center defining the Γ-point at $k = 0$. Since the bands are free-electron-like, Luttinger's theorem tells us that each band holds two electrons per unit cell (one for spin up and one for spin down). Thus, the area of the Brillouin zone corresponds to two electrons per atom. Each of the two closely-spaced bands occurring in most Au chain structures is about half-filled and thus contains one electron if unpolarized (1/2 electron if spin-split, see Subsect. 5.3.7 and [70, 77]). The surfaces tilted towards the [11$\bar{2}$] azimuth (lower half of Fig. 6.16) contain an extra single band with a filling of about 1/3, i.e. 2/3 of an electron per unit cell. Together with the half-filled doublet the filling comes out to be very close to 8/3 of an electron per unit cell (5/3 for a spin-split doublet).

The fractional band filling of some of the gold chain structures raises interesting questions about possible connections to the fractional quantum Hall effect in 2D. However, the photoemission measurements are performed without a magnetic field, and even with an applied field one might question whether the analog of a two-dimensional Landau orbit exists for a one-dimensional chain. There is a less exotic, but nevertheless intriguing explanation of the fractional filling. The Si(553)-Au surface exhibits a tripling of the unit cell near defects [11, 68, 69, 71, 72]. In fact, the structural model with the lowest total energy (among about 50 models tested [12]) has 3×1 periodicity along the chains, as shown in Fig. 6.20. Two extra Si atoms are incorporated at the step edge per 3×1 unit cell. The Au chain at the center of the terrace, on the other hand, does not see much of the 3×1 periodicity at the step edge. It exhibits 1×1 periodicity in STM at room temperature and distorts into 3×1 only near defects and at low temperature [71,72]. Judging from the role of extra Si atoms as dopants (discussed below), it is likely that the extra Si atoms dope the Au chains. If the two Si atoms donated all their 8 valence electrons, that would make 8/3 of an electron per Au atom. This is the upper limit. It is more likely that only two of the four Si valence electrons dope the chains, while the other two electrons are needed for covalent bonds (compare the calculations for Si(111)5×2-Au [51] discussed below). In that case, one would have an electron count of 4/3 of an electron per Au atom, which is close to the 5/3 expected in a scenario with spin-split bands. In any case, a fractional filling seems to be closely tied to the existence of extra Si atoms at the step edge that act as dopants for the gold chains in the middle of the terrace. This type of indirect doping allows for high mobility of electrons near the Au chains, which are not disturbed by dopant atoms. The 1D Au chains are very much the 1D analog of the 2D CuO planes in high-temperature superconductors. In both cases the dopants reside in the next higher dimension.

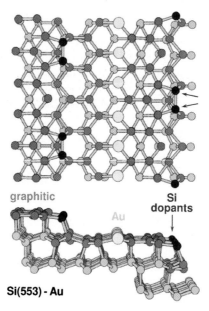

Fig. 6.20. Structural model of the Si(553)-Au surface from total energy minimization (from [12]). Extra Si atoms at the step edge dope the Au chain at the center of the terrace and lead to fractional band filling by tripling the unit cell

6.3.6 Doping of Chains and Nanoscale Phase Separation

The observation of fractional band filling suggests that doping of 1D gold chains is rather peculiar. They cannot be doped continuously like their two-dimensional counterparts in Sec. 6.3.3. Each of them has a well-defined optimum doping, which can be shifted only by a few percent with excess Au or Si. The role of dopant atoms has been investigated closely for the Si(111)5×2-Au structure, which is shown in Fig. 6.21. On top of faint atom chains with 5×2 periodicity one finds additional atoms in regular 5×4 lattice sites. However, the extra atoms occupy only half of the available 5×4 sites. They do this in seemingly random fashion, not by forming the regular 5×8 lattice expected at that coverage. Closer inspection reveals that the extra atoms form a complete 5×4 lattice for typically five sites, followed by an empty stretch of comparable length. This behavior can be quantified by determining the pair correlation function for STM images with thousands of atoms, and the result can be modeled by a nearest neighbor repulsion together with an oscillatory potential with 5×4 periodicity [75]. These interatomic potentials are in the meV regime, and they correspond to mini-gaps in the Fermi surface of the order 10–100 meV when assuming an electronic origin. The additional atoms can be identified as single Si atoms, not the deposited Au atoms: Deposition of an extra 1/40 of a monolayer of Si and annealing below 300° creates a metastable structure where all 5×4 sites are filled [74].

Fig. 6.21. Si dopants adsorbed on top of the chains of the Si(111)5×2-Au structure, forming sections of a filled 5×4 lattice alternating with empty 5×2 sections (from [73]). The filled (*doped*) sections are semiconducting, and the empty sections metallic (see Fig. 6.22). This is a one-dimensional analog of nano-scale phase separation ("stripes") observed in high-temperature superconductors. This structure can be used to fabricate an atomic scale memory, where a bit is stored by the presence or absence of a single Si atom [74]

Particularly interesting is the electronic structure of these alternating segments of filled and empty 5×4 sites. As shown in Fig. 6.22, the empty segments are metallic, while the filled segments are semiconducting with a 0.5 eV gap and E_F located near the bottom of the gap [13]. This behavior is reminiscent

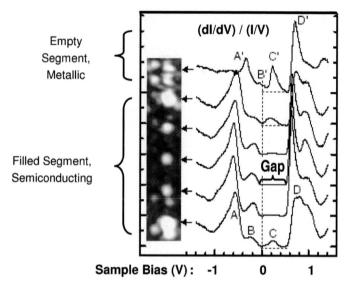

Fig. 6.22. Scanning tunneling spectroscopy data from the phase-separated Si(111)5×2-Au surface (see Fig. 6.21), providing evidence for alternating metallic and semiconducting electronic structure. From [13]

of the "stripes" in high-temperature superconductors, where metallic, doped regions alternate with gapped, undoped regions on a nanometer length scale. These static stripes, and their dynamical counterparts are viewed as a special mechanism of bringing two incompatible ingredients of high-temperature superconductivity together, magnetism and superconductivity. The segments of doped and undoped Si chains may be viewed as one-dimensional analog of stripes. The only difference is that metallic segments are undoped, and vice versa. Thus, the Si(111)5×2-Au surface may serve as prototype for the formation of stripes in general.

For 1D chains there is a model explaining the formation of doped and undoped segments [52], and it is again a compromise between two conflicting requirements. The Si(111)5 × 2-Au structure has an optimum doping of one Si atom per 5 × 8 cell according to total energy calculations [51]. However, a 5 × 4 surface periodicity is dictated by the Fermi surface measured with angle-resolved photoemission [52]. It consists of ideally-nested Fermi lines located at the zone boundaries of the 5 × 4 lattice. A 5 × 4 reconstruction opens a mini-gap at these points, which reduces the energy of the occupied states. This is similar to a Peierls gap, except for a quadrupling of the period instead of a doubling. The 5 × 4 periodicity of the inter-atomic potential [75] has probably the same origin. Likewise, a connection has recently been made between the Fermi surface and the chain length distribution for Si(553)-Au [69].

A set of alternating metallic and insulating segments brings up the conceptual question whether such a surface is a metal or an insulator. In fact, there has been a debate in the photoemission literature about this question. This could be settled by angle-resoled photoemission: If one finds two separate band structures for the metallic and semiconducting segments, the answer would be that the surface is both metallic and insulating. The scanning tunneling spectroscopy result suggests such a scenario. If the wave functions were delocalized over both segments, one would expect a single band structure with weakly metallic character.

There is an interesting application of the Si(111)5 × 2-Au structure as atomic scale memory, where a bit is stored by the presence or absence of an extra Si atom [74]. The chains are used as self-assembled tracks that are precisely five atom rows wide. The storage density is comparable to that of DNA (1 bit per $5 \times 4 = 20$ atoms versus 32 atoms per bit for DNA). If one wants to apply Moore's law to estimate the arrival of such a technology, one can scale the density up to 250 Terabits/inch^2 or the track width down to 1.6 nm and finds that this is expected to happen around 2040. Moore's law would have to remain valid twice as long as it has been in operation. Such a memory is highly impractical today, but allows testing the fundamental limits of data storage (see [74] for details). The density limit is actually due to the repulsive interaction between extra Si atoms discussed above, which makes it difficult putting two Si atoms into adjacent 5 × 2 cells and thereby doubling the density.

6.3.7 The Puzzle of Two Half-Filled Bands: Spin-Splitting?

The observation of two half-filled bands brings up an intriguing question: Why does the surface choose two half-filled bands (corresponding to two broken bonds) instead of one completely-filled band (corresponding to a covalent bond). An early explanation associated the splitting with spin-charge separation [45], one of the exotic phenomena predicted for 1D electrons. However, this option is ruled out by the fact that the splitting does not vanish at E_F (see [76] and Fig. 6.23, right).

This leaves several other potential explanations for the splitting, some of them rather exotic as well, such as a spin-splitting at a non-magnetic surface. First, consider the orbitals surrounding each Au atom. Fig. 6.20 gives a structural model for the Au chain, which applies to both Si(557)-Au and Si(553)-Au. Although Au is coordinated with three Si atoms underneath, it is monovalent and forms only a single covalent bond. By symmetry, one can assume this to be the bond to the left in Fig. 6.20. The two remaining Si atoms on the right point towards the Au with broken bond orbitals. These have $p_x + p_y$, $p_x - p_y$ character and thus form a doubly-degenerate state. The degeneracy can be removed by the formation of bonding and antibonding combinations. In fact, a local density calculation [70] predicts such a band (full circles in Fig. 6.23 left). The bonding-antibonding splitting creates a gap of about 0.2 eV at the zone boundary $ZB_{1\times 2}$.

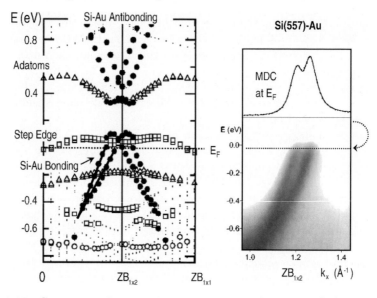

Fig. 6.23. Comparison between band calculations (left, from [70]) and angle-resolved photoemission (right, from [76]) for the Si(557)-Au chain structure. In both cases, a nearly-degenerate doublet of bands crosses the Fermi level at $ZB_{1\times 2}$, the Brillouin zone boundary of a 1×2 cell (doubled along the chain direction by Si adatoms). The calculation shows back-folding at the 1×2 boundary, which is very weak in the experiment due to the weakness of the 1×2 lattice potential at the center of the terrace, where the Au chain resides

The actual splitting of the band in Fig. 6.23, however, is a spin splitting created by the spin-orbit interaction. This model reproduces the two bands seen in photoemission remarkably well when taking into account that the back-folded part of the band between $ZB_{1\times 2}$ and $ZB_{1\times 1}$ is hardly visible in photoemission. Spin-orbit interaction with heavy elements, such as Au, is able to produce spin polarization at the surface of a non-magnetic material. While the net spin polarization integrated over the Brillouin zone remains zero, individual parts of the Brillouin zone become 100% spin-polarized. Figure 6.24 shows a a schematic for a well-studied two-dimensional case [33, 34] where a single surface state band on Au(111) splits into two spin-split bands at the Fermi surface. These exhibit circular spin structures rotating in opposite directions. These two-dimensional Fermi circles can be converted into one-dimensional Fermi lines by cutting them and deforming them into lines. The resulting one-dimensional spin pattern for the spin-split band on Si(557)-Au contains opposite spins running up and down the Fermi lines. The splitting is actually larger for Si(557)-Au than for Au(111). A critical test of this spin-orbit model for the band splitting would be spin-polarized, angle-resolved photoemission or the detection of dichroism with circularly-polarized light. The recently observed pattern of band crossings at the $ZB_{1\times 2}$ supports the spin-polarized

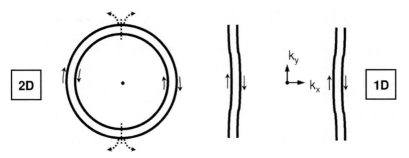

Fig. 6.24. Schematic of the spin splitting of the Fermi surface in 2D and 1D, induced by spin-orbit interaction. The prototypical 2D case is Au(111) (left, see [33, 34]), where the Rashba Hamiltonian shows how the lack of inversion symmetry about the surface lifts the spin-degeneracy. The 2D Fermi surface can be converted into 1D by cutting it in two places and deforming it into the 1D Fermi surface observed on Si(557)-Au and Si(553)-Au (compare Fig. 6.19). This explains the spin splitting predicted for Si(557) [70] (see the left panel of Fig. 6.23)

model (crossings between equal spins are avoided, but not between opposite spins [77]).

The band model in Fig. 6.23 left is rather promising, but several loose ends remain. The band filling of the spin-split bands is odd and leads to an odd overall electron count per Au atom, a problem already encountered with the spin-charge separation model. The electron count cannot be changed by adding or removing Si atoms from the surface since each Si has four valence electrons. Bringing in electrons from the bulk would require the Fermi level to lie at or below the valence band maximum, which is not observed experimentally ($E_F \approx$ VBM + 0.1 eV at these surfaces). A second unsolved question is the existence of a flat band that straddles $E_F = 0$ in Fig. 6.23 and becomes occupied near the Γ-point. It has not been observed in photoemission, but it would explain spectroscopic STM imaging, where metallic states are found near the step edge [78–80].

The empty part of the calculated surface band structure is completely different from the occupied part. Instead of a steep, free-electron-like band with a small effective mass of about 0.5, one finds a flat band near E_F. At slightly higher energy there is even a band gap due to the large bonding/antibonding splitting of the Si-Au orbitals. These states are just beginning to be explored by two-photon photoemission (see Sect. 6.4).

6.3.8 Peierls Gap and Charge Density Waves

A characteristic feature of one-dimensional systems is the Peierls distortion. Using rather generic arguments of trading off strain energy cost and electronic energy gain, one can show that a one-dimensional system with a half-filled, metallic band generally doubles its period at low enough temperatures. The doubled lattice period induces a mini-gap at the Fermi level, which lowers the energy of the occupied states and leads to a metal-insulator transition. Although there may be exceptions from this generic case (for example in multiple chains or carbon nanotubes), the threat of a Peierls transition prevents searching for exotic 1D phenomena at very low temperatures. A better strategy consists of raising the transition temperature for the interesting phases, such as superconductivity and spin-charge separation, such that they can compete with the Peierls transition. This may be achieved by tailoring the interactions such that the electrons become increasingly correlated (for the case of spin-charge separation see the Chap. 4 by R. Claessen, J. Schäfer, and M. Sing in this volume).

Despite the fact that atom chains at semiconductor surfaces are firmly embedded into the substrate lattice, they seem not to be completely immune against the formation of Peierls-like gaps. Angle-resolved photoemission finds that the metallic bands gradually open up mini-gaps at E_F in the temperature range between 300 K and 100 K (see Fig. 6.25 and [72, 81]). Even the fractionally-filled bands show such a gap. The formation of mini-gaps is accompanied by period doubling (tripling for the fractionally-filled band). These effects have been studied extensively by comparing angle-resolved photoemission and scanning tunneling spectroscopy [43, 44, 71, 72, 81].

Observing period doubling by scanning tunneling microscopy does not necessarily imply a lattice distortion, as in the Peierls scenario. Charge density waves are able to modulate STM images without any atomic displacements. They are driven purely by the instability of parallel ("nested") Fermi surfaces towards the opening of a gap at E_F. One-dimensional Fermi surfaces are straight lines and, thus, always perfectly nested.

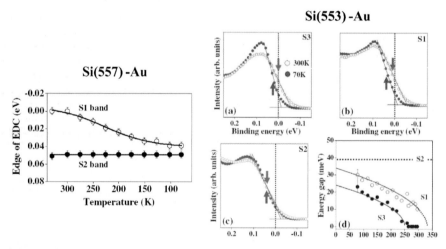

Fig. 6.25. Formation of a Peierls-like gap in the one-dimensional, metallic bands of Si(557)-Au and Si(553)-Au. From [72, 81]

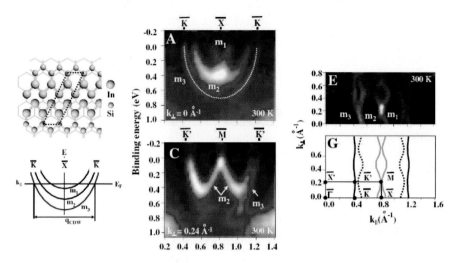

Fig. 6.26. Multiple metallic bands and interacting charge density waves in the four-chain structure Si(111)4×1-In. From [44]

Multiple chain structures can give rise to multiple charge density waves with different periods. That opens up a rich area of interacting atomic chains and the charge density waves associated with them. An example is the four-chain structure of the Si(111)4×1-In surface in Fig. 6.26. The phase transitions leading to these charge density waves are rather sluggish and lead to dynamical fluctuations between two phases over a large temperature range, another characteristic of low-dimensional systems [84, 85]

6.4 Summary and Future Avenues

There is one key message to be taken away from this overview: Atomic chains at surfaces provide a new playground for low-dimensional physics. These structures can be tailored systematically, which makes it possible to explore a large part of the parameter space of one-dimensional phase diagrams. The dimensionality, for example, is varied via the step spacing from 2D to 1D. Metallic chains on semiconducting substrates can be completely de-coupled from the bulk electronically, while locking the chain atoms firmly in place by backbonds. Spin chains can be created by transition metal and rare earth atoms. Magnetic Co chains exhibit a dramatic increase of the orbital moment.

One-dimensional states on noble metal surfaces tend to derive their electronic structure from free-electron-like 2D surface states, which makes them prototypes for demonstrating quantum confinement. One can distinguish states confined to terraces, step states, and superlattice states which propagate across steps and terraces. It has become possible to determine these wave functions directly from a Fourier transformation of the momentum distribution observed by photoemission, using phase recovery methods from x-ray optics.

All kinds of unexpected, sometimes counter-intuitive phenomena appear in chains on semiconductor substrates. Chains of metal atoms do not attach themselves to step edges, as expected from step-flow growth and observed on metal surfaces. They stay right in the middle of the terrace. The step edge is taken over by the honeycomb chain at Si(111) surfaces, a graphitic form of silicon stabilized by π-bonding, which is highly unstable in Si compounds. In fact, this graphitic strip of silicon may be the driving force for stabilizing 1D over 2D structures. Fractional band filling is seen for Si(553)-Au and explained by indirect doping of the chains at the center of the terrace by Si dopants at the step edge. Nanoscale phase separation takes place by bunching of dopants and leads to alternation sections of metallic and semiconducting chain segments, a one-dimensional prototype for the two-dimensional stripes in high temperature superconductors.

This field is still wide open. Many of the unusual phenomena are still unexplained, and most of the structures have yet to be determined. Even the origin of the metallicity of atom chains on semiconductors is still in question. Do the metallic electrons come from the metal chains at the center of the terrace or from the graphitic Si chain at the edge?

Angle-resolved photoemission needs to be pushed to its limits by these structures, as far as momentum resolution is concerned. The unit cells in real space are rather large, in order to de-couple the chains, and that leads to tiny Brillouin zones. In the future, the angle-resolved photoemission with high spatial resolution may become an important tool for zooming in on small, but highly-perfect regions of a surface. For example, optical spectra from self-assembled quantum dots on semiconductors become increasingly sharper when focusing a laser down to the micron regime [82]. The ultimate goal would be angle-resolved photoemission from single nano-objects, such as a quantum dot or a nanotube. The size distribution would be eliminated completely. The wave-function reconstruction techniques developed for confined terrace states on metals are well-matched to such systems.

Isolating the photemission signal of such nano-objects from the substrate background is going to be quite a challenge. The cross section can be optimized for specific wave functions by varying the photon energy and the polarization using synchrotron radiation. For example, the one-dimensional states on Si surfaces become most pronounced in the magic photon energy range 34–40 eV, while being rather weak at the He I resonance line. Resonant photoemission enhances a specific element, while a Cooper minimum can be used to suppress the substrate.

Another challenge for angle-resolved photoemission is the connection to transport measurements, such as conductivity and superconductivity of surfaces. These measurements become very difficult to apply to atomic chains, let alone a single atom chain. Photoemission measures the complete set of quantum numbers for surface electrons, and the resulting set of energy bands and linewidths could conceivably be used to calculate transport from photoemission data. That will require a very high energy resolution, with Δ_E small compared to the thermal energy $k_B T$.

While photoemission does an excellent job with occupied electronic states, it does not provide information about unoccupied states between the Fermi level and the vacuum level – apart from a small energy slice above the Fermi edge that is populated by thermal electrons. Inverse photoemission is the natural counterpart to photoemission, but the extremely low cross section of this technique has not allowed it to reduce the energy resolution below thermal energies. There is another option for accessing unoccupied states with high energy resolution, as long as they are long-lived: That is two-photon photoemission, where the first photon pumps electrons into an unoccupied state and the second photon ionizes them. This technique is just beginning to be applied to atomic chain structures, as shown in Fig. 6.27 [83]. The schematic diagram on the right illustrates the process. By combining high intensity infrared and low-intensity UV photons ($h\nu$ and $3h\nu$) one avoids swamping the analyzer with low-energy electrons from the thermal tail above E_F excited by single UV photons. The sequence of events can be determined by delaying the UV with respect to the IR pulse or vice versa. This type of spectroscopy does not only provide energy bands, it also follows the relaxation of the excited electrons in real time. These initial results suggest that there will be a lot more information available in the near future to solve the many riddles of atomic chains at surfaces.

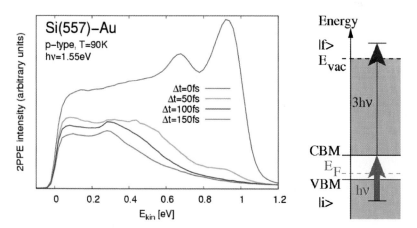

Fig. 6.27. Two-photon photoemission as future avenue for probing the unoccupied states of low-dimensional surface structures and determining the dynamics of hot electrons. The electrons with high kinetic energy disappear with 50 fs and then accumulate over several hundred fs at the bottom of a band. From [83]

References

1. E. H. Libe and D. C. Mattis (eds.) in: *Mathematical Physics in One Dimension: Exactly Soluble Models of Interacting Particles* (Academic, New York, 1966)
2. T. Giamarchi in: *Quantum Physics in One Dimension* (Oxford University Press, New Nork, 2004).
3. J. Solyom: Adv. Phys. **28**, 201 (1979)
4. K. Schönhammer in: *Strong Interactions in Low Dimensions*, ed. by D. Baeriswyl and L. Degiorgi (Klumer Academic Publishers, 2003) Ch.1 Section 5.2.
5. J. Voit: Rep. Prog. Phys. **58**, 977 (1995)
6. G. Gruner: *Density Waves in Solids* (Perseus Publishing, Cambridge, Massachusetts 1994)
7. P. M. Chaikin et al: J. Phys.-Condes. Matter **10**, 11301 (1998)
8. H. Ohnishi et al: Nature **395**, 780 (1998)
9. V. Rodrigues et al: Phys. Rev. Lett. **85**, 4124 (2000)
10. J. N. Crain and F. J. Himpsel, Appl. Phys. A **82**, 431 (2006)
11. J. N. Crain et al: Phys. Rev. Lett. **90**, 176805 (2003)
12. J. N. Crain et al: Phys. Rev. B **69**, 125401 (2004)
13. H. S. Yoon et al: Phys. Rev. Lett. **92**, 096801 (2004)
14. A. Vindigni et al: Appl. Phys. A **82**, 385 (2006)
15. A. Kirakosian et al: Surf. Sci. **498**, L109 (2002)
16. J. N. Crain et al: Phys. Rev. B **66**, 205302 (2002)
17. J.-L. Lin et al: Appl. Phys. Lett. **78**, 829 (2001)
18. N. Papageorgiou et al: Appl. Phys. Lett. **82**, 2518 (2003)
19. E. D. Williams: Surf. Sci. **299/300**, 502 (1994)
20. A. R. Bachmann et al: Phys. Rev. B **64**, 153409 (2001)
21. J. Lobo et al: Phys. Rev. Lett. **93**, 137602 (2004)
22. J. E. Ortega et al: Phys. Rev. B **72**, 195416 (2005)
23. J. E. Ortega et al: New Journal of Phys. **7**, 101 (2005)
24. A. Mugarza and J. E. Ortega: J. Phys. Cond. Mat. **15**, S3281 (2003)
25. A. Mugarza et al: J. of Phys. C **18**, S27 (2006)
26. L. Bürgi et al: Phys. Rev. Lett. **81**, 5370 (1998)
27. A. Mugarza et al: Phys. Rev. B **67**, 081404 (2003)
28. J. Miao et al: Nature (London) **400**, 342 (1999)
29. M. Giesen et al: Phys. Rev. Lett. **82**, 3101 (1999)
30. K. Morgenstern et al: Phys. Rev. Lett. **89**, 226801 (2002)
31. F. Baumberger et al: Phys. Rev. Lett. **92**, 16803 (2004)
32. A. Mugarza et al: Phys. Rev. B **66**, 245419 (2002)
33. S. LaShell et al: Phys. Rev. Lett. **77**, 3419 (1996)
34. F. Reinert et al: Phys. Rev. B **63**, 115415 (2001)
35. F. Schiller et al: Phys. Rev. Lett. **95**, 066805 (2005)
36. M. Roth et al: Phys. Rev. Lett. **88**, 096802 (2002)
37. S. Smadici and R. M. Osgood: Phys. Rev. B **71**, 165424 (2005)
38. F. Schiller et al: Phys. Rev. Lett. **94**, 016103 (2005)
39. R. Eder and H. Winter: Phys. Rev. B **70**, 085413 (2004)
40. X. Y. Wang et al: Phys. Rev. B **56**, 7665 (1997)
41. S. Hasegawa et al: Prog. Surf. Sci. **60**, 89 (1999)
42. J. N. Crain et al: Phys. Rev. B **72**, 045312 (2005)

43. H. W. Yeom et al: Phys. Rev. Lett. **82**, 4898-4901 (1999)
44. J. R. Ahn et al: Phys. Rev. Lett. **93**, 106401 (2004)
45. P. Segovia et al: Nature **402**, 504 (1999)
46. A. Kirakosian et al: Appl. Phys. Lett. **79**, 1608 (2001)
47. J. Kuntze et al: Appl. Phys. Lett. **81**, 2463 (2002)
48. S. C. Erwin and H. H. Weitering: Phys. Rev. Lett. **81**, 2296 (1998)
49. D. Y. Petrovykh et al: Surf. Sci. **512**, 269 (2002)
50. R. Losio et al: Phys. Rev. Lett. **85**, 808 (2000)
51. S. C. Erwin: Phys. Rev. Lett. **91**, 206101 (2003)
52. J. L. McChesney et al: Phys. Rev. B **70**, 195430 (2004)
53. J. Viernow et al: Appl. Phys. Lett. **72**, 948 (1998)
54. J.-L. Lin et al: J. Appl. Phys. **84**, 255 (1998)
55. D. Sánchez-Portal and R. M. Martin: Surf. Sci. **532**, 655 (2003)
56. S. C. Erwin: unpublished.
57. I. K. Robinson et al: Phys. Rev. Lett. **88**, 096104 (2002)
58. D. Sánchez-Portal et al: Phys. Rev. B **65**, 081401 (2002)
59. R. I. G. Uhrberg et al: Phys. Rev. B **65**, 081305(R) (2002)
60. Y. G. Ding et al: Phys. Rev. Lett. **67**, 1454 (1991)
61. H. Aizawa and M. Tsukada: Phys. Rev. B **59**, 10923 (1999)
62. R. Losio et al: Phys. Rev. B **61**, 10845 (2000)
63. J. Ortega et al: Phys. Rev. B **58**, 4584 (1998)
64. F. Flores et al: Surf. Rev. Lett. **4**, 281 (1997)
65. J.E. Demuth et al: Phys. Rev. Lett. **51**, 2214 (1983)
66. R. Schillinger et al: Phys. Rev. B **72**, 115314 (2005)
67. I. Barke et al: Phys. Rev. Lett., **96**, 216801 (2006)
68. J. N. Crain and D. T. Pierce: Science **307**, 703 (2006)
69. J. N. Crain et al: Phys. Rev. Lett. **96**, 156801 (2006)
70. D. Sánchez-Portal et al: Phys. Rev. Lett. **93**, 146803 (2004)
71. P. C. Snijders et al: Phys. Rev. Lett. **96**, 076801 (2006)
72. J. R. Ahn et al: Phys. Rev. Lett. **95**, 196402 (2005)
73. F. J. Himpsel et al: J. Phys. Chem. B **108**, 14484 (2004)
74. R. Bennewitz et al: Nanotechnology **13**, 499 (2002)
75. A. Kirakosian et al: Phys. Rev. B **67**, 205412 (2003)
76. R. Losio et al: Phys. Rev. Lett. **86**, 4632 (2001)
77. I. Barke, Fan Zheng, T. K. Rügheimer, and F. J. Himpsel, Phys. Rev. Lett. **97**, 226405 (2006)
78. H. W. Yeom et al: Phys. Rev. B **72**, 035323 (2005)
79. M. Schöck et al: Europhys. Lett. **74**, 473 (2006)
80. M. Krawiec et al: Phys. Rev. B **73**, 075415 (2006)
81. J. R. Ahn et al: Phys. Rev. Lett. **91**, 196403 (2003)
82. D. Gammon et al: Appl. Phys. Lett. **67**, 2391 (1995)
83. T. K Rügheimer et al: Phys. Rev. B, submitted
84. G. Lee et al: Phys. Rev. Lett. **95**, 116103 (2005)
85. C. Gonzales et al: Phys. Rev. Lett. **96**, 136101 (2006)

Part III

Ultimate Resolution

7

High-Resolution Photoemission Spectroscopy of Low-T_c Superconductors

T. Yokoya[1], A. Chainani[2], and S. Shin[2,3]

[1] The Graduate School of Natural Science and Technology, Okayama University, 3-1-1 Tsushima-naka, Okayama 700-8530 Japan
yokoya@cc.okayama-u.ac.jp
[2] RIKEN SPring-8 Center, Sayo-gun, Hyogo 679-5148, Japan
chainani@spring8.or.jp
[3] The Institute for Solid State Physics, University of Tokyo, Kashiwa, Chiba 277-8581, Japan
shin@issp.u-tokyo.ac.jp

Abstract. The high-resolution photoemission spectroscopy of conventional superconductors is reviewed. It is shown that with the presently available resolution (0.360 meV using laser excitation) the gap structure (like two gaps in MgB_2 or an anisotropic gap in $CeRu_2$) can be resolved with a high degree of accuracy. It is pointed out that the use of low-photon-energy laser excitation (10 eV or less) is an alternative way (as compared to the use of high photon energies) to make more bulk-sensitive measurements with photoemission spectroscopy.

7.1 Introduction

Bardeen, Cooper, and Schrieffer's (BCS) theory of superconductivity is based on the condensation of Cooper pairs into a spin singlet induced by an isotropic (or momentum(k)-independent) electron–phonon coupling [1]. In a simple description, the superconducting transition temperature (T_c) is expressed as $T_c = \theta_D \cdot exp(-1/N(0) \cdot g)$, where θ_D is the Debye temperature, $N(0)$ is the electronic density of states at the Fermi level (E_F), and g is the effective attraction energy between a spin-up electron and a spin-down electron. Formation of a Cooper pair gives rise to an excitation energy gap, namely the superconducting energy gap (SC gap) in the density of states (DOS) across E_F (Fig. 7.1). The magnitude of the SC gap thus reflects the strength of the pairing interaction of Cooper pairs, and the reduced gap value $2\Delta/k_B T_c$ compared to the mean-field value of 3.51, is used to classify a superconductor as a weak-coupling or strong-coupling superconductor. The pairing function, which is also called the superconducting order parameter, has an isotropic s-wave symmetry for the phonon-mediated case (Fig. 7.2(a)). In the 1960's, soon after the formulation of the BCS theory, tunneling spectroscopy revealed

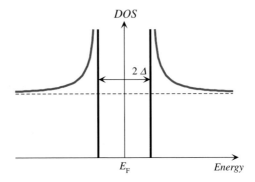

Fig. 7.1. BCS superconducting density of states

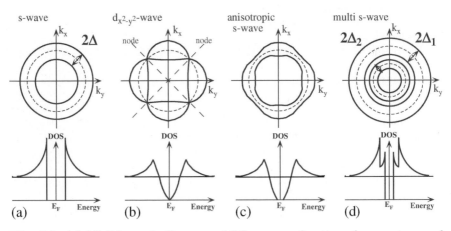

Fig. 7.2. (a)-(d) Schematic diagrams of SC gap as a function of momentum and superconducting DOS: (a) k-independent (isotropic) s-wave; (b) $d_{x^2-y^2}$-wave; (c) k-dependent (anisotropic) s-wave, and (d) FS sheet dependent s-wave

the superconducting electronic structures of elemental superconductors. The experimentally obtained electronic-structure changes across T_c provided an important and direct proof of the weak-coupling BCS theory. However, along with the confirmation of BCS theory, the experimental results also indicated deviations from the BCS theory in the form of larger gaps compared to the mean-field value and additional structure in the measured DOS, which defined the need of the strong-coupling theory [2]. Nonetheless, the simple BCS theory that assumes a k-independent electronic structure and electron–phonon coupling has succeeded to explain a large set of superconducting, thermodynamic and spectroscopic results [3].

A low-T_c superconductor in the present review is used to mean a superconductor other than the cuprate high-temperature superconductors (HTSCs), although in early work it has been used to represent conventional superconductors which are described by the weak-coupling BCS or the strong coupling theory. The HTSCs clearly belong to a special class, where a highly anisotropic order parameter of the $d_{x^2-y^2}$-wave symmetry (Fig. 7.2(b)) has been established and different mechanisms of superconductivity other than phonons are actively discussed, although no consensus has been achieved [4]. In addition to the conventional superconductors, the new families of superconductors discovered after HTSCs also exhibit anomalous superconducting, transport and magnetic properties that cannot be explained by a simple BCS theory [5]. In particular, we include them here because observations of multiple gaps or anisotropic gaps suggest $\boldsymbol{k}$ and Fermi surface (FS) sheet dependent superconducting DOSs, as schematically illustrated in Figs. 2(c) and (d).

Photoemission spectroscopy is an experimental probe which provides the electronic structure of solids. The energy resolution of photoemission spectroscopy drastically improved in the late 1980's [6,7], driven by the motivation to measure the SC gap of the HTSCs. Because of advances in energy resolution, together with its capability of measuring k-resolved electronic structure, angle-resolved photoemission spectroscopy (ARPES) played a crucial role in establishing $d_{x^2-y^2}$-wave symmetry of the HTSCs. It also provided evidence for a pseudogap in the normal state of the HTSCs, with the same symmetry as the SC gap [8,9]. These results have provided important experimental information to understand the mechanism of superconductivity in HTSCs. However, results for conventional superconductors showing a clear SC gap and the expected pile-up in DOS have been limited [10,11] due to the energy resolution and sample cooling technique of spectrometers. The absence of reliable reference spectra of conventional superconductors with low-T_c was considered as a weak point for the justification of photoemission results of HTSCs. From an experimental point of view, the limitations of photoemission spectroscopy for studying very low energy scales has been discussed [12]. Therefore, confirmation of the superconducting transition of low-T_c superconductors by photoemission spectroscopy had been an important and challenging issue.

In the late 1990's, photoemission results showing the superconducting transition of element superconductors like Nb and Pb could be obtained, and thereby, the reliability and powerfulness of photoemission spectroscopy for studying superconducting transitions has been established [13]. Moreover, extremely high resolution as well as the $\boldsymbol{k}$-dependent capability of ARPES has revealed importance of $\boldsymbol{k}$-dependent electronic structure and electron–phonon coupling, which were not taken into consideration in early work. In the following sections, we review high-resolution photoemission studies of the superconducting transitions for various low-T_c superconductors.

7.2 High-Resolution and Low-Temperature Photoemission Spectroscopy

In Fig. 7.3, we illustrate a schematic diagram of the spectrometer we used to obtain most of the photoemission results discussed in this article [14]. It is equipped with a Gammadata-Scienta SES2002 electron analyzer, a Gammadata high-flux discharging lamp with a toroidal grating monochromator, a flowing liquid-He cryostat for sample cooling, and a newly designed double thermal shield which also works as a cryopump. A simple idea of using an additional thermal shield from an independent cryostat is essential to achieve the lowest sample temperature down to 4 K. As for the photoemission studies of conventional superconductors, which normally have a T_c less than 10 K and SC gap energy scales of ~1 meV, both the requirements, lower sample temperature and higher energy resolution, are essential. The sample temperature is measured using a silicon-diode sensor mounted near the sample and the temperature can be controlled from 4 K (by pumping of liquid He) to 300 K. The main chamber is pumped with two turbo molecular pumps (TMP) connected in series, a Ti getter pump, and the cryopump. The base pressure is better than 5×10^{-11} Torr. To increase the lifetime of fresh surfaces, a thin-film filter for improving vacuum during measurements can be used [15].

Figure 7.4 shows photoemission spectrum of an evaporated gold film. As determined from the Fermi edge spectrum, the highest energy resolution we achieve is 1.4 meV, which includes the energy width of the He Iα resonance line of 1.1–1.2 meV. This indicates that the resolution of the analyzer is about 0.8 meV and therefore one can obtain μeV resolution by using a

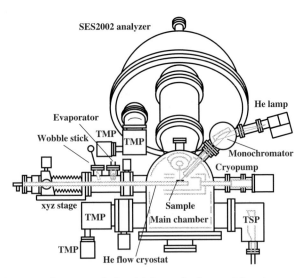

Fig. 7.3. Schematic diagram of ultrahigh-resolution and low-temperature photoemission spectrometer

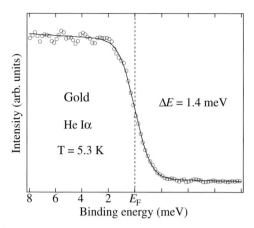

Fig. 7.4. Ultrahigh-resolution photoemission spectra of gold (*open circles*) together with a FD function of 5.3 K convolved by a Gaussian of FWHM of 1.4 meV (*solid line*) [14]

higher resolution phonon source. Recently, laser sources, with a much narrower linewidth than a discharging lamp, have been used to perform photoemission studies and will be briefly discussed in the future prospects section.

7.3 Superconducting DOS

7.3.1 SC Gap Functions

In the weak-coupling regime, BCS theory predicts that the superconducting DOS is $\omega/(\omega^2 - \Delta^2)^{1/2}$, where ω is the binding energy and Δ is the SC gap value [1]. The Dynes function is the modified BCS function including a superconducting energy gap value Δ and a thermal broadening parameter Γ, where the superconducting density of states $N(\Delta, \Gamma)$ is expressed as $N(\omega, \Delta, \Gamma) = (\omega - i\Gamma)/(\omega - i\Gamma^2 - \Delta^2)^{1/2}$. This was first introduced by Dynes et al. for explaining the temperature-dependent electronic-structure broadening of Pb [16], and has been widely used to deduce SC gap values from tunneling spectra of weak- and strong-coupling superconductors [17]. For anisotropic superconductors, k-dependent $\Delta(k)$s suitable for obtained spectra are used. For HTSC, $\Delta(k) = \Delta_{max}\cos(2\theta)$, where Δ_{max} is the maximum values of gap and θ is the polar angle in k-space [18]. For borocarbides, $\Delta(k) = \Delta_{min} + (\Delta_{max} - \Delta_{min})\cos(2\theta)$ ($0 \le \theta \le \pi/4$), where Δ_{min} is the minimum values of gap [19]. For a two-gap superconductor MgB_2, a weighted sum (N_{L+S}) of two Dynes functions (N_L for the larger gap and N_S for the smaller one), with $N_{L+S}(\omega, \Delta, \Gamma) = 1/(1+R) \cdot N_L(\omega, \Delta, \Gamma) + R/(1+R) \cdot N_S(\omega, \Delta, \Gamma)$, where R is an amplitude ratio of the smaller gap to the larger one, was found to reproduce the obtained spectra [20]. For obtaining $\Delta(k)$ from ARPES spectra, a phenomenological spectral function

$\pi A(\boldsymbol{k},\omega) = \Sigma''(\boldsymbol{k},\omega)/[(\omega - \epsilon_{\boldsymbol{k}} - \Sigma'(\boldsymbol{k},\omega))^2 + \Sigma''(\boldsymbol{k},\omega)^2]$ with $\Sigma(\boldsymbol{k},\omega) = -i\Gamma_1 + \Delta^2/[(\omega + i0^+) + \epsilon(\boldsymbol{k})]$, as is used for HTSCs [9], was applied also for low-T_c superconductors.

7.3.2 Eliashberg Equations

While the gap functions described above explain a wide range of experimentally measured gaps, the strong-coupling superconductors not only show an enhanced gap value compared to the BCS mean-field result, but also show additional features at higher energy scales compared to the gap. Based on the Eliashberg phonon interaction between electrons, the energy gap function $\Delta(\omega)$ of such a superconductor can explain these additional features, and the complete method is described in detail in early work [2,3,21,22]. Specifically, the features occur at the energy gap plus multiples of the phonon frequencies and are thus derived from the phonon density of states of the superconductor.

The complex gap function is given by

$$\Delta(\omega) = \Phi(\omega)/Z(\omega) \quad (7.1)$$

where,

$$\Phi(\omega) = \int_0^\infty d\omega' Re[\Delta'/(\omega'^2 - \Delta'^2)^{1/2}] \int d\omega_q \alpha^2 F(\omega_q)[D_q(\omega' + \omega) + D_q(\omega' - \omega) - U]$$

$$\Psi(\omega) = 1 - Z(\omega)\omega = \int_0^{\omega_c} d\omega' Re[\omega'/(\omega'^2 - \Delta'^2)^{1/2}]$$
$$\times \int d\omega_q \alpha^2 F(\omega_q)[D_q(\omega' + \omega) - D_q(\omega' - \omega)]$$

are coupled integral equations for the self-energies with $D_q(\omega) = (\omega + \omega_q - i0^+)^{-1}$, and U is the Coulomb potential.

The Eliashberg function is $\alpha^2 F(\omega)$, with $F(\omega)$ being the phonon density of states and α^2 is an effective electron–phonon coupling function. The coupled integral equations can be solved iteratively if we know $\alpha^2 F(\omega)$, i.e. for a phonon DOS obtained from an independent experiment(such as neutron scattering), or using a trial function, and the reduced electronic DOS is given by $N_s(\omega)/N(0) = Re[|\omega|/(\omega^2 - \Delta^2)^{1/2})]$ where, $N_s(\omega)$ is the superconducting-phase DOS and $N(\omega)$ is the normal-phase DOS which are measured by tunneling or photoemission spectroscopy. Thus, the calculated reduced electronic DOS can be compared with the experimentally measured reduced DOS. The dimensionless electron–phonon coupling parameter is obtained as $\lambda = 2 \int_0^{\omega_m} d\omega' (\alpha^2 F(\omega'))/\omega'$.

There also exists the inverse theory [2,3] in which one uses the experimentally obtained DOS to calculate the Eliashberg function $\alpha^2 F(\omega)$. While both, the direct and the inverse methods have been used for tunneling spectra extensively [2,3], recent work has demonstrated its applicability even for photoemission spectra of low-T_c superconductors as well as the HTSCs [23–25]. While the method has been applied to HTSCs, it is done under the assumption that the electrons couple to a bosonic mode whose origin is a topic of active present-day research [23–25].

7.3.3 Comparison with Photoemission Spectrum

Since photoemission measures the occupied DOS, the quantification of the gaps requires the theoretical functions to be multiplied by the Fermi–Dirac distribution function of the measurement temperature and convolved with a Gaussian corresponding to the experimental energy resolution. Photoemission spectroscopy succeeded to measure the SC gaps of the HTSCs before that of the low-T_c superconductors, as the required resolutions were lower for measuring the larger SC gaps of HTSCs. With the improvement in resolution, the quantification of the SC gaps in a variety of materials has led to novel results for the low-T_c superconductors.

7.4 Photoemission Results of Superconducting Gap and Strong-coupling Line Shape

7.4.1 Elemental Metals: Pb and Nb

Pb and Nb are elemental metal superconductors having T_c of 7.1 and 9.2 K, respectively. Among the elemental metal superconductors, Nb exhibits the highest T_c, while Pb has the highest electron–phonon coupling constant. Figures 7.5(a) and (b) show the first angle-integrated photoemission spectra of

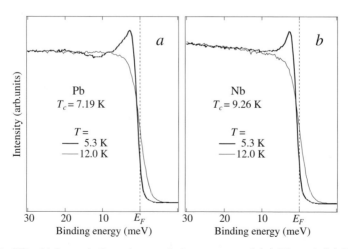

Fig. 7.5. Ultrahigh-resolution photoemission spectra of (**a**) Pb and (**b**) Nb, measured at 5.3 K (superconducting state) and 12.0 K (normal state). Redistribution of spectral weight and the opening of SC gaps are observed for Pb and Nb [13]

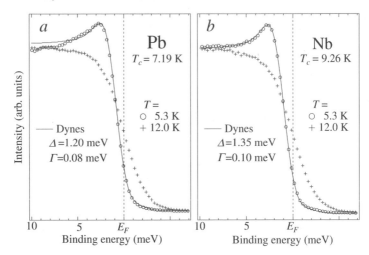

Fig. 7.6. Dynes function fits to the peak in the DOS for (**a**) Pb and (**b**) Nb are used to estimate the SC gap values [13]

Pb and Nb measured at 5.3 K and 12.0 K across the superconducting transitions [13]. As for Nb spectra, while the 12.0 K spectrum shows a Fermi-edge structure, the spectrum at 5.3 K shows a sharp peak at 2.7 meV binding energy with a shift of the leading edge to higher binding energy and with redistribution of spectral weight upto 15 meV as compared to the normal state spectrum. The observed change in the spectral shape represents opening of a SC gap. Similarly, Pb spectra show gap formation, a sharp peak at 2.5 meV, followed by fine structures at higher energies (discussed in detail later). The values of the gaps were estimated by using a Dynes function fit (Figs. 7.6(a) and (b)) to the peak and the leading edge [$\Delta(5.3\,\mathrm{K}) = 1.35$ meV and a $\Gamma = 0.10$ meV for Nb, and $\Delta(5.3\,\mathrm{K}) = 1.20$ meV and a $\Gamma = 0.08$ meV for Pb]. Using the measured values of $\Delta(5.3\,\mathrm{K})$ and the known dependence of the reduced energy gap $\Delta(T)/\Delta(0)$ versus reduced temperature (T/T_c) from strong-coupling theory [2] (the reduced energy gap versus reduced temperature is known to be very similar to the BCS weak-coupling result, as well as other experiments), $2\Delta(0)/k_B T_c$ for Nb is found to be 3.7 and for Pb to be 4.9. These values of $2\Delta(0)/k_B T_c$ are in good agreement with values known from thermodynamic measurements – 3.8 for Nb and 4.5 for Pb [3].

The Dynes function fit deviates from experiment at binding energies beyond the peak, particularly for Pb. This is more clearly seen when we enlarge the superconducting-state spectra as shown in Fig. 7.7. We see in the Pb spectrum : (i) the peak in the superconducting spectrum is itself asymmetrically broadened on the higher binding energy side (4–6 meV), (ii) a weak feature at about 9 meV, (iii) a dip at 10–15 meV, and a hump around 20 meV. It is known from neutron-scattering studies [26] and from a strong-coupling analysis of the tunneling spectra [2, 21, 22] that Pb exhibits a transverse phonon at

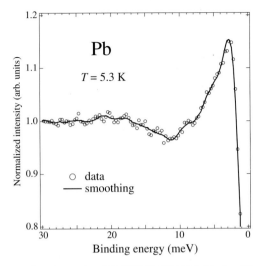

Fig. 7.7. Photoemission intensity I(5.3 K)/DOS(12.0 K) for Pb, plotted in order to see the details in the superconducting-phase spectra in analogy to tunneling conductance experiments [13]

4.4 meV and a longitudinal phonon at 8.5 meV, and we attribute the features seen in photoemission also to the same origin. Recently, an Eliashberg analysis of the spectral shape of Pb in the superconducting phase was carried out using the known phonon DOS, and a consistent value of the electron–phonon coupling parameter, $\lambda = 1.55$ was deduced [27]. A momentum distribution curve analysis has also been applied to deduce the real and imaginary parts of the self-energy of quasi-particles in Pb [27]. In Fig. 7.8, we show ARPES spectra at k_F of Pb measured below and above T_c. We find that the energy distribution curve shows a peak at E_F in the normal state, a clear quasi-particle peak in the energy distribution curve which sharpens below T_c, giving rise to a leading edge shift and SC gap formation. Such a spectral shape change is in contrast to underdoped and optimal doped HTSCs, where normal-state spectra have a broad structure and a sharp coherent peak emerges only below T_c [8,9]. The overdoped HTSCs, however, do show a quasi-particle peak in the normal state [28] and this is considered a characteristic of a Fermi liquid, as is also observed in the present case for Pb. The observation of the peak-dip-hump structure in a low-T_c system shows that the peak-dip-hump structure is indeed a characteristic of strong-coupling superconducting transitions.

7.4.2 A15 Superconductors

A15 superconductors consist of more than 70 different materials and were known to exhibit the highest transition temperature (Nb_3Ge, $T_c = 23\,\mathrm{K}$) and highest critical fields, before the discovery of HTSC [29]. First photoemission

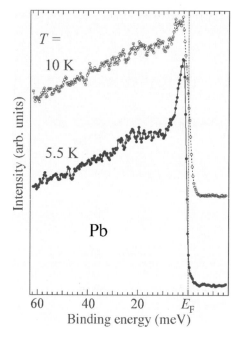

Fig. 7.8. ARPES spectra of Pb at k_F measured below and above T_c

results on the observation of SC gap in A15 compounds was reported in 1991 by Grioni et al. [10] for Nb$_3$Al ($T_c = 18.6$ K) with the world record high energy resolution at that time (energy resolution of 13 meV and sample temperature of 10 K). But they could only observe a slight shift of the leading edge below T_c compared with that above T_c. In 2001, Reinert et al. reported SC gap opening for V$_3$Si ($T_c \sim 17$ K) [30] (Fig. 7.9) and a peak above E_F, which was ascribed to thermally excited electrons across SC gap. They concluded that the observed superconducting spectral shape is consistent with BCS theory.

Figure 7.10 shows high-resolution photoemission spectra of Nb$_3$Al ($T_c = 17.6$ K) measured at 5.5 K using He Iα resonance line, but normalized with 18.0 K (normal state) data. The spectra exhibits a sharp peak just below E_F followed by a dip and a hump. A Dynes analysis indicates a gap value $\Delta = 3.0$ meV and a thermal broadening parameter $\Gamma = 0.005$ meV (curve), which reproduces the experimental result very well [31]. The obtained $\Delta = 3.0$ meV corresponds to a reduced gap value of 4.1, classifying Nb$_3$Al into a medium- to strong-coupling superconductor, consistent with the previous thermodynamic and tunneling measurements [3]. In addition, the small value of Γ may imply that the SC gap of Nb$_3$Al is very isotropic.

The dip and hump structures are better seen in the inset to Figure 7.10, where we plot the reduced photoemission DOS $= N_{exp}/(N_{BCS} - 1)$ as a function of (binding energy-Δ), where N_{BCS} is the BCS DOS broadened by the experimental resolution. The reduced DOS plot was originally used

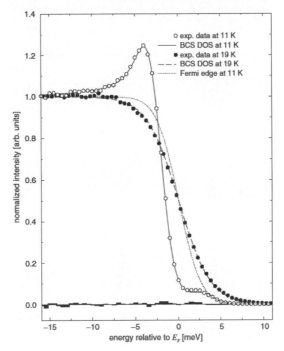

Fig. 7.9. Comparison of experimental data of V_3Si at $T = 11$ and $19\,K$ with the broadened BCS density of sates. Free parameters of the least-squares fit: energy resolution $\Delta E = 2.9\,\text{meV}$, gap width $\Delta(11\,\text{K}) = 2.5\,\text{meV}$. The *dotted line* describes a metallic Fermi edge at $T = 11\,K$. The residuum of the fit at $11\,K$ is given as *black bars* at the bottom of the figure [30]

in tunneling studies to obtain differences compared to the BCS expectation. Indeed, the reduced photoemission DOS gives a quantitative match with the reduced tunneling DOS (see inset, from [32]) and which can be well reproduced by an Eliashberg analysis using a electron–phonon coupling parameter, $\lambda = 1.7$. Photoemission spectroscopy can thus provide a microscopic insight into mechanisms of superconductivity and can be used to distinguish between weak and even moderately strong-coupling superconductivity.

7.4.3 BKBO

$Ba_{1-x}K_xBiO_3$ (BKBO) is a perovskite oxide series and shows an interesting phase diagram [33]. At $x = 0$, it is an insulator because of a three dimensional charge density wave (CDW) formation, in contrast to the simple expectation of a metal. As x increases, the system undergoes an insulator to metal transition and exhibits superconductivity [34]. High resolution photoemission studies of BKBO (x = 0.33, 0.4) shows a peak with leading edge shift below T_c, as seen in systematic temperature dependent spectra (Fig. 7.11(a)).

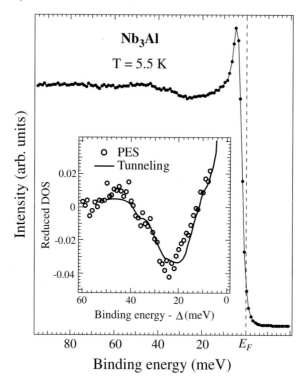

Fig. 7.10. High-resolution PE spectrum of polycrystalline Nb$_3$Al ($T_c = 17.6$ K) measured at 5.5 K (superconducting state) using He Iα resonance line (21.218 eV) with a resolution of 4.1 meV. Inset shows the intensity-enlarged reduced photoemission DOS (*open circles*), along with the reduced tunneling DOS (*curve*) from [32]

A Dynes function analysis of the results indicates a BCS type temperature evolution of SC gap (Fig. 7.11(b)) and a reduced gap value at $T = 0$ of 3.9 [35]. The spectra also show suppression of spectral weight upto a binding energy of 70 meV, which corresponds to the breathing mode phonon, and indicates formation of an electron–phonon coupling induced pseudogap.

Figure 7.12 shows an Eliashberg analysis for the $x = 0.33$ sample, carried out using the phonon DOS obtained from an inelastic neutron scattering study [36] and an iterative procedure, as is used for tunneling spectra [2]. A comparison of the experimentally obtained reduced DOS I_{sup}/I_{normal} with a converged Eliashberg calculation, shows consistency between weak structures in the reduced DOS attributable to phonons. The electron–phonon coupling parameter was obtained to be $\lambda = 1.2 \pm 0.1$, in agreement with results of an inverse Eliashberg analysis of tunneling spectra [37].

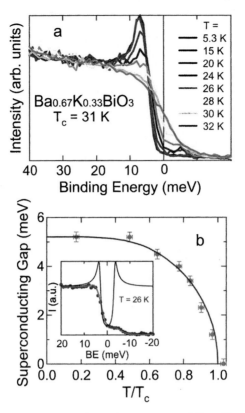

Fig. 7.11. (a) The temperature dependent spectra of $B_{0.67}K_{0.33}BiO_3$ ($T_c = 31$ K) near E_F exhibit a systematic pile up in the DOS; (b) experimental (*symbols*) and BCS (*line*) gap vs T/T_c. Inset to (b): the spectrum (*symbols*) with the fit (*gray line*) for $T = 26$ K. The small peak above E_F originates in the superconducting DOS (*black line*) [35]

7.4.4 C_{60} Fullerides

Some of the Alkali-metal-doped C_{60} fullerides exhibit superconductivity as high as 36 K (Rb_2CsC_{60}) [38]. In 1994, Gu et al., reported a photoemission study of the SC gap of Rb_3C_{60} single crystal ($T_c = 30.5$ K) with an energy resolution of 13 meV and an observed shift of the leading edge below T_c [11]. The reduce gap value was reported to be ∼ 4.1 K at 13 K with a relatively large error bars due to the absence of the SC condensation peak. Photoemission studies for K_3C_{60} ($T_c = 18$ K) and Rb_3C_{60} ($T_c = 28$ K) films have been reported by Hesper et al. in 1999 with a higher energy resolution of 7.5 meV [39], as shown in Fig. 7.13. The observed spectra show a small peak just below E_F and were found to be reproduced by the Dynes function. The reduced gap values of ≈ 3.5 have been reported, and concluded that fulleride superconductors are weak coupling BCS superconductors.

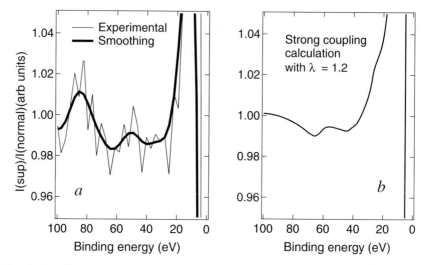

Fig. 7.12. Comparison of the experimental reduced DOS I_{sup}/I_{normal} with an Eliashberg analysis strong coupling calculation for BKBO, using the phonon DOS obtained from neutron scattering. The experimental data shows weak structures in the reduced DOS attributable to phonons, as clarified by the Eliashberg analysis

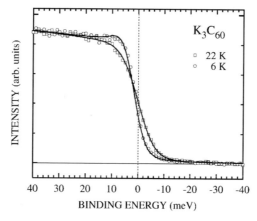

Fig. 7.13. Photoemission spectra of K_3C_{60} at 22 K (*squares*) and at 6 K (*circles*). The data are modeled (*solid lines*) with, respectively, a Fermi–Dirac function at 22 K and a BCS function with a gap $\Delta = 2.9$ meV, broadened by the resolution functions found for the Pt reference [39]

7.4.5 Silicon Clathrate

Doped Silicon clathrates belong to a new family of superconductors discovered in 1995 [40]. The crystal structure consists of Si_{20} and Si_{24} cages sharing their pentagonal faces with each other, with Ba atoms occupying these cages. Such a structure of the silicon clathrate is reminiscent of the electron-doped C_{60}

fullerides. However, the inter-cage bonding in the clathrates makes them different from the doped C_{60} fullerides where individual C_{60} molecules essentially determine its physical properties : Clathrates are covalent crystals, while fullerides are molecular crystals. The photoemission study of Ba_8Si_{46} ($T_c = 8\,K$) was reported [41] prior to tunneling spectroscopy, and thus provided first direct measurements of the SC gap. The observed spectra can be fitted with a Dynes function using $\Delta = 1.3\,meV$ and $\Gamma = 0.3\,meV$, corresponding to a $2\Delta(0)/k_BT_c = 4.38$ and it can be classified as a strong coupling superconductor [41, 42]. The difference in coupling between fullerides and clathrate may be originating in the difference in their structures.

7.5 Anomalous SC Gap Form

In the previous section, we reviewed observation of the SC gap and strong coupling features. Although the variations of reduced SC gap values and fine structures beyond the sharp condensation peak have been observed, the shape of the SC gap itself could be well-described by a modified BCS function representing an isotropic SC gap. The two boride superconductors described below have been found to have anomalous superconducting properties, which cannot be explained with an isotropic SC gap. Observed high-resolution spectra appear to be different from those observed for other superconductors, but are directly related to the observed anomalous bulk properties [5].

7.5.1 Evidence for Anisotropic s-Wave Gap in YNi_2B_2C

For Ni borocarbides, there have been a lot of experimental evidences for the anisotropy of the SC gap [43–46], leaving further experimental investigation to address whether the order parameter of YNi_2B_2C is a d-wave or a highly anisotropic s-wave. Figure 7.14 shows ultrahigh-resolution photoemission spectra in the vicinity of E_F for high quality YNi_2B_2C and $Y(Ni_{0.8}Pt_{0.2})_2B_2C$ samples measured at 6 K and 20 K. Substitution of Ni for Pt introduces impurities as well as changes T_c, which can be qualitatively explained by variations in the DOS at E_F as determined by specific-heat measurements [47]. In the spectra, besides the characteristic temperature dependent changes of SC gap opening, small enhancement of intensity around 7 meV binding energy in $x = 0.0$ spectrum can also be found, which coincides with the phonon structure reported by neutron-scattering measurements [48]. In Fig. 7.15, enlarged superconducting spectra of $x = 0.0$ and 0.2 are compared, highlighting the small differences between the two compounds. The slope is found to be more gentle in $x = 0.0$ than in $x = 0.2$.

From an anisotropic Dynes function analysis, the s-wave function fails to reproduce the shape of the experimental spectra due to its large coherent peak intensity and steep edge, while the anisotropic Dynes function gives a reasonable fit with $\Delta_{max} = 2.2 \pm 0.2$ meV, $\Delta_{min} = 0.0 \pm 0.2$ meV, and

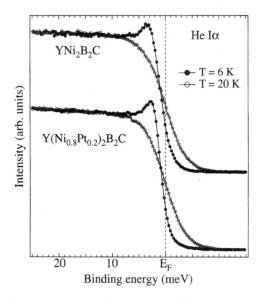

Fig. 7.14. Ultrahigh-resolution photoemission spectra in the vicinity of E_F of YNi_2B_2C (*upper panel*) and $Y(Ni_{0.8}Pt_{0.2})_2B_2C$ (*lower panel*) measured at 6 K (superconducting state) and 20 K (normal state) [19]

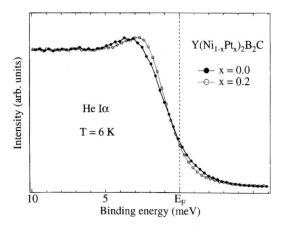

Fig. 7.15. Enlarged superconducting-state spectra of YNi_2B_2C and $Y(Ni_{0.8}Pt_{0.2})_2B_2C$. There is a small but significant difference in the slope of the leading edge [19]

$\Gamma = 0.5 \pm 0.2$ meV. For $x = 0.2$, we need to use $\Delta_{max} = 1.5 \pm 0.2$ meV, $\Delta_{min} = 1.2 \pm 0.2$ meV, and $\Gamma = 0.3 \pm 0.2$ meV. These studies indicate an anisotropic gap in $x = 0.0$ and an almost isotropic gap in $x = 0.2$, which is consistent with low-temperature specific-heat measurements under magnetic field [47]. The results provide direct evidence for a highly anisotropic s-wave gap, and not a d-wave gap, in Ni borocarbides, consistent with theoretical studies [49, 50].

7.5.2 MgB$_2$

MgB$_2$ consists of alternating stack of boron and magnesium layers and shows the highest T_c among intermetallic superconductors [51]. The physical properties of MgB$_2$ were found to exhibit anomalous behavior which cannot be explained even by an anisotropic SC gap [52]. According to band-structure calculations, it has three Fermi surface sheets: one three dimensional FS having Boron π bonding and two cylindrical FS sheets of Boron σ orbitals [53]. The first photoemission study on the SC gap of MgB$_2$ reported an isotropic gap of $\Delta = 4.5$ meV and $\Gamma = 1.1$ meV [54]. Later, high resolution study using crystals made with high-temperature and high-pressure technique showed anomalous spectral shape [20]. In Fig. 7.16, high-resolution photoemission spectra of MgB$_2$ measured at 5.4 K (superconducting state) and 45 K (normal state) are shown.

While the normal-state spectrum is characterized by the Fermi edge like structure, the superconducting-state spectrum shows a peak and a shift of the leading edge, indicative of an opening of the SC gap. An enlargement of the near-E_F region has been shown in the inset of Fig. 7.16 with open circles. One finds, besides the peak around 7 meV, a weak shoulder structure around

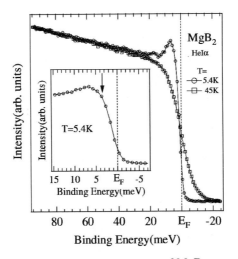

Fig. 7.16. High-resolution photoemission spectra of MgB$_2$ measured at 5.4 K (*open circles* connected with a *solid line*) and 45 K (*open squares* connected with a *solid line*) with He Iα resonance line (21.212 eV). The inset shows an expanded spectrum at 5.4 K in the vicinity of E_F. Please note that the spectrum has a peak with a shoulder as is emphasized with an arrow, which indicates a non simple isotropic gap [20]

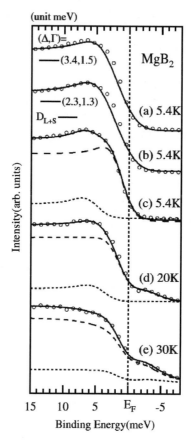

Fig. 7.17. Results of fittings (*solid lines*) and temperature-dependent experimental spectra (*open circles*)). (**a**) Single Dynes function with $\Delta = 3.4\,\text{meV}$ and $\Gamma = 1.5\,\text{meV}$ and (**b**) single anisotropic Dynes function with $a = 1$, $\Delta = 2.3\,\text{meV}$, and $\Gamma = 1.3\,\text{meV}$ for 5.4 K. The results of fittings are to reproduce the peak position; (**c**)–(**e**) the weighted sum of two Dynes functions (c) for 5.4 K with $\Delta_S = 1.7\,\text{meV}$ (*broken line*) and $\Delta_L = 5.6\,\text{meV}$ (*dotted line*) having the same $\Gamma = 0.10\,\text{meV}$; (d) for 20 K with $\Delta_S = 1.7\,\text{meV}$ (*broken line*) and $\Delta_L = 4.5\,\text{meV}$ (*dotted line*) having the same $\Gamma = 0.20\,\text{meV}$ and (e) for 30 K with $\Delta_S = 1.2\,\text{meV}$ (*broken line*) and $\Delta_L = 2.2\,\text{meV}$ (*dotted line*) having the same $\Gamma = 0.20\,\text{meV}$ [20]

3 meV. Further, the leading edge mid-point of MgB_2 is not so large compared to that of Nb_3Al in spite of the large difference in T_c. This indicates that the SC gap of MgB_2 is not a simple isotropic one.

From a fit to a weighted sum of the two Dynes functions, we can reproduce the spectral shape very nicely, as shown in (c) of Fig. 7.17, where broken and dotted lines represent the two Dynes functions: one with $\Delta = 1.7\,\text{meV}$ and $\Gamma = 0.1\,\text{meV}$ and the second with $\Delta = 5.6\,\text{meV}$ and $\Gamma = 0.1\,\text{meV}$. Here, the intensity of the Dynes function with the smaller gap is about five times larger than that for the lager gap. This analysis indicates that the SC gap is neither a simple isotropic gap nor an anisotropic gap, but rather consists of two dominant components. This is consistent with results reported from several groups using different experimental techniques [52]. The same analysis for temperature dependent spectra showed that both, the larger and smaller gaps close at the bulk transition temperature, guaranteeing that the smaller gap is not due to reduced superconducting phases which are expected to have smaller gap values and thus have smaller T_cs. From these analyses, the reduced gap values are 1.08 and 3.56 for smaller and larger gaps, respectively.

Thus advances in the energy resolution has enabled us to observe not only SC gap and strong-coupling spectral features but also anomalies in the SC gap function that suggests novel SC gap structures, as schematically shown in Fig. 7.2. To more directly observe these, one should perform angle-resolved photoemission(ARPES) studies and we describe some of the recent studies which exemplify the role of ARPES in studying low-T_c superconductivity.

7.6 Fermi Surface Sheet Dependence

Theoretically, the SC gap value dependence for different FS sheets have been studied by Suhl in terms of 'two-band superconductivity'more than 40 years ago [55]. A multiple SC gap feature in k-integrated information has been observed in Nb doped $SrTiO_3$ by tunneling studies, providing the first evidence of two band superconductivity [56]. However, more direct evidence has been provided by recent ARPES which is capable of distinguishing FS sheets and hence, SC gaps on particular FS sheets.

7.6.1 2H-NbSe$_2$

The first such evidence of Fermi surface sheet dependent superconductivity was shown for the transition-metal dichalcogenide superconductor, 2H-NbSe$_2$ [57] A very suitable condition for ARPES study is the quasi two dimensionality of 2H-NbSe$_2$ and band-structure calculations have shown that 2H-NbSe$_2$ consist of two types of Fermi surface sheets with different character: a pan-cake-like FS with dominant Se 4p character and two cylindrical FS sheets with dominant Nb 4d character [58]. Figure 7.18 shows high-resolution ARPES spectra of 2H-NbSe$_2$ measured at Fermi momentum (k_F) on different FS sheets as shown in the insets. It is evident that while the temperature dependence of spectra across T_c for the three-dimensional Se 4p-derived bands are negligible(Fig. 7.18(a)), that for quasi two-dimensional Nb 4d band shows clear indication of opening of SC gap(Fig. 7.18(b) and (c)). A spectral function analysis, as is used for studying HTSC gaps, provides an ~ 0 meV gap for the mainly Se 4p-derived 3D FS sheet and 0.9–1.0 meV for mainly Nb 4d-derived quasi-2D FS sheet, the latter of which corresponds to reduced gap values of 3.6–3.9.

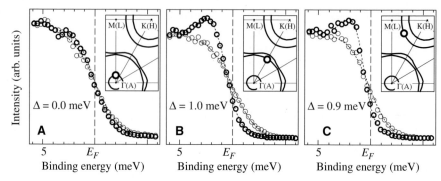

Fig. 7.18. (a), (b), and (c) temperature-dependent ultrahigh-resolution ARPES data of $2H$-NbSe$_2$ measured at FS sheets related to the Se $4p$-derived 16 th, and the Nb derived 17 th and 18 th bands as referred in the band structure calculations [58], respectively. The *black* and *gray circles* correspond to the measured ARPE spectra obtained at 5.3 K (superconducting state) and 10 K (normal state), respectively. Locations of the measured points in the Brillouin zone are shown at a *black circle* in each inset. We employ higher resolution (2.5 meV) and smaller step size (0.3 meV) to detect spectral changes as a function of temperature. The *black dotted lines* superimposed on the measured spectra are the numerical calculation results. The SC gap size Δ used for fitting the superconducting state spectrum is written in each panel [57]

7.6.2 MgB$_2$ and Ca(Al, Si)$_2$

Similarly, multiple gap behavior with different values was observed in MgB$_2$ by Souma et al. [59] and Tsuda et al. [60], the latter of whom clearly demonstrated that the SC gap exhibits a FS sheet dependence of the bulk boron $2p$ σ- and π-orbital-derived bands and inter-band coupling as discussed below. Figures 7.19(a) and (b) show high-resolution ARPES spectra measured across T_c at k_F's on FS sheets formed by the σ and π bands, respectively. Here, one can clearly distinguish that the SC gap on the σ sheets is larger than that of the π sheet by the larger leading edge shift at the lowest temperature, which corresponds to the opening of the SC gap. Comparison with first-principle band-structure calculations allows one to describe the FS sheet dependence to be due to the existence of two types of FS sheets with very different character and corresponding k-dependence in electron–phonon coupling [61].

Ca(Al,Si)$_2$ (T_c = 7.7 K) [62], which has the same crystal structure as MgB$_2$, but exhibits identical SC gaps on different FS's which possess the same character of electronic states (Figure 7.20) [63]. The observation of the FS sheet dependent gap in two different compounds ($2H$-NbSe$_2$ and MgB$_2$) provides experimental evidence that the FS sheet dependent gap is indeed an important feature for superconductors having multiple FS sheets with very different character. The similarity of the superconducting properties between 2H-NbSe$_2$ and MgB$_2$ from thermal-conductivity and heat-capacity measurements [64] indeed supports present ARPES studies.

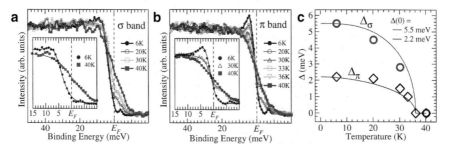

Fig. 7.19. (a) and (b), temperature-dependent high-resolution spectra of MgB$_2$ measured for the FS sheets derived from the boron σ- and π-orbital bands, respectively. Inset shows an energy-enlarged spectra (*symbols*) for selected temperatures, as compared with the fitting results (*lines*), contrasting the difference in gap size on different FSfs. (c), Plot of the SC gap size $\Delta(T)$ used for fitting the superconducting-state spectrum. *Open circles* and *diamonds* denote the gap values on the σ and π-bands, respectively, as compared with theoretical predictions [61] (*lines*) [60]

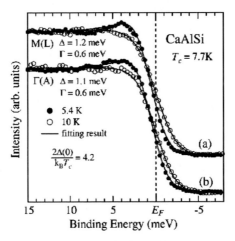

Fig. 7.20. Temperature-dependent ultrahigh-resolution spectra of Ca(Al,Si)$_2$ (T_c = 7.7 K) measured on the FS sheets at the M(L) and Γ(A) points. The *open* and *filled circles* show the experimental spectra of the normal and the SC states, respectively [63]

The temperature dependence of the small and large gap can provide a deeper insight into the inter-band coupling for the superconductivity. According to calculations by Suhl [55], the two gaps close at the same temperature for the case of finite inter-band coupling, while they close at different temperatures without inter-band coupling. In MgB$_2$, the temperature dependence shows that both the gaps close at the same temperature (Figure 19(c)), indicating finite inter-band coupling. As for the 2H-NbSe$_2$, the ARPES results suggest the possibility of smaller gap closing at temperature below T_c, and this has been confirmed from recent tunneling studies with much higher resolution and at much lower temperature [65]. These results indicate the difference of the inter-band coupling between the two compounds. The inter-band pairing interaction in the two-band superconductor MgB$_2$ has been investigated experimentally in detail by tunneling [77]. In this work also a theoretical analysis of the electron–phonon interaction in this system is presented.

7.6.3 ZrTe$_3$

The electronic structure of some low-T_c superconductors exhibit FS sheets with differing dimensionality derived from different character electronic states, and consequently, can lead to k-dependent superconductivity. ZrTe$_3$ exhibits a CDW transition around 63 K, remains metallic across a bump anomaly in the electrical resistivity and shows filamentary superconductivity below 2 K [66]. Two independent band-structure calculations of ZrTe$_3$ have predicted a 3D FS sheet around the G point , quasi-1D FS sheets along the B(A)–D(E) direction, and a vHs on B-A line where hybridization of Zr $4d$ and Te $5p$-derived states add a more 3D character to the quasi-1D sheet (thus 'dual quasi-1D+3D') [67, 68]. The calculated quasi-1D FS sheet forms a nesting vector consistent with the CDW vector obtained by the electron microscopy study. From ARPES studies at FS crossings k_F with different dimensionality, a clear difference in temperature dependence of energy distribution curves was observed: (i) a narrowing line shape (or increasing quasi-particle coherence) on the 3D FS and (ii) a simultaneous formation of a pseudogap on the 1D FS (Fig. 7.21) [69].

One may then expect that the ungapped parts on FS sheets, where increasing quasi-particle coherence was observed, can play an important role for the superconductivity.

7.7 Summary and Future Prospects

High-resolution photoemission spectroscopy studies of low-T_c superconductors were reviewed. They measure the SC gap opening and phonon-induced fine structures, as predicted by theories. Moreover, due to the advance in the energy resolution and capability of k-resolved electronic states, they have revealed the origin of anomalous superconducting electronic structure occurring in a variety of low-T_c superconductors. These results have established that high-resolution photoemission spectroscopy has now become one of the most powerful experimental techniques to study electronic structures of solids and, by increasing its energy resolution, it will be able to reveal the superconducting electronic states of unconventional superconductors, e.g. heavy-fermion, ruthenate, and organic superconductors. Indeed, this direction in photoemission spectroscopy has been demonstrated using a laser as a photon source [70, 71]. A recent study reported an extremely high-resolution of

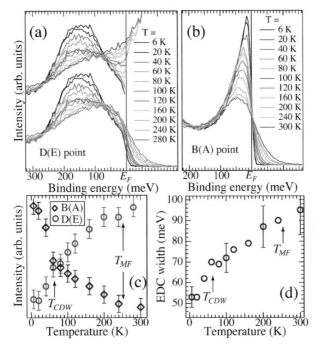

Fig. 7.21. T-dependent (6–300 K) ARPES spectra of ZrTe$_3$ normalized for intensity with scan time are shown for (**a**) the quasi-1D FS showing a fluctuation induced pseudogap at the D(E) point in k-space and (**b**) quasi-particle coherence at the B(A) point which corresponds to the vHs with dual quasi 1D and 3D character. The top panel of (**a**) is FD function subtracted spectra of the bottom panel of (**a**); (**c**) T-dependent intensities at E_F due to (i) the pseudogap (*gray circles*) at the D(E) point, and (ii) increasing quasi-particle coherence (*black diamonds*) at the B(A) point. (**d**) The quasi-particle peak FWHM showing the reduction in peak width, clearly below 200 K and across T_{CDW} [69]

360 μeV, which has been achieved using a laser as a photon source, and successfully measured the superconducting electronic structures of CeRu$_2$ [70] and MgB$_2$ [65], and an anomalous normal state pseudogap in cobaltate superconductors [72] (for valence band spectra on CeRu$_2$ see also the chapter by Sekiyama et al. in this volume). The laser PES study of CeRu$_2$ could even discuss the anisotropy of the SC gap in the compound (Fig. 7.22). Since the photon energy of laser PES is lower, the larger electron escape depth can become another advantage, which has been already indicated for elemental metals [71] and very recently reported for high-T_c superconductors [73].

Another future direction in photoemission spectroscopy was shown by soft x-ray ARPES, which can detect bulk sensitive electronic structures of solids [74–76]. Very recent soft x-ray ARPES results on heavily boron-doped diamond superconductors have shown the power and reliability of this technique to study new superconductors [76]. The study clearly resolved band

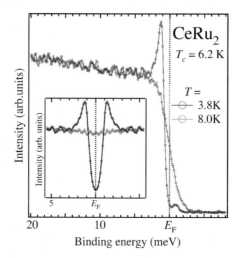

Fig. 7.22. T-dependent ultrahigh-resolution spectra near E_F of CeRu$_2$ ($T_c = 6.2$ K) with an inset showing the symmetrized spectra from the same data [70]

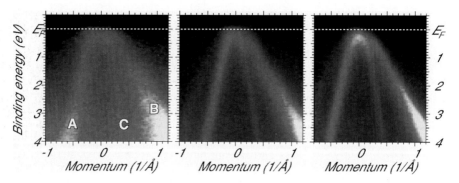

Fig. 7.23. Boron concentration dependent near E_F SXARPES intensity maps from single crystal diamond(111) using 825 eV. The shift of E_F is clearly seen [76]

dispersions and a systematic shift of E_F as a function of B-doping concentrations, as shown in Fig. 7.23. These two photoemission spectroscopies will become important techniques to study electronic structures of functional materials as well as low-T_c superconductors in near future.

Acknowledgements

The photoemission results on low-T_c superconductors described here have been obtained in collaboration with T. Kiss, S. Tsuda, T. Baba, T. Watanabe, A. Fukushima, K. Hirata, T. Nishio, H. Uwe, K. Kobayashi, K. Moriguchi, A. Shintani, H. Fukuoka, S. Yamanaka, M. Nohara, H. Takagi, Y. Takano,

K. Togano, H. Kito, H. Ihara, A. Matsushita, F. Yin, J. Itoh, M. Imai, K. Yamaya, T. Oguchi, H. Harima, I. Hase, and we sincerely thank all of them.

References

1. J. R. Schrieffer: *Theory of Superconductivity* (Preseus books, Reading, MA, 1983)
2. W. L. MacMillan and J. M. Rowell: Tnneling and Strong-Coupling Superconductivity. In: *Superconductivity*, ed by R. D. Parks (Dekker, New York, 1969), Vol. 1, Sec. V.
3. For an extensive review on conventional superconductor, see. J. P. Carbotte: Rev. Mod. Phys. **62**, 1027 (1990)
4. P. W. Anderson: *The Theory of Superconductivity in the High-T_c Cuprates* (Princeton University Press, Princeton, New Jersey, 1997)
5. B. H. Brandow: Phil. Mag. **1**, 2487 (2003)
6. J.-M. Imer et al: Phys. Rev. Lett. **62**, 336 (1989)
7. F. Patthey et al: Phys. Rev. Lett. **58**, 2810 (1987)
8. A. Damascelli et al: Rev. Mod. Phys. **75**, 473 (2003)
9. J. C. Campuzano et al: Photoemission in the High-T_c Superconducotrs. In: *The Physics of Superconductors* (Springer-Verlag, Berlin Heidelberg, 2004) pp 167–273.
10. M. Grioni et al: Phys. Rev. B **43**, 1216 (1991)
11. C. Gu et al: Phys. Rev. B **50**, 16566 (1994)
12. D. Purdie et al: J. Electron Spectrosc. Relat. Phenom. **101–103**, 223 (1999)
13. A. Chainani et al: Phys. Rev. Lett. **85**, 1966 (2000)
14. T. Yokoya et al: J. Electron Spectrosc. Relat. Phenom. **124**, 99 (2002)
15. T. Yokoya et al: Jpn. J. Appl. Phys. **43**, 3618 (2004)
16. R. C. Dynes et al: Phys. Rev. Lett. **41**, 1509 (1965)
17. For examples, see T. Ekino and J. Akimitsu: Tnuueling Spectroscopy on High-T_c Superoncutors. In: *Studies of High Temperature Superconductors*, ed by A. V. Narlikar (Nova, New York, 1992) pp 259–309
18. J. Zasadzinski: Tunneling Spectroscopy of Conventional and Unconventional Superconductors. In: *The Physics of Superconductors* (Springer-Verlag, Berlin Heidelberg, 2004) pp 591–646.
19. T. Yokoya et al: Phys. Rev. Lett. **85**, 4952 (2000)
20. S. Tsuda et al: Phys. Rev. Lett. **87**, 177006 (2001)
21. D. J. Scalapino: The Electron–Phonon Interacton and Strong-Coupling Superconductors. In: *Superconductivity*, ed by R. D. Parks (Dekker, New York, 1969), Vol. 1, Sec. IV.
22. W. L. MacMillan and J. M. Rowell: Phys. Rev. Lett. **14**, 108 (1965)
23. E. Schachinger et al: Phys. Rev. B **67**, 214508 (2003)
24. S. V. Dordevic et al: Phys. Rev B **71**, 104529 (2005)
25. X. J. Zhou et al: Phys. Rev. Lett. **95**, 117001 (2005)
26. B. N. Brockhouse et al: Phys. Rev. **128**, 1099 (1962)
27. F. Reinert et al: Phys. Rev. Lett. **91**, 186406 (2003)
28. Z. M. Yusof et al: Phys. Rev. Lett. **88**, 167006 (2002)
29. Y. A. Izyumov et al: Sov. Phys-Usp. **17**, 356 (1975)
30. F. Reinert et al: Phys. Rev. Lett. **85**, 3930 (2001)

31. T. Yokoya et al: J. Phys. Chem. Solids **63**, 2141 (2002)
32. J. Kuo and T. H. Geballe: Phys. Rev. B **23**, 3230 (1981)
33. S. Pei et al: Phys. Rev. B **41**, 4126 (1990)
34. L. F. Mattheiss et al: Phys. Rev. B **37**, 3745 (1988)
35. A. Chainani et al: Phys. Rev. B **64**, 180509(R) (2001)
36. C. K. Loong et al: Phys. Rev. B **45**, 8052 (1992)
37. Q. Huang et al: Nature (London) **347**, 369 (1990).
38. For review, see O. Gunnarrson: Rev. Mod. Phys. **69**, 575 (1994)
39. R. Hesper et al: Phys. Rev. Lett. **85**, 1970 (2000)
40. H. Kawaji et al: Phys. Rev. Lett. **74**, 1427 (1995)
41. T. Yokoya et al: Phys. Rev. B **64**, 172504 (2001)
42. T. Yokoya et al: Phys. Rev. B **70**, 159902(E) (2004)
43. G.-Q. Zheng et al: J. Phys. Chem. Solids **59**, 2169 (1998)
44. T. Kohara et al: Phys. Rev. B **51**, 3985 (1995)
45. M. Nohara et al: J. Phys. Soc. Jpn. **66**, 1888 (1997)
46. G. Wang and K. Maki: Phys. Rev. B **58**, 6493 (1998)
47. M. Nohara et al: J. Phys. Soc. Jpn. **68**, 1078 (1999)
48. H. Kawano et al: Phys. Rev. Lett. **77**, 4628 (1996)
49. L. S. Borkowski and P. K. Hirschfeld: Phys. Rev. B **49**, 15404 (1994)
50. R. Fehrenbacher and M. R. Norman: Phys. Rev. B **50**, 3495 (1994)
51. J. Nagamatsu et al: Nature (London) **410**, 63 (2001)
52. C. Buzea and T. Yamashita: Supercond. Sci. Technol. **14**, R115 (2001)
53. A. Y. Liu et al: Phys. Rev. Lett. **87**, 087005 (2001)
54. T. Takahashi et al: Phys. Rev. lett. **86**, 4915 (2001)
55. H. Shul et al: Phys. Rev. Lett. **3**, 552 (1959)
56. G. Binnig et al: Phys. Rev. Lett. **45**, 1352 (1980)
57. T. Yokoya et al: Science **294**, 2518 (2001)
58. R. Corcoran et al: J. Phys. -Condens. Matter **6**, 4479 (1994)
59. S. Souma et al: Nature (London) **423**, 65 (2003)
60. S. Tsuda et al: Phys. Rev. Lett. **91**, 127001 (2003)
61. H. J. Choi et al: Nature (London) **418**, 758 (2002)
62. M. Imai et al: Appl. Phys. Lett. **80**, 1019 (2002)
63. S. Tsuda et al: Phys. Rev. B **69**, 100506(R) (2004)
64. E. Boaknin et al: Phys. Rev. Lett. **90**, 117003 (2003)
65. S. Tsuda et al: Phys. Rev. B **72**, 064527 (2005)
66. S. Takahashi et al: J. Physique Coll. **44**, 1733 (1983)
67. C. Felser et al: J. Mater. Chem. **8**, 1787(1998)
68. K. Stöwe and F. W. Wagner: J. Solid State Chem. **138**, 160-168 (1998)
69. T. Yokoya et al: Phys. Rev. B **71**, 140504 (2005)
70. T. Kiss et al: Phys. Rev. Lett. **94**, 057001 (2005)
71. T. Kiss et al: J. Electron Spectrosc. Relat. Phenom. **144**, 953, (2005)
72. T. Shimojima et al: Phys. Rev. B **71**, 020505(R) (2005)
73. J. D. Koralek et al: Phys. Rev. Lett. **96**, 017005 (2006)
74. N. Kamakura et al: Europhys. Lett. **67**, 240 (2004)
75. T. Claesson et al: Phys. Rev. Lett. **93**, 136402 (2004)
76. T. Yokoya et al: Nature (London) **438**, 647 (2005)
77. J. Geerk et al: Phys. Rev. Lett. **94**, 227005 (2005)

Part IV

Molecules

8

Very-High-Resolution Laser Photoelectron Spectroscopy of Molecules

K. Kimura

Institute for Molecular Science, Okazaki 444-8585, Japan and Japan Advanced Institute of Science and Technology, Nomi 923-1292, Japan
k-kimura@ims.ac.jp

Abstract. A very-high-resolution molecular photoelectron spectroscopy based on 'resonantly enhanced multiphoton ionization' (REMPI) is described, in which zero-kinetic-energy (ZEKE) photoelectrons are measured as a function of laser wavelength in two-color experiments with tunable pulsed UV/visible lasers. The REMPI-based photoelectron technique provides both 'cation spectroscopy' and 'excited-state spectroscopy' for gaseous molecular species. The following topics are mainly described. The principles, characteristics, and advantages of the REMPI-based photoelectron spectroscopy are described in Sect. 8.2; several types of compact cm^{-1}-resolution and high-brightness ZEKE photoelectron analyzers in Sect. 8.3; and its typical applications to jet-cooled van der Waals molecules in Sect. 8.4.

8.1 Introduction

Molecular photoelectron spectroscopy has been developed since the early 1960s and may be divided into the following three fields, as mentioned in a recent review article [1]: Namely, (1) photoelectron spectroscopy with a 58.4-nm He I resonance source; (2) laser photoelectron spectroscopy associated with electron kinetic-energy measurements on the basis of 'resonantly enhanced multiphoton ionization' (REMPI); (3) REMPI-based photoelectron spectroscopy associated with zero-kinetic-energy (ZEKE) photoelectron measurements in a very high resolution. The development of these three kinds of molecular photoelectron spectroscopy is illustrated in chronological order, together with their photoelectron energy resolution in Fig. 8.1.

1. Molecular photoelectron spectroscopy with a single VUV photon provides direct spectroscopic information about the ionization transitions of molecules, subsequently providing their ionization energies and the corresponding ionic states produced immediately after ionization transitions. Molecular photoelectron spectroscopy with a He(I) resonance source has been developed originally by Turner et al. [2] in the 1960's. Since then, the valence electronic structures for a number of organic and inorganic molecules

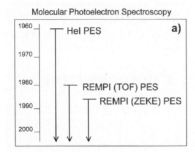
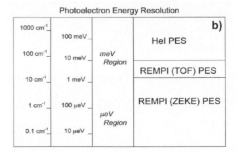

Fig. 8.1. (a) Development of molecular photoelectron spectroscopy in chronological order; (b) its progress in the photoelectron energy resolution. Here, PES means 'photoelectron spectroscopy', REMPI 'resonantly enhanced multiphoton ionization, TOF 'time-of-flight', and ZEKE 'zero kinetic energy'

have been studied on the basis of the vertical ionization energies obtained from their He(I) photoelectron spectra. A handbook of He(I) photoelectron spectra of many molecules has been published first by Turner et al. [2]. Another comprehensive handbook on He(I) photoelectron spectra has been published by the author's group, including about 200 fundamental organic molecules with their vertical ionization energies as well as their *ab initio* molecular orbital assignments in the full He(I) region [3].

2. When gaseous molecules are irradiated with a UV/visible laser, resonant ionization is remarkably enhanced, as first indicated by Johnson et al. [4] from their multiphoton ionization (MPI) ion-current measurements. On the basis of 'resonantly enhanced multiphoton ionization' (REMPI), therefore, it has been possible to observe photoelectron spectra due to molecular resonant excited states in one- and two-color experiments with a tunable pulsed UV/visible laser, as seen from some earlier review articles [5,6]. The REMPI photoelectron kinetic-energy spectroscopy has been developed in the early 1980s by the author's group [7], independent of other groups [8–11], mainly with a time-of-flight (TOF) electron analyzer to measure photoelectron kinetic-energy spectra of jet-cooled molecules. Such a REMPI-based photoelectron spectroscopy provides a dynamic aspect of various molecular excited states from the photophysical and photochemical points of view [6]. Therefore, it may be called dynamic photoelectron spectroscopy. It is also possible to study nonradiative electronic states with this technique.

3. REMPI-based photoelectron spectroscopy with a ZEKE photoelectron technique have been developed originally by Müller-Dethlefs et al. [12] and also independently by Achiba, Sato, and Kimura [13]. They have extended their earlier REMPI photoelectron technique to their ZEKE photoelectron technique, with which two-color ZEKE photoelectron spectra have been observed with jet-cooled aniline and benzene molecules [13]. Since then,

a series of several compact cm^{-1}-resolution high-brightness ZEKE photoelectron analyzers have been developed in the author's laboratory, as mentioned in Sect. 3.

The photoelectron energy resolutions attained in the three kinds of fields associated with molecular photoelectron spectroscopy are also illustrated in Fig. 8.1, showing a dramatic progress in molecular photoelectron spectroscopy.

8.2 REMPI Photoelectron Spectroscopy

8.2.1 Principles and General Features

The multiphoton ionization (MPI) of a molecule by a pulsed UV/visible laser is remarkably enhanced at a laser frequency at which the photon energy is in exact resonance with one of its specific excited states. Such enhancement is widely observed when the total ion current is measured as a function of the laser wavelength, giving rise to an MPI ion-current spectrum [4].

This MPI process is called 'resonantly enhanced multiphoton ionization' (REMPI). In other words, photoelectron emission takes place at each resonant excited state of the molecules. A REMPI process emitting photoelectrons in two-color experiments with two pulsed lasers (ν_1, ν_2) is described by

$$M + nh\nu_1 \rightarrow (M^*)_i \tag{8.1}$$

$$(M^*)_i + h\nu_2 \rightarrow (M^+)_j + (e^-)_j \tag{8.2}$$

Here, M means a molecule at the electronically ground state, $(M^*)_i$ its i-th resonant excited state, and $(M^+)_j$ the j-th ionic state produced after photoelectron emission.

An MPI ion-current spectrum is shown schematically in Fig. 8.2, in general consisting of many peaks, which correspond to n-photon allowed resonant excited states. From an MPI ion-current spectrum, one can obtain spectroscopic information about the resonant excited states, but no information is provided about the resulting ionic states. As shown in Fig. 8.2, at each MPI ion-current peak we measure a REMPI photoelectron spectrum by means of a photoelectron kinetic-energy analysis, this providing new information about the ionic states produced by ionization transition from the resonant excited states.

From a technical point of view, we must record first an MPI ion-current spectrum prior to REMPI photoelectron experiments. MPI ion-current spectra primarily provide information about the energy levels and population of excited states of a neutral molecule, while REMPI photoelectron spectra contain information about the energy levels and populations of ionic states produced from a selected excited state, as illustrated schematically in Fig. 8.3.

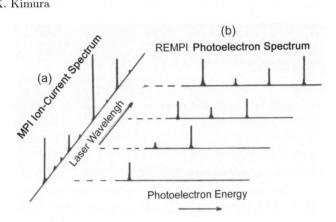

Fig. 8.2. Schematic drawing showing the relationship between (**a**) an MPI ion-current spectrum and (**b**) REMPI photoelectron spectra observed at the individual ion-current peaks

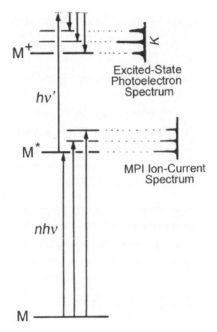

Fig. 8.3. Schematic drawing showing an energy level diagram, which is relevant to both the MPI ion-current and the REMPI photoelectron spectrum. The ion-current peaks correspond to the excited states (M^*), while the photoelectron peaks correspond to the ionic states (M^+). The photoelectron peak intensity depends on the ionization transition probability between the excited and the ionic state

In the two-color process, the wavelength of the second laser can be selected independently of the first one, so that it is always possible to carry out single-photon ionization of M^*, if $h\nu_2$ has an enough energy. Such a two-color ($n + 1'$) REMPI experiment is especially important for studying an excited-state photoelectron spectrum. Otherwise, another resonance might takes place at a higher excited state of the molecule, making it difficult to analyze the resulting photoelectron spectrum.

In two-color REMPI experiments of molecules, in general, the following four kinds of measurements are carried out:

1. At the first step, total-ion-current measurements are carried out as a function of laser wavelength to study the energy levels of resonant excited states.
2. Mass spectroscopic measurements are carried out to identify a molecular species associated with an observed MPI ion-current spectrum.
3. Measurements of photoelectron kinetic-energy spectra are carried out as a function of electron kinetic energy to survey ionic states in a wide energy region.
4. Measurements of ZEKE photoelectron spectra are carried out as a function of laser wavelength for studying photoelectron spectra in very high resolution.

8.2.2 REMPI Photoelectron Kinetic-Energy Spectra

Laser photoelectron spectroscopy based on REMPI has been originally developed mainly with a time-of-flight electron analyzer in the early 1980s [7–11], as seen from earlier review articles [5,6]. A time-of-flight (TOF) photoelectron analyzer has been used in the earlier measurements of REMPI photoelectron kinetic-energy spectra with jet-cooled molecules, by combining a tunable UV/visible laser system [5].

Photoelectron kinetic-energy measurements with a TOF electron analyzer simply provide excited-state photoelectron spectra in a wide energy region, if an appropriate UV/VUV laser system is available. The primary information deduced from an excited-state photoelectron spectrum relates to the ionic states produced by the optical selection rule from the resonant excited states in the ionization transition, as mentioned before. In this sense, an REMPI photoelectron spectrum gives rise to fingerprint identification for excited states.

Figure 8.4 shows a schematic drawing of both the two-photon resonant ionization and the single-photon ionization, together with their experimental examples obtained in the case of the NO molecule. The one-photon resonant two-photon ionization through the $\nu' = 0$ vibrational level of the Rydberg $A^2\Sigma^+$ state of NO shows only a single vibrational peak due to the $\nu^+ = 0$ level of the $NO^+(X)$ ion, because the NO bond distance of this Rydberg excited state is almost the same as that of the ionic state [5](b). On the other hand, a He(I) photoelectron spectrum of NO shows the first ionization band consisting of several vibrational peaks [2](a) and [3].

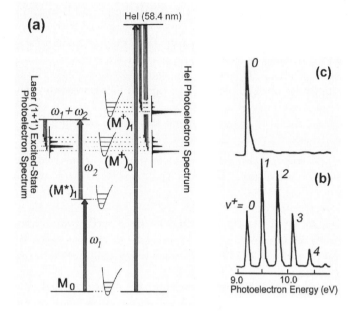

Fig. 8.4. (a) The single- and two-photon ionization processes associated with photoelectron spectroscopy are compared schematically with each other; (b) a He(I) photoelectron spectrum of NO showing the first ionization band, which consists of several vibrational peaks $(\nu^+ = 0\text{--}4)$ [3]; (c) a REMPI photoelectron spectrum of NO showing only a single peak due to $\nu^+ = 0$, obtained by $(1 + 1')$ resonant ionization via the excited A state $(^2\Sigma^+)$ at the $\nu' = 0$ level [5](b)

Comparison of these two kinds of photoelectron spectra demonstrates a dramatic difference in spectral pattern. In general, an excited-state photoelectron spectrum essentially differs from a VUV ground-state photoelectron spectrum, reflecting more or less a difference in molecular geometry between the excited state and the ground state.

REMPI-based excited-state photoelectron spectroscopy has the potential for observing the dynamic behavior of excited states. Since a molecular excited electronic state may undergo photophysical or photochemical phenomena, this REMPI-based photoelectron technique should provide new information about the photophysics and photochemistry of electronically excited molecules in the gas phase. The time evolution of the excited state can be studied in terms of the change of the photoelectron spectral pattern under various laser conditions. Because of this capability of observing the time evolution of the excited states, excited-state photoelectron spectroscopy may be called 'dynamic photoelectron spectroscopy' [6]. This is in striking contrast to a single-photon VUV photoelectron spectroscopy associated with the static aspect of ground-state molecules. Furthermore, molecular non-radiative electronic states, for

8 Very-High-Resolution Laser Photoelectron Spectroscopy of Molecules   221

which direct observation is difficult by fluorescence spectroscopy, can also be studied by this technique.

The photoelectron spectrum of a molecular excited electronic state is similar in quality of information to the corresponding fluorescence spectrum, when fluorescence is emitted from that state. However, as far as transition probability is concerned, the two kinds of electronic processes are quite different. Ionization is always allowed for one-electron transition for any excited state, and the ionization transition probabilities are of the same order of magnitude. However, fluorescence transition probabilities vary by several orders of magnitude. Observations of molecular fluorescence spectra in the gas phase are very limited. Different electronic states of ions can be produced as the final states in ionization by an appropriate laser. This situation also differs from that in molecular fluorescence spectroscopy.

The REMPI photoelectron technique with a TOF electron analyzer has been widely applied to study photochemical dynamics in the following several topics by using mainly one-color $(2+1)$ scheme in the author's laboratory: 1) One-photon forbidden excited state of O_2 [22](a). 2) Autoionization of NO [22](b), and its super-excited states [22](c). 3) Small van der Waals molecules Ar-NO and $(NO)_2$ in the $(2+1)$ scheme via the Rydberg C state [22](d–f). 4) Simple molecules NO and NH_3 [22](g), and the rare gas atoms Ar and Xe [22](h). 5) The intramolecular vibrational relaxation of benzene in the $^1E_{1u}$ states [22](i) and naphthalene in the excited S_1 and S_2 states [22](k), and the channel-three problem of benzene in the $^1B_{2u}$ state [22](j). 6) A total of thirteen low-lying electronic states of Fe atoms have been identified in the photodissociation of $Fe(CO)_5$ [22](l – n).

8.2.3 Two-Color ZEKE Photoelectron Spectroscopy and its Features and Advantages

The two-color ZEKE photoelectron technique provides a very-high-resolution excited-state photoelectron spectroscopy as well as a cation spectroscopy in which the molecular ionization energy can be obtained in the wavenumber resolution or even in the sub-wavenumber resolution.

In general, two-color ZEKE photoelectron spectroscopy is carried out in the following scheme:

$$M \xrightarrow{h\nu} M^* \xrightarrow{h\nu'} \left(M^+ \cdots e^-\right)_{\text{Rydberg}} \xrightarrow{\text{PFI}} (M)^+ + e^-_{\text{ZEKE}} \quad (8.3)$$

Namely, high lying Rydberg states $(M^+ \cdots e^-)_{\text{Rydberg}}$ converging into a specific ionization threshold are initially produced by a two-color REMPI process, and then the Rydberg electrons are collected as 'ZEKE electrons' by applying an appropriate delayed pulsed electric field during scanning the second laser wavelength $(h\nu')$ [15,17]. This is pulsed field ionization (PFI) and the technique is called ZEKE-PFI or PFI-ZEKE photoelectron spectroscopy, or simply ZEKE photoelectron spectroscopy.

Several important features and advantages of two-color REMPI-based ZEKE photoelectron spectroscopy may be summarized as follows [1].

1. Even for a mixture of different molecular species, it is possible to select a specific molecular species as well as a specific resonant excited state during two-color REMPI photoelectron experiments. This is especially useful for studying various molecular vdW (van der Waals) complexes and molecular clusters formed in a supersonic jet. Prior to the photoelectron experiments, it is necessary to carry out mass-selected MPI ion-current measurements in order to identify a specific molecular species.
2. A ZEKE photoelectron spectrum can be obtained in a very high resolution, by combining the following 'pulse' techniques; namely, the pulsed sampling, the pulsed laser, and the pulsed electric field.
3. Due to the laser tunability, a ZEKE photoelectron spectrum can be obtained as a function of laser wavelength by scanning it at an appropriate speed.
4. The energy resolution in ZEKE photoelectron spectroscopy is very high; namely, it is an order of a wavenumber or sub-wavenumber.
5. The collection efficiency of ZEKE photoelectrons is in general very high, comparable to ion-current detection, although it depends on the solid angle of collecting electrons as well as its device.

The intensity distribution of a ZEKE photoelectron spectrum is very reliable: For example, a series of vibrationally resolved ZEKE photoelectron bands is interpreted in terms of Franck-Condon factor on the basis of a theoretical one.

8.3 Compact cm^{-1}-Resolution ZEKE Photoelectron Analyzers

Let us first mention briefly the background of the development of our compact cm^{-1}-resolution ZEKE photoelectron analyzers in our laboratory. In the early 1980s, we have developed one- and two-color REMPI–based photoelectron spectroscopic techniques, in which excited-state photoelectron spectra were observed as a function of the photoelectron kinetic energy by using a time-of-flight (TOF) electron analyzer. This has soon been extended to REMPI-based ZEKE photoelectron experiments in the middle 1980s. The first ZEKE photoelectron analyzer that we designed and constructed is a cylindrical-type electrostatic analyzer, which is shown schematically in Fig. 8.5, consisting of a set of three electrodes (A, B, and D) and a disk (D) with a small hole located at an off-axis position [13]. By applying a set of appropriate voltages on the three electrodes, ZEKE photoelectrons coming though the small hole were detected, providing a ZEKE spectrum as a function of laser wavelength during two-color REMPI experiments.

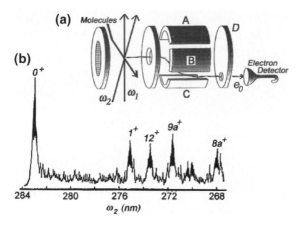

Fig. 8.5. (a) A ZEKE photoelectron analyzer consists of three electrostatic electrodes (A, B, and C) and a thin disk (D) with a hole at an off-center position. A set of appropriate voltages is applied on A, B, and C so as to pull out ZEKE photoelectrons through the hole toward the electron detector; (b) a ZEKE photoelectron spectrum due to (aniline)$^+$ is also shown, obtained in a supersonic jet by two-color $(1 + 1')$ excitation via the S_1 origin

This analyzer has been applied to aniline and benzene molecules in supersonic jets for their cation spectroscopy [13]. A ZEKE photoelectron spectrum thus obtained with a two-color $(1 + 1')$ REMPI scheme for aniline via the S_1 state is also shown in Fig. 8.5, clearly indicating several vibrational peaks due to (aniline)$^+$. The energy resolution was about $20\,\text{cm}^{-1}$ in this case.

On the other hand, in the case of benzene, several ZEKE photoelectron spectra have been obtained via its three S_1 vibronic levels ($6_0^1 1_0^n$, $n = 1 - 3$) at delay times of 0.0, 0.5 and 1.0 ns, providing an interesting dynamical information that the intensity distribution obtained at $n = 3$ clearly differs from those observed at $n = 0$ and 1 [13]. In other words, this result suggests that fast IVR (intramolecular vibrational relaxation) takes place above the vibrational excess energy of about $3000\,\text{cm}^{-1}$, as expected from the so-called channel-three problem [23].

Since then, we have developed a series of several compact cm^{-1}-resolution and high-brightness ZEKE electron analyzers with a short flight distance as follows: Namely, (1) a capillary type [16](a–d), (2) a deflection type [16](d), (3) a pulsed-electric-field ionization (PFI) type [16](f), and (4) a more improved two-pulsed-field ionization (PFI) type [17](a,b). The initial ideas behind these analyzers are to use a short flight distance as well as to apply a pulsed-electric-field technique for carrying out time-resolved discrimination after an appropriate delay time (typically 500–700 ns) at each laser shot [16](a–f). As a result, this has made it possible to obtain ZEKE photoelectron spectra with a resolution of a few cm^{-1}. For this purpose, it is important to ZEKE photoelectrons as quickly as possible at a shortest possible flight distance, because the brightness of the analyzer is especially important from an experimental point of view. Furthermore, by introducing the second pulsed electric field to remove a background signals, the energy resolution has been more improved [16](d,e). In the cases of the third and fourth analyzers, PFI electrons emitted from high-lying Rydberg states are collected by applying the collection field, always mixed with optically prepared ZEKE photoelectrons.

8.3.1 Capillary Type and Deflection Type

In Fig. 8.6, the capillary- and deflection-type ZEKE photoelectron analyzers earlier developed in our laboratory are schematically shown, together with a profile of the pulsed electric field used [16](a–d). Photoelectrons with various kinetic energies are initially produced at the ionization point (Q) at each laser shot, and then quickly disperse from the ionization region. Only ZEKE photoelectrons are collected at several hundred nanoseconds after each laser shot, by using the following two techniques; namely, 'time-resolved discrimination' and 'angular discrimination'.

Let us consider a sphere of 10 mm in diameter surrounding the ionization point (Q). Electrons with energies lower than a few cm^{-1} should remain in the sphere for as long as 500 ns after each laser shot. Only ZEKE photoelectrons can be collected typically at 500 ns after each laser shot. Very low energy electrons can be therefore collected with such a small-size compact analyzer by applying a pulsed electric field (a few V/cm) across P_1 and P_2.

In the capillary type, some ZEKE photoelectrons reaching the capillary plate were removed by applying a pulsed electric field [16](a–d). The

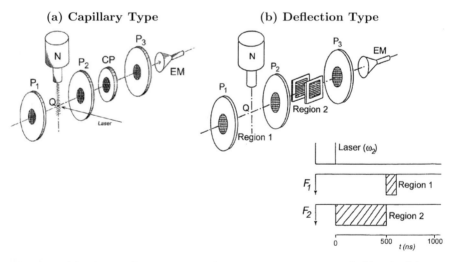

Fig. 8.6. (a) The capillary-type ZEKE photoelectron analyzer [16](a–d); (b) the deflection-type ZEKE photoelectron analyzer and the pulse profiles used for the laser (ω_2) and the electric fields (F_1 and F_2) [16](d,e)

deflection-type analyzer, on the other hand, is very bright and very high resolution ($1-2\,\mathrm{cm}^{-1}$), and very compact for the following reasons: (1) All the ZEKE photoelectrons ejected in the whole space (4π) are collected; (2) The detector is located only at 5 cm from the ionization point; and (3) the analyzer is quite simple in structure and small in size [16](d,e). A series of seven rotational peaks of NO^+ have been obtained with bandwidths of $1-3\,\mathrm{cm}^{-1}$ (Fig. 8.7) [16](a). The deflection type has been used in studies of several van-der-Waals complexes: NO-Ar [16](d,e), aniline-$Ar_{1,2}$ [18](a), *cis and trans* p-dimethoxybenzene-$Ar_{1,2}$ [18](b), and styrene-Ar and phenylacetylene-Ar [18](c).

A rotationally-resolved ZEKE photoelectron spectrum has been obtained with the deflection-type analyzer in the $(1+1')$ scheme via the well-defined rotational level of NO $A^2\Sigma^+$ state, ($\nu' = 0$, $N' = 7$). Here, the R_1 rotational band ($J'' = 6.5$) in the $A^2\Sigma^+ - X^2\Pi_{1/2}(0,0)$ transition was chosen, because the ion-current peak is sufficiently isolated and the cation spectrum is also well resolved. The NO^+ rotational spectrum consisting of seven peaks shown in Fig. 8.7(a) has been interpreted in terms of $\Delta\nu = 0$ and $\Delta N = N^+ - N' = 0, \pm 1, \pm 2, \pm 3$ [16](a). It should be mentioned that the most intense rotational band ($N^+ = 7$) in Fig. 8.7(a) has been observed with a resolution of $1.1\,\mathrm{cm}^{-1}$ by using an improved 2PFI analyzer mentioned later [17](a). The resulting

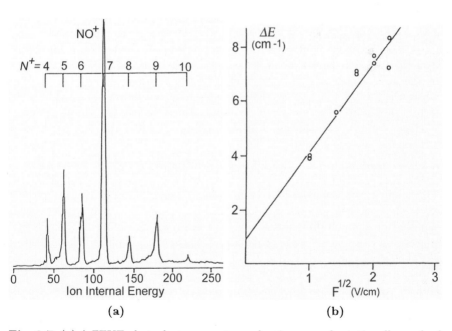

Fig. 8.7. (a) A ZEKE photoelectron spectrum showing several rotationally resolved peaks of NO^+, obtained with the deflection-type analyzer in the two-color $(1+1')$ scheme via the one of the rotational level of NO $A^2\Sigma^+$ state ($\nu' = 0, N' = 7$); (b) the bandwidths (ΔE) are plotted against the square root of $F(V/cm)$ [16](d)

energy resolution is evaluated to be $2\,\text{cm}^{-1}$ (FWHM), although it depends on the delay time of the electric field as well as on the field strength. In an earlier study with the capillary-type analyzer, the energy resolution was $4\,\text{cm}^{-1}$ without using a pulsed field technique [16](c).

The electric-field (F) dependence may be explained as follows. In addition to the ZEKE photoelectrons, PFI electrons are also emitted from its highly excited Rydberg states in the two-color REMPI experiments of NO. Consequently, the ZEKE photoelectrons should mix with the PFI electrons. The field ionization gives rise to a red shift in the ionization energy. The resulting energy shift ΔE should be taken into account in very-high-resolution photoelectron spectroscopy.

The $\Delta N = 0$ bandwidth gradually increases with F_1 as plotted in Fig. 8.7(b), which shows a linear relationship; ΔE is proportional to the square root of F. This linear relationship indicates that the field ionization of highly excited Rydberg states takes place to emit electrons. Consequently the field-ionization electrons more or less mix with the ZEKE photoelectrons at a delay time of 500 ns.

The intercept ($0.9\,\text{cm}^{-1}$) of the linear line shown in Fig. 8.7 corresponds to the energy resolution expected in the two-color REMPI experiments under the field-free conditions. Since the effective diameter of the capillary plate (CP) in Fig. 8.6(a) is 10 mm, the ZEKE photoelectrons remaining in the effective volume after 500 ns should have kinetic energies lower than $2.3\,\text{cm}^{-1}$. The intercept of $0.9\,\text{cm}^{-1}$ might correspond to the width of the ionizing laser (ω_2).

It should also be mentioned that energy resolution of our ZEKE photoelectron analyzer at a delay time of 500 ns may be divided into two terms governed by the laser wavelength and the field strength: Namely, $\Delta E = \Delta E_\text{L} + \Delta E_\text{F}$, where the first term is due to the laser wavelength resolution, and the second term is proportional to the square root of the electric field (F) [16](a). In order to achieve a higher resolution, it is desirable to lower the applied electric field.

8.3.2 1- and 2-Pulsed-Field-Ionization Type

The analyzer of the field-ionization type consists of basically two parallel electrodes set 3 cm apart, as shown in Fig. 8.8 [16](f). The application of a pulsed field inevitably causes field ionization of high-lying Rydberg states converging to the ionic state of interest. Consequently, pulsed-field-ionization (PFI) electrons are mixed with ZEKE photoelectrons. Discrimination against PFI electrons in favor of ZEKE photoelectrons is a rather difficult process, and for most purposes it makes more sense to collect PFI electrons at the expense of ZEKE photoelectrons.

The solution to this problem is to remove all the photoelectrons prior to the detection of PFI electrons, and to apply a pulsed 'extracting field' (V_1) immediately after the laser shot. Since this field causes field ionization of high-lying Rydberg states, it is essential to ensure that the field strength of the collection pulse (V_2) is higher than that of the discrimination pulse (V_1).

8 Very-High-Resolution Laser Photoelectron Spectroscopy of Molecules   227

The balance between these two field settings influences the spectral width thus providing a convenient way to control the resolution. A positive discrimination pulse (V_1) is applied to the extracting plate, while a negative collection pulse (V_1) is applied to the repelling plate (see Fig. 8.8). Using this technique, a study of anthracene-argon vdW complexes has been carried out, and signals due to the possible vdW isomers have been resolved [16](f).

Using the PFI analyzer shown in Fig. 8.8, the delay between the laser shot and the discrimination pulse was set to 25 ns with a pulse width of 250 ns and a height of +2.7 V which translates to a field strength of $V_1 = 0.9\,\mathrm{V/cm}$. The delay between the laser shot and the collection pulse V_2 was set to 400 ns with a pulse width of 800 ns and a height of 6.9 V (field strength of $V_1 = 2.3\,\mathrm{V/cm}$). These settings resulted in a typical spectral resolution of $7\,\mathrm{cm}^{-1}$ (FWHM) using two dye lasers as pump and probe (typical linewidth of $1\,\mathrm{cm}^{-1}$) [16](f).

A further improved version of our compact ZEKE photoelectron analyzer is described below. Figure 8.9 schematically illustrates a remarkable difference

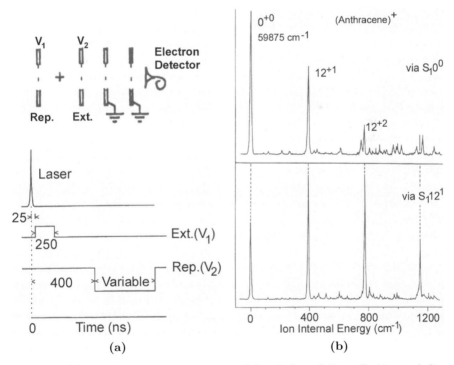

Fig. 8.8. (a) A schematic representation of the timing of the collection and discrimination pulses. The discriminating pulsed extracting field V_1 is applied to the extraction electrode (Ext) and the pulsed repelling field, V_2, used to collect the ZEKE electrons is applied to the repeller electrode (Rep); (b) the two-color $(1+1')$ REMPI ZEKE photoelectron spectra of anthracene recorded via the $S_1 0^0$ and $S_1 12^1$ vibrational levels [16](f)

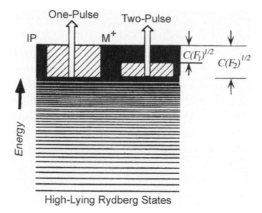

Fig. 8.9. Comparison between the one-pulsed-field (1PFI) and the two-pulsed-field ionization (2PFI) techniques detecting the high-lying Rydberg states produced by two-color laser excitation at near the photoionization threshold

in the principle between the 1PFI and 2PFI method. The 1PFI technique was used to detect high-lying Rydberg states below the ionization threshold within the range ΔE (the shaded part) in Fig. 8.9, which is given by $\Delta E = C(F)^{1/2}$, where F (V/cm) is the field strength, and C is the constant in the range 4-6 depending on the field ionization mechanism, i.e., diabatic or adiabatic [24]. In the 2PFI technique, the first pulsed field (F_1) was applied to remove shallow Rydberg states, and then the second pulsed field ($-F_2$) was applied to collect deeper Rydberg states (the shaded part in Fig. 8.9). The energy width is ideally given by $\Delta E = C\{(F_2)^{1/2} - (F_1)^{1/2}\}$, broadened by some relaxation processes such as a collisional relaxation occurring between the two pulses.

The ZEKE analyzer mentioned above has been improved in brightness as well as in resolution, by modifying the combination of the collection pulse with the discrimination pulses used [80]. A schematic drawing of the analyzer is shown in Fig. 8.10, together with a timing chart of time-delayed pulsed electric fields with respect to a laser shot. The analyzer consists of three electrodes, a repelling electrode (P_1), an extracting electrode (P_2), and a shielding electrode (P_3). All the optically prepared electrons are extracted in the direction opposite to an electron detector by a positive discrimination pulse V_1 applied on P_1. This pulse also field-ionizes high-lying Rydberg states. The electrode P_2 is always at ground potential, and a pulsed electric field is applied only on P_1. This configuration makes it possible to remove kinetic energy electrons effectively from reaching the detector. When this technique was employed in measurements of the ZEKE rotational spectra of NO in a supersonic jet, a rotational band due to NO$^+$ was observed with a bandwidth of 1.1 cm^{-1}; this was the $\Delta N^+ = 0$ transition via the $A^2\Sigma^+$ state ($\nu' = 0$, $N' = 10$) [17](a).

Let us mention the whole experimental setup used for our ZEKE photoelectron spectroscopy briefly [17](a), which is shown schematically in Fig. 8.11.

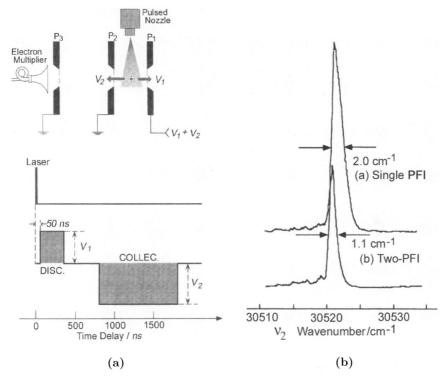

Fig. 8.10. (a) Schematic drawing of an improved 2PFI analyzer and a timing chart of the pulsed electric field in relation to a laser shot. P_1 and P_2 are the repelling and extracting electrodes, respectively. The electrode P_3 is used to shield a high voltage applied on an electron multiplier. The discrimination field and the collection field are indicated by DISC and COLLEC, respectively; (b) two ZEKE photoelectron rotational bands due to NO$^+$($^1\Sigma^+$) ($v^+ = 0$, $N^+ = 10$), obtained with one and two pulses of electric field, showing a difference in the bandwidth [17](a)

The outputs of two dye lasers pumped by a Nd:YAG laser (5 ns, 10 Hz) were frequency-doubled with linewidths of about 0.6 cm^{-1} by nonlinear crystals (KD*P or BBO) mounted on an autotracking system. The visible outputs of two dye lasers were 10 mJ/pulse for ν_1, and 15 mJ/pulse for ν_2. The ν_1 UV output without any focusing lens was attenuated to less than 10 µJ/pulse by a neutral density filter to avoid any MPI process. The ν_2 UV output of a few hundreds µJ/pulse was loosely focused with a 40 cm focal lens. The wavelengths were calibrated with a Fabry-Perot interferometer to an accuracy of +0.002 nm. The ZEKE/MATI (mass analyzed threshold ionization) analyzer consists of two grids spaced by 30 mm, across which a pulsed electric field was applied. An ion flight tube of 20 cm long was used in MATI experiments.

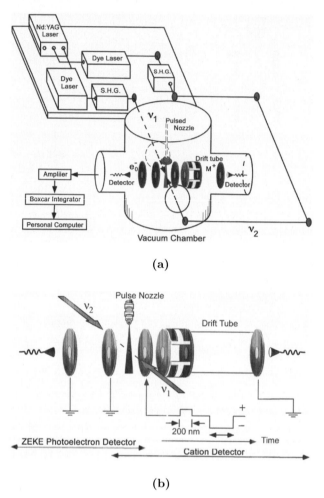

Fig. 8.11. (a) A block diagram of the experimental setup, consisting of a two-color laser system and a vacuum chamber; (b) schematic drawing of an analyzer detecting both ZEKE/MATI signals in an expanded scale. MATI (mass analyzed threshold ionization) experiments as well as the detection of mass-selected ions were also carried out with the drift tube shown here

8.4 Application

In this Section, several topics of REMPI photoelectron spectroscopic studies are first summarized, in which the cm^{-1}-resolution ZEKE photoelectron analyzers were used to jet-cooled molecules in our laboratory, and then its application to the van-der-Waals (vdW) molecules of aniline with argon atoms is described somewhat in detail as a typical example.

The compact ZEKE analyzers mentioned in Sect. 8.3 have been applied mainly to the following photochemical and photophysical topics studied with jet-cooled molecules in the $(1 + 1')$ REMPI scheme: namely, (1) rotational spectra of NO^+ [16](a); (2) vibrational coupling of (naphthalene)$^+$ [18](a); (3) large-amplitude torsional motion of (tolane)$^+$ [18](b); (4) rotational isomers of (*cis and trans n*-propylbenzene)$^+$ and structural isomers of (2-hydroxy-pyridine)$^+$ [18](b,f–h); (5) proton tunneling of (tropolone)$^+$ and (9-hydroxyphenalenone)$^+$ [18](i,j); (6) intramolecular vibrational redistribution (IVR) of *trans* stilbene [18](k); and (7) aromatic van der Waals complexes with argon atoms.

Molecular van der Waals (vdW) complexes have attracted much interest, because of their low binding energies, large intermolecular equilibrium distances, and very low frequency vdW vibrations. The REMPI ZEKE photoelectron spectroscopy has the following advantages for molecular vdW complexes; namely, (1) only a specific vdW species among many analogous species is selectively ionized, (2) the adiabatic ionization energy is accurately determined as well as the change in dissociation energy upon the vdW complex formation, (3) a low-frequency vibrational progression due to the vdW vibration is often observable.

The first observation of vdW vibrations in ZEKE photoelectron spectra is for $(NO-Ar)^+$ and $(aniline-Ar_{1,2})^+$ [16](d,e), [18](a). Since then, many dvW complexes have been studied by the author's group with the compact cm^{-1}-resolution ZEKE photoelectron technique, as shown in Table 8.1.

Table 8.1. Adiabatic ionization energies (I_a) obtained by ZEKE photoelectron spectroscopy. The data shown here in this table are those obtained by the author's group. Also, see the data shown in the footnotes, which have been reported by other workers.

Molecular Species	I_a (cm^{-1})	ΔI_a	Reference
Aniline (C_6H_5–NH_2)	$62\,268 \pm 4$		[16](a)*1
Aniline-Ar	$62\,157 \pm 4$	-111	[16](a)*2
Aniline-Ar$_2$	$62\,049 \pm 4$	-219	[16](a)*3
Anisole (C_6H_5–OCH_3)	$66\,396 \pm 6$		[19](l)
Anisole-Ar	$66\,200 \pm 6$	-196	[19](l)
Anisole-Ar$_2$	$66\,023 \pm 6$	-373	[19](l)
Anthracene ($C_{14}H_{10}$)	$59\,872 \pm 5$		[16](f)
Anthracene-Ar (isomer I)	$59\,807 \pm 5$	-65	[16](f)
Anthracene-Ar (isomer II)	$59\,825 \pm 5$	-47	[16](f)
Anthracene-Ar$_2$ (isomer I)	$59\,757 \pm 5$	-115	[16](f)
Anthracene-Ar$_2$ (isomer II)	$59\,774 \pm 5$	-98	[16](f)
Anthracene-Ar$_3$ (isomer I)	$59\,695 \pm 5$	-177	[16](f)

continued on next page

continued from previous page

Molecular Species	I_a (cm^{-1})	ΔI_a	Reference
Anthracene-Ar$_4$ (isomer II)	59 606 ± 5	−212	[16](f)
Anthracene-Ar$_4$ (isomer I)	59 660 ± 5	−266	[16](f)
Anthracene-Ar$_5$ (isomer I)	59 565 ± 5	−307	[16](f)
Azulene (C$_{10}$H$_8$)	59 781 ± 5		[19](i)
Azulene-Ar	59 708 ± 5	−73	[19](i)
Benzene-N$_2$	74 528 (lower limit)		[19](g)
Benzonitrile (C$_6$H$_5$–C≡N)	78 490 ± 2		[19](a)
Benzonitrile-Ar	78 241 ± 4	−249	[19](a)
Benzonitrile-Ar$_2$	78 007 ± 4	−483	[19](a)
m-Chlorophenol (cis)	69 810 ± 10		[18](g)
m-Chlorophenol (trans)	70 027 ± 10		[18](g)
p-Dimethoxybenzene (cis)	60 774 ± 7		[18](b)
p-Dimethoxybenzene (trans)	60 563 ± 7		[18](b)
p-Dimethoxybenzene-Ar (cis)	60 687 ± 7	−87	[18](b)
p-Dimethoxybenzene-Ar (trans)	60 479 ± 7	−84	[18](b)
p-Dimethoxybenzene-Ar$_2$ (trans)	60 295 ± 7	−268	[18](b)
p-Dimethoxybenzene-Ar$_2$ (cis)	60 509 ± 7	−265	[18](b)
Ethylbenzene	70 762 ± 6	−128	[19](k)
Ethylbenzene-Ar	70 634 ± 6		[19](k)
Fluorobenzene (C$_6$H$_5$–F)	74 238 ± 4		[19](d)*4,5
Fluorobenzene-Ar	74 011 ± 4	−227	[19](d)*6,7
Fluorobenzene-Ar$_2$	73 816 ± 4	−422	[19](d)
Fluorobenzene-N$_2$	74 172 (lower limit)		[19](g)
Indole (C$_9$H$_7$N)	62 592 ± 4		[19](c)
Indole-Ar	62 504 ± 6	−88	[19](c)
Naphthalene (C$_{10}$H$_8$)	65 692 ± 3		[19](b)
Naphthalene-Ar	65 607 ± 3	−85	[19](b)
Phenylacetylene (C$_6$H$_5$–C≡CH)	71 175 ± 5		[18](c)
Phenylacetylene-Ar	71 027 ± 5	−148	[16](c)
p-Phenylenediamine	54 640 ± 8		[16](b)
n-Propylbenzene (trans)	70 278 ± 8		[18](f)
n-Propylbenzene (gauche)	70 420 ± 8		[18](f)
n-Propylbenzene (trans)	70 267 ± 6		[19](k)
n-Propylbenzene (trans)-Ar	70 155 ± 6	−112	[19](k)
Pyrimidine (C$_4$H$_4$N$_2$)	75 261 ± 6		[19](j)
Pyrimidine-Ar	75 000 ± 6	−261	[19](j)
Pyrimidine-Ar$_2$	74 745 ± 6	−516	[19](j)
Styrene (C$_6$H$_5$–CH=CH$_2$)	68 267 ± 5		[18](c)
Styrene-Ar	68 151 ± 5	−116	[18](c)

continued on next page

continued from previous page

Molecular Species	I_a (cm^{-1})	ΔI_a	Reference
Thioanisole (C$_6$H$_5$–SCH$_3$)	63 906 ± 3		[19](e)
Thioanisole-Ar	63 789 ± 3	−117	[19](e)
Thioanisole-Ar$_2$	63 675 ± 3	−231	[19](e)
Toluene	71 203 ± 5		[19](f)*8
Toluene-Ar	71 037 ± 5		[19](f)*9

The following data have been reported:
*1 62 281 cm^{-1} [Ref. [21](d)]   *2 62 168 cm^{-1} [Ref. [21](d)]   *3 62 061 cm^{-1} [Ref. [21](d)]
*4 74 229 cm^{-1} [Ref. [21](b)]   *5 74 222 cm^{-1} [Ref. [21](c)]   *6 74 004 cm^{-1} [Ref. [21](b)]
*7 74 000 cm^{-1} [Ref. [21](c)]   *8 71 199 cm^{-1} [Ref. [21](a)]   *9 71 033 cm^{-1} [Ref. [21](a)]

A series of vibrational progressions of (aniline-Ar)$^+$ and (aniline-Ar$_2$)$^+$ have been clearly resolved in ZEKE photoelectron spectra, as shown in Fig. 8.12, which were obtained via the S$_1$ origins [18](a). Their first ionization bands are also shown in an expanded scale in Fig. 8.13. Each vibrational band is divided into several peaks upon the complex formation. The adiabatic ionization energies (I_a) and their shifts (ΔI_a) have been determined from the cation origin bands (D$_0$ 0$^+$): namely, I_a = 62268 ± 4 cm^{-1}, and ΔI_a(aniline-Ar) = −111 cm^{-1}, and ΔI_a(aniline-Ar$_2$) = −219 cm^{-1}. The I_a shift of aniline-Ar$_2$ is almost twice that of aniline-Ar. The shifts in the S$_1$ ori-

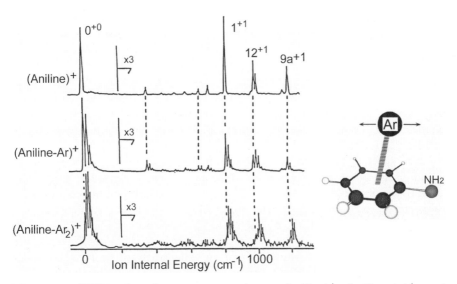

Fig. 8.12. ZEKE photoelectron spectra due to (aniline)$^+$, (aniline-Ar)$^+$, and (aniline-Ar$_2$)$^+$, which were obtained in the (1+1′) scheme via the S$_1$ origins, showing the sharp vibrational peaks due to the cation origin (D$_0$ 0^{+0}) and the low-frequency vibrational progressions due to the vdW bending modes [18](a)

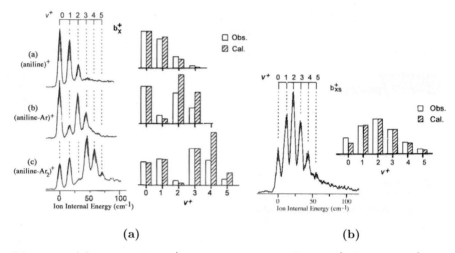

Fig. 8.13. (a) The bending b_x^+ progressions due to (aniline)$^+$, (aniline-Ar)$^+$ and (aniline-Ar$_2$)$^+$, which were obtained via (1) the S$_1$ origin, (2) S$_1$ b_x^1, and (3) S$_1$ b_x^2. Their observed Franck-Condon factors are compared with the calculated ones; (b) the bending b_{xs}^+ progression due to (aniline-Ar$_2$)$^+$. The observed Franck-Condon factors are also compared with the calculated ones

gins of aniline-Ar and $-$Ar$_2$ are -54 cm^{-1} (reported by Amirav et al. [25](a)) and -109 cm^{-1} (reported by Bieske et al. [25](b)), respectively. Therefore, the shifts in I_a are almost twice as much as those in the S$_1$ states.

The aniline-Ar cation has the following three vdW vibrational modes, if C$_s$ symmetry is assumed: Namely, the totally symmetric bending b_x^+(a'), the non-totally symmetric bending b_y^+(a''), and the 'stretching' s_z^+(a'). On the other hand, the aniline-Ar$_2$ cation has a total of six vdW vibrational modes, if C$_{2v}$ symmetry is assumed; namely, the two in-phase symmetric bending modes b_{xs}^+(a$_1$) and b_{ys}^+(b$_2$), the two out-of-phase antisymmetric bending modes b_{xa}^+(b$_1$) and b_{ya}^+(a$_2$), and the symmetric and antisymmetric stretching modes s_{zs}^+(a$_1$) and s_{za}^+(b$_1$).

The spectra due to (aniline-Ar$_{1,2}$)$^+$ in Fig. 8.13 clearly show the vibrational progressions with separations of 16 and 11 cm^{-1}, respectively, attributable to the symmetric bending (b_x^+) and the in-phase bending mode (b_{xs}^+), respectively. In (aniline-Ar)$^+$, the progressions with the same frequency have been obtained via the S$_1$ 0^0, b_x^1 and b_x^2 levels, as seen from Fig. 8.13, in which the Franck-Condon calculations are also compared with the experimental ones.

Possible structural parameters can be deduced from the Franck-Condon calculations with a simple one-dimensional harmonic oscillator model, as reported in the cases of benzonitrile-Ar$_{1,2}$ [19](a), fluorobenzene-Ar$_{1,2}$ [19](d), and azulene-Ar$_{1,2}$ [19](i). The three kinds of the first band progressions shown in Fig. 8.13(a) have well been reproduced with a value of $\Delta\phi = 8.2°$ (the difference in the angle between S$_1$ and D$_0$). The shift in I_a is given by the difference in the dissociation energy (D_0) by

$$\Delta I_a = I_a(\text{aniline-Ar}) - I_a(\text{aniline}) = D_0(\text{aniline-Ar}) - D_0(\text{aniline-Ar})^+ \quad (8.4)$$

The dissociation energies can not be determined directly from the ZEKE photoelectron spectra, but only their difference is determined. In order to determine the dissociation energy of the vdW cation, it is necessary to find the onset of the dissociation directly, as demonstrated in a mass-analyzed threshold ionization (MATI) study [26].

Finally, let us mention another example of ZEKE photoelectron spectroscopy, in which a series of vdW complexes of alkylbenzenes with Ar have been studied by Sato et al. [19](k). Figure 8.14(a) shows ZEKE spectra of the first four vdW complexes of alkylbenzenes (benzene-Ar, toluene-Ar, ethylbenzene-Ar and n-propylbenzene-Ar) in the low-energy region. In the case of benzene-Ar, in addition to the cation origin (0^+), several very weak bands appear at 24, 29, 38 and 48 cm^{-1}, which have been tentatively assigned as b_x^{+2}, b_y^{+2}, s_z^{+1}, and b_x^{+4}, respectively, as shown in Fig. 8.14(b). The benzene-Ar cation has a total of three vdW vibrational modes, which are the totally symmetric bending mode (b_x^+), the non-totally symmetric bending mode (b_y^+), and stretching mode (s_z^+). The vibrational frequencies thus deduced are 12 cm^{-1} (b_x^+), 15 cm^{-1} (b_y^+), and 38 cm^{-1} (s_z^+ mode) [19](k). The overtones of vdW bending vibration of benzene-Ar were observed with very weak intensities, since the vdW bending vibration of benzene-Ar is non-totally symmetric under C_{6v} symmetry.

The spectrum of (toluene-Ar)$^+$ shown in Fig. 8.14(a) is essentially the same as that earlier reported by Inoue et al. [19](f). A spacing of 17 cm^{-1} observed in the ZEKE vibrational progression of (toluene-Ar)$^+$ has been assigned to the vdW bending mode (b_x^+) of the vdW cation [19](f). The frequency of 17 cm^{-1} is similar to that reported for the b_x^+ mode in other aromatic-Ar cations (for example, 16 cm^{-1} for the aniline-Ar cation, 12 cm^{-1} for the benzonitrile-Ar cation, and 12 cm^{-1} for the fluorobenzene-Ar cation) [18](a), [19](a,d). The ZEKE vibrational progressions obtained via the S_1 b_{x0}^n ($n = 1 - 3$) levels have been interpreted in terms of the cation vdW bending b_x^{+n} ($n = 0 - 4$). Furthermore, the vdW bending (b_x^+) at 17 cm^{-1} and the vdW stretching (s_z^+) at 49 cm^{-1} have been identified for the toluene-Ar cation from its ZEKE photoelectron spectra obtained via the S_1 vdW stretching level s_{z0}^1 largely overlapped with b_x^3 [19](k).

All the vdW molecular complexes studied by the very-high-resolution ZEKE photoelectron techniques in the author's group are listed in Table 8.1, indicating their accurate adiabatic ionization energies (I_a) as well as their differences (ΔI_a) upon the complex formations.

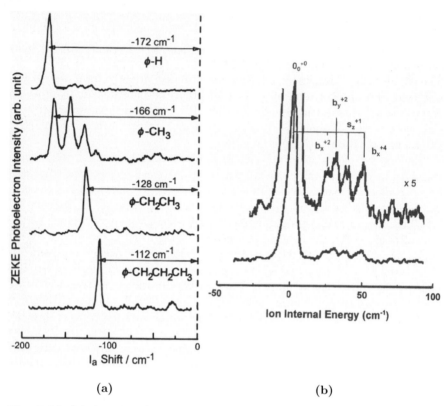

Fig. 8.14. (a) A series of ZEKE photoelectron spectra observed for benzene-Ar, toluene-Ar, ethylbenzene-Ar, and *trans* n-propylbenzene-Ar in the low-energy region are compared with one another [19](k); (b) a ZEKE electron spectrum of benzene-Ar, indicating the cation origin peak (0^{+0}) as well as several weak vdW vibrational bands with their assignments [19](k)

8.5 Concluding Remarks

It should be emphasized that the one- and two-color REMPI-based photoelectron spectroscopic techniques with tunable pulsed UV/visible lasers are capable of performing state-selective and species-selective photoionization experiments for any organic and inorganic molecular species in a supersonic jet. With the two-color REMPI technique, it is also possible to carry out time-resolved photoionization and photoelectron experiments.

Since the ionization transition is always allowed, REMPI-based photoelectron spectroscopy with an appropriate tunable pulsed laser system is a versatile technique to study dynamics of molecular excited states and their ionic states. The REMPI photoelectron spectroscopy method of measuring a photoelectron kinetic energy spectrum is complementary to that of measuring a ZEKE photoelectron spectrum. The advantage of the former is to be able to use a one-color REMPI process, while the advantage of the latter is to carry out very-high-resolution cation spectroscopy. The present lecture note is far from a complete one, but it is expected to provide some idea for future applications of the very-high-resolution REMPI-based photoelectron spectroscopy to not only single molecular species but also many kinds of vdW complexes and molecular clusters formed in a supersonic jet.

Acknowlegdgements

The author thanks his former colleagues Professors S. Katsumata, Y. Achiba, K. Okuyama, M. Takahashi, H. Ozeki, and S. Sato for their contributions to the development of our laser REMPI-based photoelectron technique and its applications to various jet-cooled molecules both at Institute for Molecular Science and at Japan Advance Institute of Science and Technology. The author also thank Professor J.M. Dyke, Dr. M.C.R. Cockett, Dr. I. Plazibat, and Dr. T. Vondrak for their contributions to the REMPI-based photoelectron work during their visits supported by the Government.

References

1. K. Kimura, J. Electron Spectrosc. Related Phenom. **100**, 273 (1999)
2. a) M. I. Al-Joboury and D. W. Turner: J. Chem. Phys. **37** 3007 (1962); b) D. W. Turner et al: *Molecular Photoelectron Spectroscopy, A Handbook of He 584 Å Spectra* (Interscience, London, 1970)
3. K. Kimura et al: *Handbook of HeI Photoelectron Spectra of Fundamental Organic Molecules, Ionization Energies, ab initio Assignments, and Valence Electron Structure for 200 Molecules*, (Halsted Press, New York; Japan Societies Press, Tokyo, 1981)
4. a) P. M. Johnson, Acc: Chem. Res. **13**, 20 (1980); b) P. M. Johnson and C. E. Otis: Ann. Rev. Phys. Chem. **32**, 139 (1981)
5. a) K. Kimura: Adv. Chem. Phys. **60**, 161 (1985); b) Y. Achiba and K. Kimura: Nippon Kagaku Kaishi (Chem. Soc. Japan) **1984**, 1529 (1984, in Japanese)
6. K. Kimura: International Rev. Phys. Chem. **6**, 195 (1987)
7. a) Y. Achiba et al: Annual Review, p. 100 (Institute for Molecular Science, Okazaki, Japan, 1980); b) Y. Achiba et al: J. Photochem. **17**, 53 (1981); c) Y. Achiba et al: J. Chem. Phys. **79**, 2709 (1982); d) Y. Nagano et al: Chem. Phys. Lett. **93**, 510 (1982)
8. a) R. N. Compton et al: Chem. Phys. Lett. **71**, 87 (1980); b) J. C. Miller and R. N. Compton: J. Chem. Phys. **75**, 22 (1981); c) J. C. Miller and R. N. Compton: J. Chem. Phys. **75**, 2020 (1981)
9. a) J. T. Meek et al: J. Chem. Phys. **73**, 3503, (1980); b) J. T. Meek et al: J. Chem. Phys. **86**, 2809 (1982)
10. a) J. Kimman et al: J. Phys. **B 14** L597 (1981); b) P. Kruit et al: Chem. Phys. Lett. **88**, 576 (1982)
11. S. L. Anderson et al: Chem. Phys. Lett. **93**, 11 (1982)

12. K. Müller-Dethlefs et al: Chem. Phys. Lett. **112**, 291 (1984)
13. Y. Achiba et al: Abstract of Second Symposium of Chemical Reaction (The Chemical Society of Japan) held at Okazaki, p. 24 (1985)
14. a) L. A. Chewter et al: Chem. Phys. Lett. **135**, 219 (1987); b) K. Müller-Dethlefs, E. W. Schlag: Annu. Rev. Phys. Chem. **42**, 109 (1991); c) K. Müller-Dethlefs et al: Chem. Rev. **94**, 1845 (1994)
15. K. Müller-Dethlefs: J. Elect. Spectrosc. Relat. Phenom. **75**, 35 (1995)
16. a) M. Takahashi et al: Chem. Phys. Lett. **181**, 255 (1991); b) H. Ozeki et al: J. Phys. Chem. **95**, 4308 (1991); c) M. Takahashi et al: J. Mol. Structure **249**, 47 (1991) d) K. Kimura and M. Takahashi: in *Optical Methods for Time- and State-Resolved Chemistry* ed. by C.-Y. Ng (The International Society for Optical Engineering, Washington, DC, 1992, SPIE) **1638**, p. 216; e) M. Takahashi: J. Chem. Phys. **96**, 2594 (1992); f) M. C. R. Cockett and K. Kimura: J. Chem. Phys. **100**, 3429 (1994)
17. a) S. Sato and K. Kimura: Chem. Phys. Lett. **249**, 155 (1996); b) S. Sato and K. Kimura: J. Chem. Phys. **107**, 3376 (1997)
18. a) M. Takahashi et al: J. Chem. Phys. **96**, 6399 (1992); b) M. C. R. Cockett et al: J. Chem. Phys. **97**, 4679 (1992); c) J. M. Dyke et al: J. Chem. Phys. **97**, 8926 (1992); d) M. C. R. Cockett et al: J. Chem. Phys. **98**, 7763 (1993); e) K. Okuyama et al: J. Chem. Phys. **97**, 1649 (1992); f) M. Takahashi and K. Kimura: J. Chem. Phys. **97**, 2920 (1992); g) M. C. R. Cockett et al: Chem. Phys. Lett. **187**, 250 (1991); h) H. Ozeki et al: J. Phys. Chem. **99**, 8608 (1995); i) H. Ozeki et al: J. Chem. Phys. **95**, 9401 (1991); j) H. Ozeki et al: J. Chem. Phys. **99**, 56 (1993); k) M. Takahashi and K. Kimura: J. Phys. Chem. **99**, 1628 (1995)
19. a) M. Araki et al: J. Phys. Chem. **100**, 10542 (1996); b) T. Vondrak et al: Chem. Phys. Lett. **261**, 481 (1997); c) T. Vondrak et al: J. Phys. Chem. **A 101**, 2384 (1997); d) H. Shinohara et al: J. Phys. Chem. **A 101**, 6736 (1997); e) T. Vondrak et al: J. Phys. Chem. A **A 101**, 8631 (1997); f) H. Inoue et al: J. Electron Spectrosc. Related Phenom. **88**, 125 (1998); g) S. Shinohara et al: J. Electron Spectrosc. Related Phenom. **88–91**, 131 (1998); h) S. Sato et al: J. Electron Spectrosc. Related Phenom. **88–91**, 137 (1998); i) D. Tanaka et al: Chem. Phys. **239**, 437 (1998); j) S. Sato et al: J. Electron Spectrosc. Related Phenom. **97**, 121 (1998); k) S. Sato et al: J. Electron Spectrosc. Related Phenom. **112**, 241 (2000); l) K. Tsutsumi et al: unpublished data (Master thesis by Tsutsumi at Japan Advance Institute of Science and Technology, 1996)
20. For the vdW vibrational frequencies obtained by ZEKE photoelectron spectroscopy, see the following review articles: a) K. Kimura: J. Electron Spectrosc. Related Phenom. **108**, 31 (2000); b) K. Kimura: *Zero-Kinetic-Energy Photoelectron Spectroscopic Studies of Aromatic-Argon van der Waals Complexes*, in *Photoionization and Photodetechment*, Ed. by C.-Y. Ng (World Scientific Publishing, New Jersey, 2000) Advanced Series in Phys. Chem. **10 A**, p. 246; c) K. Kimura: J. Chin. Chem. Soc. **48**, 433 (2001)
21. a) K.-T. Lu and J. C. Weisshaar: J. Chem. Phys. **99**, 4249 (1993); b) G. Lembach and B. Brutschy: J. Phys. Chem. **100**, 19758 (1996); c) Th. L. Grebner and H. J. Neusser: Int. J. Mass. Spect. Ion Processes **159**, 137 (1996); d) X. Zhang et al: J. Chem. Phys. **97**, 2843 (1992)
22. a) K. Katsumata et al: J. Electron Spectrosc. Relat. Phenom. **41**, 325 (1986); b) Y. Achiba et al: J. Chem. Phys. **82**, 3959 (1985); c) Y. Achiba and K. Kimura:

Chem. Phys. **129**, 11 (1989); d) K. Sato et al: J. Chem. Phys. **80**, 57 (1984); e) K. Sato et al: J. Chem. Phys. **85**, 1418 (1986); f) K. Sato et al: Chem. Phys. Lett. **126**, 306 (1986); g) Y. Achiba et al: J. Chem. Phys. **78**, 5474 (1983); h) K. Sato et al: J. Chem. Phys. **80**, 57 (1984); i) Y. Achiba et al: J. Chem. Phys. **79**, 5213 (1983); j) Y. Achiba et al: J. Chem. Phys. **80**, 6047 (1984); k) A. Hiraya et al: J. Chem. Phys. **82**, 1810 (1985); l) Y. Nagano et al: J. Chem. Phys. **84**, 1063 (1986); m) Y. Nagano et al: J. Phys. Chem. **90**, 615 (1986); n) Y. Nagano et al: J. Phys. Chem. **90**, 1288 (1986)
23. C. S. Parmenter: Adv. Chem. Phys. **22**, 365 (1972)
24. W. A. Chupka: J. Chem. Phys. **98**, 4520 (1993)
25. a) A. Amirav et al: Mol. Phys. **49**, 899 (1983); (b) E. J. Bieske et al: J. Chem. Phys. **94**, 7019 (1991)
26. H. J. Neusser and H. Krause: Chem. Rev. **94**, 1829 (1994)

Part V

High-Temperature Superconductors
and Transition-Metal Oxides

9

Doping Evolution of the Cuprate Superconductors from High-Resolution ARPES

K. M. Shen[1] and Z.-X. Shen[2]

[1] Laboratory of Atomic and Solid State Physics, Cornell University, Ithaca NY 14853, USA
kmshen@ccmr.cornell.edu
[2] Department of Physics, Applied Physics, and Stanford Synchrotron Radiation Laboratory, Stanford University
zxshen@stanford.edu

> "Angle resolved photoemission is, for this problem, the experiment that will play the role that tunneling played for BCS superconductivity"
>
> – P.W. Anderson on the cuprates, *Physics Today*, 1992

Abstract. Recent advances in both materials synthesis and state-of-the-art instrumentation have led to an understanding of how the electronic structure of the high-temperature cuprate superconductors first evolve away from the parent Mott insulating state. In this chapter, we describe ARPES studies of the $La_{2-x}Sr_xCuO_4$ and $Ca_{2-x}Na_xCuO_2Cl_2$ compounds which track the growth of the nodal quasiparticles, and the doping evolution of the chemical potential and Fermi surface. These results indicate that the consideration of strong electron-boson interactions are essential for an adequate description of the first doped hole states, and also suggest a relationship between the momentum anisotropy of the low-energy states and competing orders in the underdoped regime.

9.1 Introduction

In the more than ten years following Anderson's statement, ARPES has not disappointed the high-T_c community, as it has played a central role in shaping the field. The list of significant discoveries made by ARPES after this statement in 1992 is long indeed : the detection of the anisotropic d-wave superconducting gap (1993) [1,2], the dispersion of the single hole in the parent insulator (1995) [3], the detection of the normal state pseudogap (1996) [4,5], and the observation of strong electron–boson coupling in the hole-doped cuprates (2000) [6,7]. The convergence of ARPES and the high-T_c cuprates was a most fortuitous one, with their highly two-dimensional structures and stable surfaces being almost tailor-made for photoemission, and the highly k-dependent physics giving ARPES a unique advantage over momentum averaged or $q = 0$ probes. A second stroke of good fortune was that many of the salient momentum, energy, and temperature scales relevant to the high-T_c problem were already within the accessible range of the experimental ARPES systems of ten years past, allowing photoemission to directly address many of the central questions during the early stages of the field. Having established a strong foothold in this field rather early on, ARPES has now assumed a central role in the high-T_c field. The advent of the angular multiplexing analyzers, such as the Gammadata-Scienta SES200, and now sub-meV energy resolution, has allowed ARPES to address even very subtle details in the electronic structure and many-body interactions in the cuprates.

9.2 High-Temperature Superconductivity

The high-temperature superconductors are layered materials comprised of stacked, two-dimensional square CuO_2 planes separated by so-called "blocking" layers. Superconductivity in this class of materials was first discovered by Bednorz and Müller in 1986 in the $La_{2-x}Ba_xCuO_4$ compound [8], and within one year, superconducting transition temperatures (T_c's) jumped to around 100 K with the discoveries of other cuprate families, such as $YBa_2Cu_3O_{7-\delta}$. In most cuprates, the CuO_2 plane is considered to be the only electronically active component of the crystal (a notable exception being the $YBa_2Cu_3O_{7-\delta}$ family, which also has metallic Cu-O chains). The planar nature of the square CuO_2 layers results in a highly two-dimensional electronic structure (typical resistivity anisotropies, ρ_c/ρ_{ab}, are on the order of 10^4) where the electrons are largely confined within the CuO_2 planes. This layered structure also results in easily exposed cleavage planes which are highly advantageous for ARPES; the Bi-based compounds, such as $Bi_2Sr_2CaCu_2O_{8+\delta}$ are particularly micaceous and can even be tape cleaved. In Fig. 9.1, we show the high temperature (I4/mmm) structure of parent cuprate La_2CuO_4, showing the characteristic CuO_2 planes and one representative CuO_6 octahedron.

Through chemical substitution of cations in the blocking layers or through oxygen intercalation, one can alter the stoichiometry of these materials, thereby introducing mobile carriers into the CuO_2 plane in a manner similar to semiconductor doping. The materials which have undoped CuO_2 planes (i.e. stoichiometric) are commonly referred to as "parent compounds", meaning that there is one unpaired electron per CuO_2 plaquette. Within the context of band theory, this half filling would imply that these materials should be good metals. However, in the case of the parent cuprates, these compounds are antiferromagnetic insulators with a gap of ~ 2 eV. This insulating gap arises due to strong Coulomb repulsions between electrons on the Cu $3d$ and O $2p$ orbitals, and are commonly known as charge-transfer

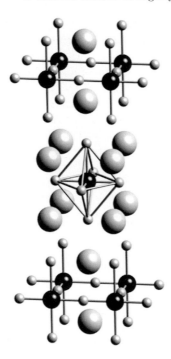

Fig. 9.1. Crystal structure of the prototypical parent cuprate La_2CuO_4, with one CuO_6 octahedron shown in the center. Cu is shown in black, O as small grey spheres, and La as large grey spheres

insulators [10], where the Coulomb repulsion energy dominates over the gain in kinetic energy associated with delocalization. Formally speaking, the cuprates are charge-transfer insulators, although here we will use that term interchangeably with the commonly used "Mott insulator" terminology. The presence of this gap, and the fact that non-interacting band theory predicts a fundamentally incorrect ground state, is a direct and dramatic example of the strong electron–electron correlations in the cuprates.

Once the parent compounds are doped with carriers (in this chapter, we will focus on hole doping, not electron doping), the average number of electrons per unit cell will deviate from 1. The phase diagram for $La_{2-x}Sr_xCuO_4$ and electron doped $Nd_{2-x}Ce_xCuO_4$ is shown as a function of temperature and hole doping in Fig. 9.2. The phase diagram for $La_{2-x}Sr_xCuO_4$ can be considered as representative of the hole doped cuprates. Even a fairly small concentration of doped holes (∼0.03 holes / unit cell) is enough to completely suppress long-range antiferromagnetic Néel order, although it is well known that short-range antiferromagnetic correlations remain robust and persist to much higher doping levels (for an overview of cuprate properties, see for instance, Kastner et al. [11]). Typically, around a hole concentration of $x \sim 0.07$, although this varies between material families, superconductivity arises. The

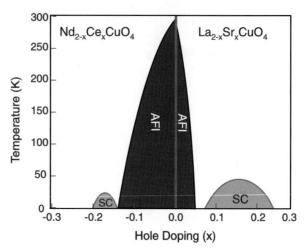

Fig. 9.2. Phase diagram of the hole-doped ($La_{2-x}Sr_xCuO_4$) and electron-doped ($Nd_{2-x}Ce_xCuO_4$) high-temperature superconducting cuprates

superconducting transition temperature, T_c, continues to increase with hole doping up until a maximum transition temperature which usually occurs around a hole concentration of $x \sim 0.16$, a value known as "optimal doping". This optimal doping value also varied slightly from family to family among the cuprates. Above this optimal doping concentration, T_c declines with further hole doping until superconductivity is lost, usually around $x \sim 0.25$. Samples with hole doping levels below optimal doping are typically referred to as "underdoped", while samples above optimal doping are referred to as "overdoped". The maximum transition temperature (T_c at optimal doping) ranges from approximately 25 K for the lowest T_c families and to up to 160 K (under pressure) for the highest. This is in obvious contrast to nearly all conventional BCS superconductors which typically have T_c's below 10 K, with a few notable exceptions. Another unusual aspect of the high-T_c superconductors is that instead of the typical s-wave gap ($l = 0$) found in virtually all conventional superconductors, the high-T_c superconductors exhibit a gap with a $d_{x^2-y^2}$ symmetry ($l = 2$), which results in ungapped electronic excitations along the Brillouin zone diagonal, even at $T = 0$ [12].

In addition to the unusually high transition temperatures and the unconventional symmetry of the superconducting gap, the high-T_c compounds also exhibit highly anomalous properties above T_c, in the normal state. In a conventional superconductor, such as lead, the material exhibits metallic properties at temperatures above T_c. However, the underdoped cuprates do not behave like ordinary metals, or in fact any other known materials, above T_c, but continue to exhibit highly unusual normal-state properties. For instance, a gap in the single-particle electronic channel persists above T_c, even though this gap can no longer be representative of the superconducting order parameter.

This normal-state gap is typically referred to as a "pseudogap", whose microscopic origin is still a topic of great debate [13]. One possibility which has been postulated is that the pseudogap is representative of preformed pairs which achieve global phase coherence only below T_c, as suggested by Emery and Kivelson [14, 15]. An alternative possibility is that the pseudogap represents the order parameter for another phase which competes against d-wave superconductivity, thereby driving down T_c in the underdoped regime. Examples of theoretically proposed competing states include various charge ordered states, orbital-current states, or d-density-wave states [16–19]. Since understanding the origin of the pseudogap likely has significant bearing on the ultimate microscopic theory of high-temperature superconductivity, this remains a topic of great importance.

9.3 Photoemission Studies of the Lightly Doped Cuprates

9.3.1 Previous ARPES Studies of High Temperature Superconductors

As discussed in the introduction, a wide variety of issues have already been studied in depth using ARPES (see also the contributions by Johnson and Valla, by Takahashi et al. and Fink et al. in this volume). Not surprisingly then, a number of excellent review articles have already been published in the past couple of years [20–23]. To date, the vast majority of existing results in this field, and those covered in the aforementioned reviews, have still come from the $Bi_2Sr_2CaCu_2O_{8+\delta}$ family, due to its particularly favorable surface properties, high transition temperature, and relative ease of synthesis. However, because of certain idiosyncratic material characteristics (superlattice modulations, strong cation disorder, interlayer coupling), it is important to understand which particular results from $Bi_2Sr_2CaCu_2O_{8+\delta}$ can be considered generic to the cuprates or are specific to that material family. With these considerations in mind, and to avoid excessive overlap with other recent reviews, we will focus this chapter primarily on materials other than $Bi_2Sr_2CaCu_2O_{8+\delta}$ and covering results which have been published after earlier reviews. In particular, we will concentrate on one particular topic where significant recent progress has made: the evolution of the electronic structure of the cuprates as they are hole doped away from the parent insulating state.

Apart from the high superconducting transition temperatures, the defining characteristic of the cuprates is the fact that the parent compounds are charge-transfer insulators, a direct consequence of strong electron–electron repulsions. How such insulators evolve with carrier doping is not only a central intellectual problem in the field of high-T_c superconductivity, but is one of the fundamental issues in the general field of strongly correlated systems [24]. Despite this fact, ARPES studies of the carrier doping evolution of the cuprates have been rather sparse over the past decade, largely due to the lack of suitable materials for such studies. Fortunately, recent advances in materials synthesis have now provided unprecedented access to the lightly carrier doped region of the phase diagram where one can study the crossover from the Mott insulator to the superconducting state. For instance, developments in high-pressure crystal growth have led to the recent synthesis of single crystals of $Ca_{2-x}Na_x$-CuO_2Cl_2 [25], which has proven to be an ideal material for ARPES studies of the undoped and very lightly doped cuprates. Together with the continued development of high-resolution photoemission spectroscopy, one can now achieve a much deeper understanding of the Mott insulator-to-superconductor transition in the cuprates, which we will detail in this chapter.

The fact that ARPES provides direct information on the many-body interactions makes it the ideal tool for studying correlation effects. In particular, how these many-body correlations evolve as a function of doping is one of key questions in high-T_c superconductivity. In tracking the insulator-to-superconductor transition, we can directly address a number of important issues through ARPES. In the remainder of this chapter, we will cover 4 such issues, as follows :

1. Behavior of the chemical potential, μ, as a function of hole doping away from half-filling
2. Evidence for Franck–Condon broadening and polaron formation in the undoped and lightly doped cuprates
3. Evolution of quasiparticle-like excitations with hole doping
4. Momentum dependence of near-E_F spectral weight

In order to address these particular subjects, we will focus on lightly doped compositions of the two particular cuprate families, $La_{2-x}Sr_xCuO_4$ and $Ca_{2-x}Na_xCuO_2Cl_2$, where ARPES has been reported. $La_{2-x}Sr_xCuO_4$ is often considered as the prototypical cuprate and is one of the most comprehensively studied among all families. It has the unique advantage that its doping composition can be varied over the entire phase diagram from very lightly doped to heavily overdoped. Although very-high-quality ARPES data has been taken on $La_{2-x}Sr_xCuO_4$, its surface characteristics are nevertheless somewhat ambiguous due to the curious lack of atomically resolved scanning tunneling microscopy (STM) images. Nevertheless, the recent successes in high-resolution ARPES coupled with the consistency of the data from other families would suggest that the ARPES results from $La_{2-x}Sr_xCuO_4$ are very likely reliable. On the other hand, $Ca_{2-x}Na_xCuO_2Cl_2$ can be grown at very low hole concentrations and meanwhile has also yielded very high quality STM/STS data [26, 27]. On the other hand, $Ca_{2-x}Na_xCuO_2Cl_2$ cannot yet be grown in single crystal form at compositions above optimal doping, limiting ARPES studies to the undoped-to-underdoped region of the phase diagram. To date, no extensive ARPES studies of other common cuprates, such as Bi_2-$Sr_2CaCu_2O_{8+\delta}$ or $YBa_2Cu_3O_{7-\delta}$, have yet been reported in this lightly doped regime. This is primarily due to constraints of crystal chemistry which make the synthesis of such compounds at such low dopings rather difficult, although future advances along these lines will be much anticipated.

9.3.2 Doping Evolution of the Chemical Potential

Understanding the doping evolution of the chemical potential, μ, is one of the critical issues in the field of high-T_c, since μ is a fundamental thermodynamic quantity, and could thus potentially allow one to clearly discriminate between different theoretical models of high-T_c superconductivity. Moreover, the nature of the electronic states occupied by these doped carriers is a non-trivial matter due to the presence of strong electronic correlations. Photoemission can directly probe the occupied electronic states relative to the position of the Fermi energy, E_F, thus making it uniquely suited to studying the doping evolution of μ. In light of this, it is somewhat surprising that over the past decade no consensus had been reached over the doping behavior of the chemical potential in the cuprates. In fact, the doping evolution of μ has remained a controversial and unresolved topic. In this section, we will focus on some recent work which may contribute to a better understanding of this critical issue.

In the past decade, there have been two dominant pictures for describing the carrier doping evolution of the chemical potential. The first scenario was that μ was pinned in the middle of the gap, approximately halfway between the lower and upper Hubbard bands. New electronic states then form in midgap, analogous to doping small concentrations of impurity states into a semiconductor, instead of moving to the top of the valence band/bottom of the conduction band. Such a claim had been made based on early valence band photoemission studies of $La_{2-x}Sr_xCuO_4$ and $Nd_{2-x}Ce_xCuO_4$ by Allen et al. [29], as well as from core-level x-ray photoemission studies (XPS) of $La_{2-x}Sr_xCuO_4$ by Ino et al. [30]; a simple picture for this is shown in Fig. 9.3(c). In addition, ARPES studies of the near-E_F states by Ino et al. and Yoshida et al. also appear to be in accordance with this picture, showing that the ostensible lower-Hubbard-band states apparently remain well below E_F upon light hole doping [31–33], as shown in Fig. 9.4. This pinning of μ has been viewed as consistent with pictures involving electronic phase separation in the cuprates (for instance, [34]). The countervailing view was that μ shifts rapidly upon carrier doping, similar to what one would expect within a simple rigid band picture. Therefore, upon hole doping, μ would drop to the top of the lower Hubbard band, and would continue to empty out the states in the lower Hubbard band, as shown in Fig. 9.3(b). This type of scenario was originally suggested by photoemission studies of near-optimally doped $Bi_2Sr_2CaCu_2O_{8+\delta}$, where the entire valence band was shown to shift considerably upon doping [35] and also supported by later core level measurements of lightly hole doped Pr and Er-substituted $Bi_2Sr_2CaCu_2O_{8+\delta}$ by Harima et al. [36]. Another important distinction between the mid-gap and chemical-potential-shift scenarios is that upon switching from hole to electron doping,

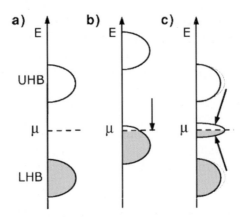

Fig. 9.3. (a) Chemical potential pinned in mid-gap of the undoped parent insulator; (b) a rigid band shift of μ, and (c) a pinned μ with the formation of mid-gap states. From Ronning et al. [28]

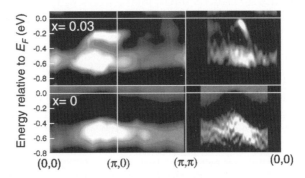

Fig. 9.4. Second derivative plots of the ARPES spectral intensity for $La_{2-x}Sr_x$-CuO_4 $x = 0$ and 0.03, showing the lower Hubbard band pinned at approximately $-0.6\,\mathrm{eV}$, and the formation of a low energy nodal quasiparticle branch at $x = 0.03$. From Yoshida et al. [33]

in the former picture, μ would remain approximately in mid-gap, while in the latter, μ would jump across the gap from the top of the valence band to the bottom of the conduction band.

Now we describe recent work which has quantified the doping evolution of μ in the $Ca_{2-x}Na_xCuO_2Cl_2$ compound. A large shift in μ in $Ca_{2-x}Na_x$-CuO_2Cl_2 was first reported by Ronning et al. [28] and Kohsaka et al. [37] when going from the undoped parent compound to the $x = 0.10$ composition. In particular, Ronning et al. clearly showed a significant overall shift of all valence band states extending up to 6 eV binding energy, when comparing the $x = 0$ and $x = 0.10$ compositions. Later detailed valence band measurements of $Ca_{2-x}Na_xCuO_2Cl_2$ were followed up by Shen et al., who were able to precisely quantify the shift in μ in $Ca_{2-x}Na_xCuO_2Cl_2$ as a function of hole concentration spanning from the undoped parent compound, $Ca_2CuO_2Cl_2$, all the way to $x = 0.12$ [9]. This analysis was performed by using particular peaks in the valence band that do not hybridize with the Zhang–Rice singlet states [38], so that these peaks could be used as a reference for μ. The data is shown in Fig. 9.5, where a fairly rapid shift of μ with hole doping can be observed ($d\mu/dx \sim -1.8 \pm 0.5$ eV / hole), and would be in support of a rapidly shifting chemical potential in $Ca_{2-x}Na_xCuO_2Cl_2$. Nevertheless, the authors argue that a simplistic rigid band picture still fails to capture the essential physics of the hole doping evolution of the electronic structure due to additional effects such as Franck–Condon broadening, as will be later detailed. In $Bi_2Sr_2CaCu_2O_{8+\delta}$, it was found that μ also shifted upon doping, but not as rapidly as measured in $Ca_{2-x}Na_xCuO_2Cl_2$. Since the $O2p_\pi$ band is comprised purely of in-plane O $2p$ states at $\bm{k} = (\pi,\pi)$, it would interesting in the future to also quantify and compare the different values of $d\mu/dx$ obtained from different families using this same marker state, to see if the results are also consistent with those from core level spectroscopies.

Despite this apparent difference in $d\mu/dx$ between $La_{2-x}Sr_xCuO_4$ and $Ca_{2-x}Na_xCuO_2Cl_2$, current work still suggests that even in $La_{2-x}Sr_xCuO_4$, the real situation may well be more complicated than a simplistic mid-gap state scenario. As will be discussed in the next section, the possibility of

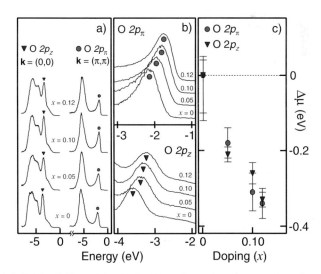

Fig. 9.5. (a) Shift of $Ca_{2-x}Na_xCuO_2Cl_2$ valence bands as a function of x at $\bm{k} = (0,0)$ and (π,π), respectively; (b) expanded plots of the shifts of the $O2p_z$ peak at $(0,0)$ (*triangle*) and the $O2p_\pi$ peak at (π,π); (c) The doping dependent shift of the $O2p_\pi$ and $O2p_z$ peaks summarized from multiple samples. Data are plotted relative to μ_0 determined from the lower bound of the pinned chemical potentials in undoped $Ca_2CuO_2Cl_2$. Adapted from Shen et al. [9]

Franck–Condon broadening in the ARPES spectra of all undoped cuprates (including La_2CuO_4 and $Ca_2CuO_2Cl_2$) may make the distinction between a pinned versus shifting μ somewhat nebulous. In undoped $Ca_2CuO_2Cl_2$ and La_2CuO_4, μ is pinned at least 0.4 to 0.5 eV above the broad peak maximum of the lower Hubbard band [31, 33]. On the other hand, recent ARPES studies of undoped Nd_2CuO_4 and very lightly electron doped $Nd_{2-x}Ce_xCuO_4$ by Armitage et al. demonstrate that μ is pinned ≈ 1.2 eV above the top of the lower-Hubbard-band peak [39]. A slight oxygen nonstoichiometry likely exists in both nominally undoped La_2CuO_4 and Nd_2CuO_4, which would result in a small concentration of holes and electrons, respectively. The fact that μ appears to jump approximately ~ 0.8 eV upon switching between a very small concentration of holes or electrons, appears inconsistent with the simple model of mid-gap states. It is important to note that this estimated value of ~ 0.8 eV is subject to error, since the absolute value of the charge transfer gap can vary significantly between different materials. This jump is also consistent with recent combined x-ray absorption and resonant photoemission studies of $La_{2-x}Sr_xCuO_4$ and $Nd_{2-x}Ce_xCuO_4$ by Steeneken et al. who used the $3d^8$ ^1G peak as an internal electronic reference to demonstrate a jump in μ of ~ 1 eV upon changing between hole and electron doping [40]. These recent studies clearly intimate that the chemical potential jumps across the charge-transfer gap upon changing between hole and electron doping. However, the reason for the substantially different values of $d\mu/dx$ in the different cuprates is still an open issue. A strong body of evidence still suggests that $d\mu/dx$ is significantly smaller in $La_{2-x}Sr_xCuO_4$ than in $Bi_2Sr_2CaCu_2O_{8+\delta}$ or $Ca_{2-x}Na_xCuO_2Cl_2$ upon hole doping, but the reasons for this difference are not yet well understood. It has been proposed by Tanaka et al. that the next-nearest neighbor hopping integral, $|t'|$, which is highly dependent on the crystal structure and can vary significantly between cuprates, and was suggested that this may have some effect on $d\mu/dx$ [41]. In addition, it has also been proposed that $d\mu/dx \approx 0$ in $La_{2-x}Sr_xCuO_4$ could be representative of tendencies towards stripe instabilities or phase separation which might be weaker in other materials [30, 36]. On the other hand, charge ordering has recently been observed by STM on the surface of $Ca_{2-x}Na_xCuO_2Cl_2$, while $d\mu/dx$ appears to be large as $x \to 0$.

9.3.3 Franck–Condon Broadening

A second outstanding problem in understanding the lightly doped cuprate materials has been the anomalously broad spectral lineshapes in the parent Mott insulators. These were first observed in the parent cuprate $Sr_2CuO_2Cl_2$ in 1995 by Wells et al. [3], but the many-body interactions giving rise to these lineshapes have still eluded explanation for the past decade. As the detailed lineshape of the ARPES spectra provides direct information about the single-electron excitations and their interactions with various many-body processes (such as electron–electron or electron–boson interactions), understanding the lineshape of the parent Mott insulator is critical to understanding the exact nature of the strong correlations in these materials. Performing photoemission experiments on the parent Mott insulator is analogous to studying the dynamics of a single hole in the Mott insulator. From a theoretical perspective, addressing this single hole problem is the most tractable scenario when considering the strongly correlated many-body system, and would provide the natural starting point for understanding the lightly doped materials and the relevant many-body processes in the high-T_c cuprates [42].

Thanks to the extremely high resolution of modern ARPES systems, Fermi-liquid-like quasiparticle lineshapes can now be directly observed in materials such as Sr_2RuO_4 [43–45], a transition-metal oxide that is nearly isostructural to $Ca_2CuO_2Cl_2$, sharing the same K_2NiF_4 structure. On the other hand, the lineshapes observed in the parent cuprates such as $Sr_2CuO_2Cl_2$, $Ca_2CuO_2Cl_2$, La_2CuO_4, or Nd_2CuO_4 are well over an order of magnitude broader than those observed in Sr_2RuO_4 and do not resemble Fermi-liquid-like quasiparticle poles whatsoever. During the first studies of the parent insulators a decade ago, little attention was paid to the details of the actual lineshape itself, which was natural given the comparatively poorer energy and momentum resolutions of such earlier systems. However, the fact that ARPES spectra from the parent insulators remained comparatively similar over the past decade, while the quality of data from the doped compounds improved dramatically, suggest an intrinsic nature to these broad spectra.

Furthermore, calculations based on the $t-J$ model (see Dagotto and references therein [46]), along with simple phase-space scattering constraints, would predict a sharp QP peak at the top of the lower Hubbard band, $(\pi/2, \pi/2)$, which is completely at odds with all experimental findings. If a reasonable comparison between theory and experiment could not be made even in this simplest-case scenario, then attempting to tackle the physics of the many-hole problem, i.e. the doped compounds, would be futile. In addition, it is generally believed that the relevant interactions in the parent Mott insulator should also be important to the doped cuprates, and might ultimately give rise to high-temperature superconductivity. This discrepancy between theory and experiment was highlighted by recent measurements on very lightly doped $La_{2-x}Sr_xCuO_4$, which showed very sharp, quasiparticle-like nodal excitations at doping levels of $x < 0.03$ [33, 47].

The work reviewed in this section draws a parallel between the behavior of μ in the previous section and the unusual lineshape of the insulator, through a model based on Franck–Condon broadening. This approach has already proven successful in describing ARPES spectra from a wide range of other materials, as will be detailed later. Until very recently, this analysis had not been directly applied to ARPES data on the cuprates, although there had been earlier theoretical suggestions along these lines (for instance, see Sawatzky [48], Alexandrov and Ranninger [49], or Hirsch [50]). In addition, a heuristic picture based on multiple initial and final states was proposed by Kim et al. regarding data on $Sr_2CuO_2Cl_2$ [51]; although this scenario did not explicitly discuss polaron formation or Franck–Condon broadening, it did

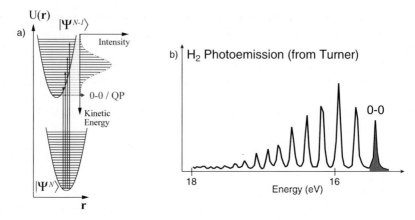

Fig. 9.6. (a) Illustration of the Franck–Condon principle, showing the transition from the $\psi_0^N \to \psi_m^{N-1}$ states; (b) photoemission spectrum from $H_2 \to H_2^+$ with the 0-0 transition filled. Adapted from Turner [52]

possess certain qualitatively similar aspects. The basic premise of the Franck–Condon broadening scenario is that an electron or hole interacts strongly with and is heavily dressed by the bosonic field. This occurs to the point where the new eigenstates of this system have very little wavefunction overlap with the single-electron state (i.e. the new quasiparticles of the system are polarons, not single-electron-like states). We should note that this bosonic dressing occurs in any real system, so there is no sharp rigorous distinction between polaronic versus Fermi-liquid-like quasiparticles. As a result of this strong coupling, the spectral function of the injected photohole consists of a manifold of bosonic shakeoff excitations (i.e. a hole + n virtual bosons). In this case, the actual electronic quasiparticle is simply the lowest-energy transition coupled to zero bosons. However, the projection of this lowest-energy electronic state onto the polaron may have vanishingly small spectral weight, and be essentially nondispersive, meaning that one can reasonably think of such polaronic excitations as effectively localized in real space.

This scenario is shown in Fig. 9.6, where a very simple example of Franck–Condon broadening is presented : the photoemission spectrum of $H_2 \to H_2^+$ is shown (from Turner [52]). Even in this simplest of cases, the $\psi_0^N \to \psi_0^{N-1}$ transition (the "0-0 transition" or the "zero-phonon line") has only a ~ 0.10 probability, and the overlap to the $N-1$ states with vibrational quanta excited, $\psi_{m>0}^{N-1}$, is about ~ 0.90. This implies that a photoemitted electron has a high probability of leaving the system in a vibrationally excited state when it leaves the molecule. In fact, the $\psi_0^N \to \psi_{1,2,3,4}^{N-1}$ transitions all have a higher probability than 0-0, and the overlap will depend strongly on the nature of the chemical bond (i.e. whether it is a bonding, antibonding, or nonbonding orbital). This picture can be quite naturally extended to the solid state by considering the intramolecular vibrations as phonons (or some generalized bosonic field). A general characteristic of polaron formation is that the spectral weight of the quasiparticle pole, Z, is reduced and transferred into a manifold of multi-boson excitations. In the simplest-case scenario (a single electronic state interacting with a single Einstein mode), the spectral function can be calculated easily and the form of the envelope for this multi-boson manifold is well known and corresponds to a Poisson distribution [53]. Although it had been well established for a number of years that the linewidth of the lower Hubbard band in the undoped cuprates was unusually broad, there had been no attempts to quantitatively model the lineshape until very recently [9]. This new work demonstrates that the ARPES lineshape of the lower Hubbard band fits very closely to a Gaussian form, as shown in Fig. 9.7. Since a Poisson distribution reduces to a Gaussian in the large-n limit, the ARPES lineshape of $Ca_2CuO_2Cl_2$ is consistent with a scenario based on small polaron formation.

This polaronic model may have a number of advantages over the conventional Fermi-liquid quasiparticle picture usually employed in the ARPES field. The first potential benefit is the excellent agreement of the Gaussian envelope (as opposed to a Lorentzian-like quasiparticle pole) to the ARPES spectra. The second is in explaining why the minimum pinned position of μ in the parent compounds never approaches closer than ~400 meV to the broad peak maximum position of the lower Hubbard band, a situation which appears to be true for all measured parent cuprates ($Sr_2CuO_2Cl_2$, $Ca_2CuO_2Cl_2$, La_2CuO_4, Nd_2CuO_4). In a simple quasiparticle picture, the peak maximum would correspond to a quasiparticle pole at the top of the valence band, and therefore μ could approach arbitrarily close to the peak maximum. The third is in reconciling the doping evolution of spectral weight. The integrated peak intensity of the broad lower Hubbard band peak in the undoped system encompasses much greater spectral weight than the weight of the sharp near-E_F peak

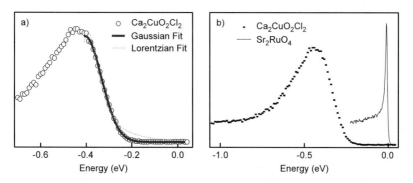

Fig. 9.7. (a) Fits to the experimental data from $Ca_2CuO_2Cl_2$ at the top of the valence band, $\boldsymbol{k} = (\pi/2, \pi/2)$, at 200 K . The fit to a Lorentzian is shown as a *dashed line*, while the fit to a Gaussian is shown as a *solid line*; (b) comparison of the data from $Ca_2CuO_2Cl_2$ in (a) with Sr_2RuO_4, taken at 15 K at $\boldsymbol{k} = (\pi, 0)$. Data adapted from [9, 44, 54]

found in lightly doped $La_{2-x}Sr_xCuO_4$ or $Ca_{2-x}Na_xCuO_2Cl_2$ [9,33]. Within a Fermi-liquid picture, this would paradoxically imply that Z is much larger in the undoped parent Mott insulator than at higher hole dopings, where one would expect to recover more Fermi-liquid-like behavior. The polaronic scenario could reconcile this problem by proposing that the observed peak in the undoped system represents incoherent spectral weight, while the true quasiparticle weight, and hence Z, is invisible and vanishingly small. This would then allow Z to evolve continuously from nearly zero in the insulator to a larger finite value in the superconductor.

There have also been recent theoretical efforts at incorporating strong electron–boson coupling into the $t-J$ model framework. A major reason that the extended $t-J$ model had long been considered successful in describing the behavior of the single hole in the antiferromagnetic insulator was the agreement of the calculated $\varepsilon(\boldsymbol{k})$ dispersion with experiment [3,42,55]. When polaronic effects are added, it is not necessarily obvious how this would affect the dispersion of the observable spectral weight, since the coherent weight has vanishing intensity. Recent calculations by Mishchenko and Nagaosa on a $t-J$ + phonons model using a diagrammatic Monte Carlo technique clearly demonstrate that although the electronic quasiparticle state is vanishingly weak and effectively nondispersive, the first moment (or centroid) of the broad Franck–Condon envelope nevertheless exactly tracks the original dispersion, $\varepsilon(\boldsymbol{k})$, in the absence of electron–boson coupling [56]. This correspondence between the first moment of spectral weight and the initial pole position was well known in the very simple case of a single electron coupled to a single Einstein mode [53], but had not been confirmed for Hamiltonians as complex as the $t-J$ model. Additional work along these lines was also performed by Rösch and Gunnarsson using a different approach, which arrived at qualitatively similar conclusions [57]. These crucial theoretical studies would then, in principle, allow one to reconcile the agreement of the dispersion with the $t-J$ model with the broad polaronic features which manifest themselves in the lineshape and position of μ.

If this Franck–Condon broadening does arise from electron–phonon coupling, one might expect that such behavior would also be observed in a wide array of systems, especially in conventional insulators or poor conductors, where the photohole could be poorly screened from the ionic charges. Such polaronic broadening has indeed been observed in such systems, starting with the broadening of core levels in the ionic alkali halides by Citrin and coworkers [58,59] in the 1970's. More recent work has focused on systems with strong electron–phonon coupling, such as the 1D charge density wave systems by Perfetti et al. [60,61], or various other insulating or poorly conducting transition metal oxides such as the colossal magnetoresistive manganites [62–64], vanadates [65,66], and magnetite [67]. It is interesting to note that studies of transition metal oxides in the 1950's [68,69] provided the stimulus for much of Holstein's early work on the concept of polarons.

For the case of the cuprates, the situation is more complex, since the photohole can in principle couple both to magnons as well as phonons. In order to determine the nature of this electron–boson interaction (i.e. whether the photohole couples primarily to bosons of lattice or magnetic origin), photoemission studies have compared the lineshape of the lower Hubbard band to particular states in the valence band which have no hybridization with the correlated Cu $3d_{x^2-y^2}$ orbital and thus no coupling to the magnetic system. Therefore, one should be able to isolate the electron–phonon contribution, an approach that was employed by Pothuizen et al. in $Sr_2CuO_2Cl_2$ [38], where it was found that linewidths of the lower Hubbard band and the $O2p_\pi$ state appeared to be very similar. By comparison to a tight-binding model, it was determined that the $O2p_\pi$ state at $k = (\pi, \pi)$ was orthogonal to the Cu $3d_{x^2-y^2}$ orbital and Zhang–Rice singlet. This $O2p_\pi$ state was used in by Shen et al. to determine the shift of μ [9], and following the approach of Pothuizen et al., those authors have also reexamined the lineshape of the $O2p_\pi$ peak in greater detail [70] and found that it possessed the same Gaussian lineshape found in the lower Hubbard band. The fact that this apparent Franck–Condon broadening was also observed in the $O2p_\pi$ state, which is decoupled from the spin system, would imply the injected photohole forms a lattice polaron state in the undoped parent cuprate. Rösch, Gunnarsson, and coworkers have also calculated the electron–phonon coupling of a single hole to the lattice in La_2-CuO_4, in a realistic model [71]. Their calculations utilize a known shell model for the phonon modes (derived earlier from neutron-scattering experiments) and a $t - J$ model for the electronic degrees of freedom, in order to best simulate the real material. Their calculations indicate that the photohole-lattice coupling is easily strong enough to form self-trapped small polarons ($\lambda = 1.2$), lending substantial support to this lattice polaron scenario.

Despite the apparent suitability of this strong-coupling/polaronic scenario in describing many aspects of the ARPES data, it should be emphasized that this does not undermine the significance of strong electron–electron correlations. The presence of the charge-transfer gap and the dispersion of the observable spectral weight are all dominated by electron correlations, not electron–phonon interactions. Therefore, one could view the interplay between electron–lattice and electron–electron interactions more as complementary, rather than competitive or an exclusive "either-or" scenario. Along these lines, the work of Mishchenko and Nagaosa in fact suggests that t-J physics enhances the propensity towards self-trapping behavior [56]. Furthermore, it is plausible that while lattice-polaron formation may account for a substantial amount of the broadening observed in the undoped Mott insulator, electronic correlation/magnetic effects should also play an important role in the lineshape of the lower Hubbard band, and cannot be excluded [70].

9.3.4 Emergence of Nodal Quasiparticle States

Recent studies of the emergence of the nodal quasiparticle states in $La_{2-x}Sr_xCuO_4$ and $Ca_{2-x}Na_xCuO_2Cl_2$ have also demonstrated how these first hole doped states emerge from the parent antiferromagnetic insulator. Here we loosely use the term "quasiparticle" to refer to the sharp, well-defined, near-E_F electronic excitations, although we cannot conclusively determine whether these are quasiparticles in the exact Landau Fermi-liquid sense. The first work on the emergence of quasiparticle states from the Mott insulator was performed by Ino et al. on lightly doped $La_{2-x}Sr_xCuO_4$ [31,32], where it was found that sharp, near-E_F peaks only began to emerge around optimal doping in the region near $(\pi, 0)$. Later, after an extensive study of the photoelectron matrix element effects in $La_{2-x}Sr_xCuO_4$ by Zhou and Yoshida, it was revealed that distinct nodal quasiparticle states began to emerge at dopings even as low as $x \approx 0.03$ [33,47] when using σ-polarized photons with $h\nu \approx 55\,eV$. In Fig. 9.8, we show the doping evolution of the nodal quasiparticle states in $La_{2-x}Sr_xCuO_4$ from Yoshida et al. [33]. Studies of the evolution of quasiparticle excitations in $Ca_{2-x}Na_xCuO_2Cl_2$ were first reported by Kohsaka et al. and Ronning et al. [28,37], demonstrating a clear nodal quasiparticle peak at $x = 0.10$, followed by a doping dependent study by Shen et al. shown in Fig. 9.9 [9]. Here, we will focus only on the excitations on the nodal directions, and other k directions, including the antinode, will be discussed later. When comparing the data in Figs. 9.8 and 9.9, one can see striking similarities between $La_{2-x}Sr_xCuO_4$ and $Ca_{2-x}Na_xCuO_2Cl_2$. In both materials, a sharp peak emerges near E_F whose intensity grows proportionally to the hole doping, as shown in Fig. 9.15(c). As discussed by Zhou et al., the quasiparticle velocity at E_F, v_F, appears to be approximately universal across cuprate

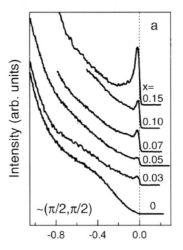

Fig. 9.8. Doping dependence of the EDC at k_F along the nodal direction for $La_{2-x}Sr_xCuO_4$. Taken from Yoshida et al. [33]

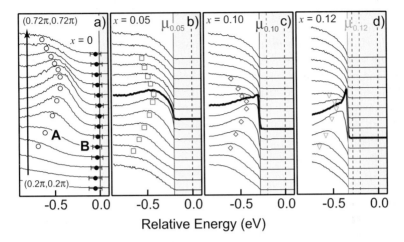

Fig. 9.9. Doping dependence of EDCs of $Ca_{2-x}Na_xCuO_2Cl_2$ along the nodal direction from $(0.2\pi,0.2\pi)$–$(0.72\pi,0.72\pi)$ for $x = 0$ (**a**), 0.05 (**b**), 0.10 (**c**), and 0.12 (**d**). Hump maxima (*symbols*) are determined from a combination of fitting and second derivatives. Data are plotted on an absolute scale relative to μ_0, but E_F for each individual sample is marked above. Taken from Shen et al. [9]

families and doping levels, and this is also the case for $La_{2-x}Sr_xCuO_4$ and $Ca_{2-x}Na_xCuO_2Cl_2$ [73]. At higher binding energies, a broad hump persists whose shape and dispersion is reminiscent of the lower Hubbard band in the undoped parent compound, clearly underscoring the continued importance of strong electronic correlations in the lightly doped materials.

This increase in the sharp near-E_F peak could be taken as consistent with a strongly doping-dependent quasiparticle residue, Z. This is much different from what one would anticipate from a weakly interacting picture, where Z would remain roughly unchanged, and it would be the Luttinger volume ($\mathbf{k}_F$) which would increase with carrier doping. However, this increasing Z might be consistent with the proposed polaronic picture discussed above, assuming that the effective electron–boson coupling strength depends strongly on doping. However, it has not yet been firmly established by other experimental techniques how strongly the electron–phonon (or electron–magnon) coupling varies as a function of hole concentration, as this would be an important test for the validity of such a scenario and determining the relevance of polaron formation to the evolution of the nodal quasiparticles. In addition, strong electronic correlations should certainly play a critical role in determining the evolution of the quasiparticle states away from half-filling. Various theoretical works, including the model for "gossamer" superconductivity by Laughlin [74] as well as calculations based on a variational projected resonating valence bond (RVB) framework by Randeria, Paramekanti, and Trivedi [75,76], also obtain a doping dependent Z which approaches zero in the undoped limit. At this point, it is difficult to unequivocally determine whether purely electronic models are sufficient for explaining the doping evolution of the nodal quasiparticle states away from half-filling, and what role strong electron–phonon coupling would play. However, the apparent importance of electron–lattice interactions in the parent insulator would seem to suggest that such interactions should not be neglected when considering the emergence of the quasiparticle states particularly at low dopings.

In addition to the doping-dependent intensity of this quasiparticle peak, the doping evolution of the Fermi wavevector, k_F, and the Fermi velocity, v_F in $Ca_{2-x}Na_xCuO_2Cl_2$ have also been investigated by ARPES. At least for the case of $Ca_{2-x}Na_xCuO_2Cl_2$, the evolution of these quantities also appears to agree rather well with the chemical potential shift discussed earlier. This is shown in Fig. 9.10(a), where the relative shift in μ, the change in k_F, and v_F in $Ca_{2-x}Na_xCuO_2Cl_2$ are all summarized within a single plot. At higher binding energies, the approximate dispersions of the broad hump maxima are shown as solid symbols while μ appears to slide down the low energy quasiparticle branch shown as bold lines whose dispersion was determined by MDC analysis. One interesting aspect of the doping evolution is that the nodal QP velocity appears to be largely independent of both hole doping concentration and material family, as discussed by Zhou et al. [73]. Within conventional Fermi-liquid theory, the renormalization of the QP velocity and the QP residue Z should be inversely related. The fact that Z appears to increase linearly with x while v_F remains roughly constant is rather remarkable. At this point, there are some theoretical works which could potentially describe this behavior, including the projected RVB models of Randeria and coworkers [76], and also a t-J model framework strongly coupled to phonons by Mishchenko and Nagaosa [77].

There also appear to be significant quantitative differences between $Ca_{2-x}Na_xCuO_2Cl_2$ and $La_{2-x}Sr_xCuO_4$, since the existing near-E_F (< 1 eV) and core level photoemission data on $La_{2-x}Sr_xCuO_4$ appears to suggest that $d\mu/dx \to 0$ close to the Mott insulating state, as discussed previously. If there are significant changes in k_F with x in $La_{2-x}Sr_xCuO_4$ while little change in μ, this would certainly present a clear contrast to $Ca_{2-x}Na_xCuO_2Cl_2$. Another difference is that a sharp nodal quasiparticle-like peak is evident at even $x = 0.03$ in $La_{2-x}Sr_xCuO_4$, while no QP peak is apparent in $Ca_{2-x}Na_xCuO_2Cl_2$ below $x = 0.10$ (although concentrations between $x = 0.05$ and $x = 0.10$ have not yet been measured), as shown in Fig. 9.10(b). However, in both materials, the generic trend that the low-energy spectral weight increases proportionally to doping is followed. In fact, if we interpret the results from undoped La_2CuO_4 within the simple polaronic framework, the fact that the lower Hubbard band of La_2CuO_4 is significantly broader than $Ca_2CuO_2Cl_2$ or $Sr_2CuO_2Cl_2$ would suggest that the electron–boson coupling in La_2CuO_4 is significantly larger than in the oxychloride compounds. One might then assume that the zero-boson QP peak in doped $La_{2-x}Sr_xCuO_4$ would then be even weaker than in the oxychlorides. The fact that this simple reasoning

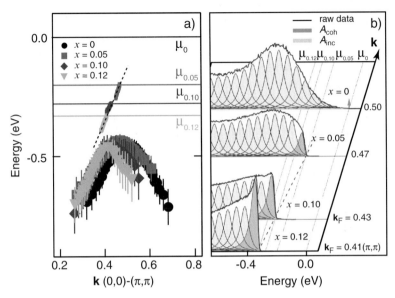

Fig. 9.10. (a) Summary of the doping dependence of the broad hump (in *symbols*, also shown in Fig. 9.9) and the low-energy MDC dispersions along the nodal direction. All data are plotted relative to μ_0, although E_F for each composition is denoted above; (b) doping dependence of the nodal EDC at k_F, combined with a cartoon schematic of the proposed distribution of coherent (*dark grey*) and incoherent (*light grey*) spectral weight, and its evolution as a function of doping. Taken from Shen et al. [9]

does not seem to hold may indicate that additional factors that have not been considered within a simple polaronic model such as electronic correlations or phase separation may also be important.

One feature that has been universally observed in the ARPES spectra of all p-type cuprates is the presence of an abrupt break or "kink" in the measured quasiparticle dispersion in the range of 40–70 meV away from E_F. This kink was first identified by Bogdanov et al. in $Bi_2Sr_2CaCu_2O_{8+\delta}$ [6], and now has also been reported in $Bi_2Sr_2CuO_{6+\delta}$, $La_{2-x}Sr_xCuO_4$, $Ca_{2-x}Na_xCuO_2Cl_2$, $Tl_2Ba_2CuO_6$, $Bi_2Sr_2Ca_2Cu_3O_{10+\delta}$ [7,9,73,78]. At this stage, there is a general consensus on a number of experimental observations including the fact that : i) this effect appears to be generic to all hole-doped cuprates, ii) the magnitude of the kink along the nodal direction appears to weaken with increasing hole doping, iii) below T_c, the strength of the kink appears to be maximal along the antinodal direction, at least for the higher transition temperature compounds such as $Bi_2Sr_2CaCu_2O_{8+\delta}$, and iv) this effect appears to be due to the strong coupling of the electrons to a bosonic mode. Despite this concordance in the experimental data, there is currently an intense debate over the origin of this kink, and particularly, whether the bosonic mode is primarily of lattice or magnetic origin (for a sampling of experimental papers and groups debating this issue, see for instance [7, 72, 78–83]). While an in-depth discussion of this controversial point is well beyond the scope of this chapter, the work on Franck–Condon broadening discussed above would appear to indicate that strong electron–phonon coupling is, at the very least, relevant to the physics of the lightly doped cuprates. Whether or not the coupling of the electrons to a magnetic mode is also of importance to describing the physics of the kink is a subject that will likely remain the source of considerable investigation and discussion for some time to come.

9.3.5 Momentum Dependence of Spectral Weight

The final topic that will be covered in this chapter is the momentum dependence of near-E_F spectral weight in the lightly doped cuprates $La_{2-x}Sr_x$-CuO_4 and $Ca_{2-x}Na_xCuO_2Cl_2$. In a conventional metal, the contour defined by the ARPES spectral weight near E_F would effectively delineate the Fermi surface. In strongly correlated materials such as the cuprates, however, it is not clear whether true Fermi-liquid quasiparticle excitations exist throughout the entire Brillouin zone. Nevertheless, these spectral weight contours are typically referred to in the literature as "Fermi surfaces" (FS), and with this caveat, we will also adopt this loose terminology. The theoretically predicted FS topology at low dopings can vary, depending on the microscopic model used. For instance, certain models predict that the Fermi surface could resemble a small hole pocket centered at $(\pi/2, \pi/2)$ with an enclosed volume of x, similar to what one would expect from a rigid band shift into the lower Hubbard band [16,84,85]. Another scenario, proposed by Furukawa and Rice [86,87] is that the FS is truncated into "hot" and "cold" spots due to strong scattering of the electrons near $(\pi, 0)$ due to antiferromagnetic fluctuations. Further, Kivelson [88], Markiewicz [34], and others have discussed the possible FS which could arise from spatially inhomogeneous charge distributions, or "stripes" which might conspire to give an apparent Fermi surface. Finally, one might also expect that upon hole doping, a large Fermi surface with a volume of $1-x$ emerges, reminiscent of the non-interacting predictions (see Pickett [89], and references within). Therefore, the topology of the Fermi surface can, in principle, carefully discriminate between different microscopic models for the cuprates.

The previous section detailed the doping evolution of the nodal quasiparticles. However, the states near $(\pi, 0)$ are also of great interest, since these states comprise the maximum, or antinode, of the d-wave superconducting gap. Therefore, it is important to determine whether the low-energy states have any momentum anisotropy along the apparent Fermi surface. This was first addressed in $La_{2-x}Sr_xCuO_4$ by Ino et al. [32] and Zhou et al. [90] who suggested that at lower hole concentrations, the antinodal states were dominant near E_F. After careful study of the role of the photoelectron matrix elements by Yoshida et al. and Zhou et al. [33,47], it was found that in fact the nodal states first emerged at low doping levels, as shown in Fig. 9.11. It

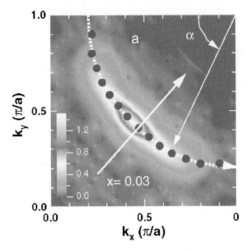

Fig. 9.11. Integrated spectral intensity within a ± 10 meV window around E_F for $La_{1.97}Sr_{0.03}CuO_4$. The angular distribution of spectral weight is remarkably similar to that found in $Ca_{2-x}Na_xCuO_2Cl_2$ as shown in Fig. 9.12. From Yoshida et al. [33]

was also found that the distribution of spectral weight in $Ca_{2-x}Na_xCuO_2Cl_2$, first studied by Kohsaka et al. and Ronning et al. demonstrated a striking similarity to the FS maps of $La_{2-x}Sr_xCuO_4$ [28, 37]. This is shown in Fig. 9.12, suggest a common picture for the momentum distribution of spectral weight in the lightly doped cuprates. The low energy spectral weight in both materials is maximum near the nodes and drops off rapidly towards the antinodes. In neither lightly doped $La_{2-x}Sr_xCuO_4$ nor $Ca_{2-x}Na_xCuO_2Cl_2$ is a small "hole pocket" topology actually observed, although this possibility is difficult to entirely rule out due to potentially weak coherence factors of the back-folded band. Instead, the apparent Fermi surface appears to extend towards the antinodes, as basically expected from the non-interacting, $1 - x$ picture, although the intensity drops to nearly zero around $(\pi, 0)$. The fact that such behavior is seen in both $La_{2-x}Sr_xCuO_4$ and $Ca_{2-x}Na_xCuO_2Cl_2$ would suggest that this type of behavior is generic to the lightly hole doped cuprates since this anisotropy was observed in different materials using a wide range of incident photon energies and polarizations, this would indicate that the effect is intrinsic and not an experimental artifact due to photoelectron matrix elements.

This anisotropy is also demonstrated in the individual EDC curves along the ostensible Fermi surface in both $La_{2-x}Sr_xCuO_4$ and $Ca_{2-x}Na_xCuO_2Cl_2$. In both compounds, the nodal excitations are most distinct, possessing clear QP-like peaks. Upon sweeping out to the antinodes, this sharp peak becomes rapidly suppressed and vanishes, resulting in a broad lineshape with a weak near-E_F shoulder in both compounds indicating a much lower QP weight. This is shown in Fig. 9.13, where selected EDCs are shown along the osten-

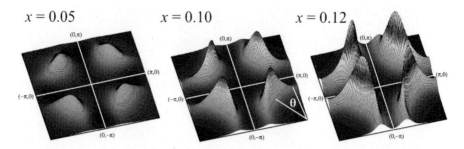

Fig. 9.12. 3D bird's eye view plots of the spectral intensity within a ±10 meV window from $Ca_{2-x}Na_xCuO_2Cl_2$, adapted from Shen et al. [54, 91]. These plots were generated by parameterizing the angular distribution of spectral weight, the width of the EDC plots, and the underlying FS. These plots demonstrate that the apparent Fermi arcs do not rigidly extend with doping, but is more representative of an overall growth in spectral weight

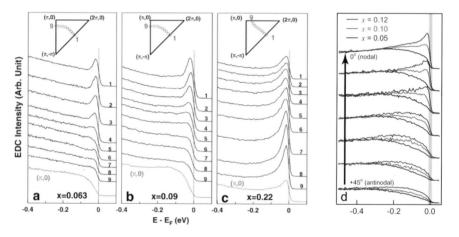

Fig. 9.13. EDCs as a function of angle around the ostensible Fermi surfaces of $La_{2-x}Sr_xCuO_4$, from Zhou et al. [47], and $Ca_{2-x}Na_xCuO_2Cl_2$ from Shen et al. [91] EDCs along the Fermi surface of $La_{2-x}Sr_xCuO_4$ $x = 0.063$ (**a**), 0.09 (**b**), and 0.22 (**c**) samples [47] and $Ca_{2-x}Na_xCuO_2Cl_2$ at $x = 0.05, 0.10$, and 0.12 as a function of angle (defined in Fig. 9.12). The *grey area* in (**d**) around E_F indicates the typical energy integration window for generating FS plots such as the ones shown in Fig. 9.12 and 9.11

sible Fermi surfaces of lightly doped $La_{2-x}Sr_xCuO_4$ and $Ca_{2-x}Na_xCuO_2Cl_2$. Despite the absence of any sharp low-energy peak around the antinodes, the spectral weight still maintains a distinct structure in momentum space. A clear intensity maximum is observed along the $(\pi, 0)$ to (π, π) direction, which would correspond to the underlying Fermi surface within a simple non-interacting picture. In $Ca_{2-x}Na_xCuO_2Cl_2$, it has been shown that the wavevector con-

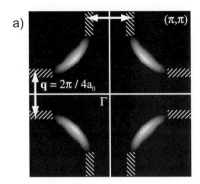 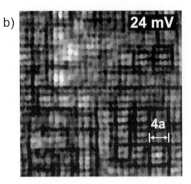

Fig. 9.14. (a) Spectral intensity for $Ca_{1.9}Na_{0.10}CuO_2Cl_2$ with approximately nested antinodal segments illustrated. Adapted from Shen et al. [9]; (b) $4a_0 \times 4a_0$ charge checkerboard pattern observed by STM, from Hanaguri et al. [27]

necting these spectral intensity maxima across the antinodes corresponds to the recently discovered $4a_0 \times 4a_0$ "checkerboard" charge modulation observed by scanning tunneling microscopy by Hanaguri et al. [27,91]. A schematic illustration of this is shown in Fig. 9.14. It is quite possible that this checkerboard modulation may represent a competing state in the lightly doped regime of the cuprates, such as a charge-density-wave (CDW) state or some other form of charge ordering. As discussed in the introduction, the existence of such a state and its competition with d-wave superconductivity could potentially be the origin of many of the unusual physical properties observed in the cuprates, and in particular, the pseudogap. This charge ordered state seems to manifest itself in k-space through the presence of faint antinodal states with a nesting wavevector corresponding to the charge modulation pattern. However, this appears to be unlike the picture for simple, nesting-driven charge density wave formation, since in conventional CDW systems the QP coherence is not necessarily lost in the nested regions of FS. This picture has been confirmed in other CDW systems [92–94], and is analogous to Bogoliubov quasiparticles in a BCS superconductor. Therefore, the fact that the antinodal states are seemingly "incoherent" (or have a much lower Z than at the antinodes) could have important implications for understanding how different regions of k-space could play special roles for disparate competing tendencies in the cuprates. Very similar behavior was also recently observed in the bilayer manganite $La_{1.2}Sr_{1.8}Mn_2O_7$ by Mannella et al. [64], demonstrating well-defined quasiparticle-like excitations only along the "nodal" (i.e. Mn-O bond diagonal) direction, and well-nested straight sections of FS away from the $(0,0)$-(π,π) line which show only broad, seemingly incoherent excitations. In addition, $La_{1.2}Sr_{1.8}Mn_2O_7$ also exhibits an apparent pseudogap as well as charge ordering phenomenon which could be related to these observed nested sections of Fermi surface.

A qualitatively similar picture to $Ca_{2-x}Na_xCuO_2Cl_2$ has also been reported in lightly doped $La_{2-x}Sr_xCuO_4$ by Zhou et al. [47] where the low energy antinodal states are also faint and apparently nested, and the EDCs near $(\pi,0)$ also fail to exhibit sharp near-E_F structures, as shown in Fig. 9.13 (a) and (b). Despite such global similarities between $Ca_{2-x}Na_xCuO_2Cl_2$ and $La_{2-x}Sr_xCuO_4$, there are a few notable differences worth discussing. The first is that the antinodal wavevector in $La_{2-x}Sr_xCuO_4$ appears to be significantly smaller than $Ca_{2-x}Na_xCuO_2Cl_2$ [47,91]. If the proposed correspondence between STM and ARPES in $Ca_{2-x}Na_xCuO_2Cl_2$ [91] can be extended to $La_{2-x}Sr_xCuO_4$, this might imply that the charge ordering pattern in $La_{2-x}Sr_xCuO_4$ could be significantly different than the $4a_0 \times 4a_0$ checkerboard. Another important difference is that the Fermi surface extracted in $Ca_{2-x}Na_xCuO_2Cl_2$ do not appear to satisfy a $1-x$ Luttinger volume, implying that "Fermi surface" is not appropriate nomenclature, while $La_{2-x}Sr_xCuO_4$ appears to come much closer to satisfying a $1-x$ volume. A particular advantage of the $La_{2-x}Sr_xCuO_4$ system is that heavily overdoped samples can also be synthesized, unlike $Ca_{2-x}Na_xCuO_2Cl_2$. The anisotropy of spectral weight can then be characterized from the lightly doped regime all the way to the overdoped regime. In the underdoped regimes of both $Ca_{2-x}Na_xCuO_2Cl_2$ and $La_{2-x}Sr_xCuO_4$, the angular distribution of spectral weight are remarkably similar (Fig. 9.15), and also remains unchanged with doping apart from a global scaling factor. However, in overdoped $La_{2-x}Sr_xCuO_4$ the distribution of spectral weight appears to change dramatically, to where the antinodes have maximal spectral weight.

This has also been recently observed and discussed in overdoped $Tl_2Ba_2CuO_{6+\delta}$ by Platé et al. [95] One interesting aspect of this change in the spectral

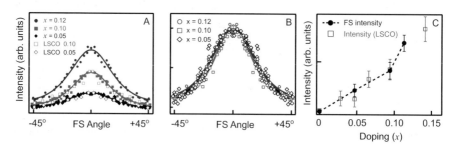

Fig. 9.15. (a) Angular distribution of spectral weight as a function of angle along the FS. Data from $Ca_{2-x}Na_xCuO_2Cl_2$ are plotted as solid symbols [91], while data from $La_{2-x}Sr_xCuO_4$, from [33], are plotted as open symbols; (b) same angular distributions from $Ca_{2-x}Na_xCuO_2Cl_2$ in (a), but normalized to the same intensity, showing the quantitatively similar angular distributions; (c) doping dependence of FS intensity which as a function of x for $Ca_{2-x}Na_xCuO_2Cl_2$ (*circles*) and $La_{2-x}Sr_xCuO_4$ (*squares*)

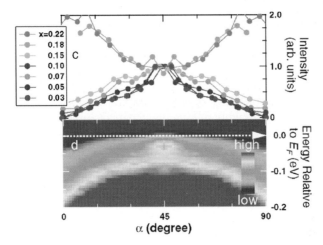

Fig. 9.16. Doping dependence of spectral weight distribution in $\text{La}_{2-x}\text{Sr}_x\text{CuO}_4$. All intensities are scaled to the same value at the node, and show an abrupt change near $x = 0.18$. From Yoshida et al. [33]

anisotropy observed in LSCO is that it occurs very abruptly, at a concentration slightly above optimal doping, as shown in Fig. 9.16.

One possibility is that this abrupt and dramatic change in spectral weight could correspond to a quantum phase transition around optimal doping [95–99]. Given the recent evidence for charge ordering/modulation in the underdoped regime, this transition could potentially correspond to the disappearance of the charge ordered state near optimal doping. While this remains an intriguing possibility, it is obvious that much more work still needs to be performed on this particular subject before any conclusive statements can be made.

9.4 Conclusions

In this chapter, we have detailed recent developments in ARPES studies of the doping evolution of the high-T_c superconductors near the Mott insulating state. This work has been made possible through the combination of recent breakthroughs in materials synthesis and high-resolution instrumentation. The work discussed in this chapter has covered a number of important recent advances in our understanding of the lightly doped region of the cuprate phase diagram. The behavior of the chemical potential with doping has been studied carefully in a number of compounds and shown to jump across the charge-transfer gap upon switching from hole to electron doping, as opposed to remaining pinned in mid-gap. Taking advantage of the high energy resolution of current ARPES systems, the dynamics of the single hole in the antiferromagnetic Mott insulator has been studied in depth, and shown to exhibit a behavior redolent of Franck–Condon broadening. Coupled with recent theoretical works, these studies indicate that a strong coupling of the photohole to the lattice, along with electron–electron interactions, may be necessary for a complete description of the dynamics of a single hole in the Mott insulator. Upon further hole doping, studies of $La_{2-x}Sr_xCuO_4$ and $Ca_{2-x}Na_xCuO_2Cl_2$ demonstrate that the first quasiparticle-like excitations emerge along the nodal direction, with a highly doping dependent Z but a fairly constant Fermi velocity, v_F. Finally, the low-energy excitations in both $La_{2-x}Sr_xCuO_4$ and $Ca_{2-x}Na_xCuO_2Cl_2$ exhibit strong k anisotropies, with the antinodal excitations heavily suppressed. At this point, the origin of this anisotropy is still not understood, but one intriguing candidate is the possibility of a competing order, such as charge ordering, which suppresses antinodal coherence in the underdoped regime.

References

1. B. O. Wells et al: Phys. Rev. B, 46:11830 (1992)
2. Z.-X. Shen et al: Phys. Rev. Lett., 70:1553 (1993)
3. B. O. Wells et al: Phys. Rev. Lett., 74:964 (1995)
4. H. Ding et al: Nature, 382:51 (1996)
5. A. G. Loeser et al: Science, 273:325 (1996)
6. P. V. Bogdanov et al: Phys. Rev. Lett., 85:2581 (2000)
7. A. Lanzara et al: Nature, 412:510 (2001)
8. J. G. Bednorz and K. A. Müller: Z. Phys. B, 64:189 (1986)
9. K. M. Shen et al: Phys. Rev. Lett., 93:267002 (2004)
10. J. Zaanen et al: Phys. Rev. Lett., 55:418 (1985)
11. M. Kastner et al: Rev. Mod. Phys., 70:897 (1998)
12. C. C. Tsuei and J. R. Kirtley: Rev. Mod. Phys., 72:969 (2000)
13. T. Timusk and B. Statt: Rep. Prog. Phys., 62:61 (1999)
14. V. J. Emery and S. A. Kivelson: Nature, 374:4347 (1995)
15. V. J. Emery and S. A. Kivelson: Phys. Rev. Lett., 74:3253 (1995)
16. I. Affleck and J. B. Marston: Phys. Rev. B, 37:3774 (1988)
17. C. M. Varma: Phys. Rev. B, 55:14554 (1997)
18. S. Chakravarty et al: Phys. Rev. B, 63:094503 (2001)
19. C. Castellani et al: Phys. Rev. Lett., 75:4650 (1995)
20. A. Damascelli et al: Rev. Mod. Phys., 75:473 (2003)
21. J. C. Campuzano et al: cond-mat/0209476, 2002
22. P. D. Johnson et al: J. Electron Spectrosc. Relat. Phenom. 117-118:153 (2001)
23. D. W. Lynch and C. G. Olson: *Photoemission Studies of High-Temperature Superconductors*. Kluwer Academic, 1st edition (1999)
24. M. Imada et al: Rev. Mod. Phys., 70:1039 (1998)
25. Y. Kohsaka et al: J. Am. Chem. Soc, 124:12275 (2002)
26. Y. Kohsaka et al: Phys. Rev. Lett., 93:097004 (2004)
27. T. Hanaguri et al: Nature, 430:1001 (2004)
28. F. Ronning et al: Phys. Rev. B, 67:165101 (2003)
29. J. W. Allen et al: Phys. Rev. Lett., 64:595 (1990)
30. A. Ino et al: Phys. Rev. Lett., 79:2101 (1997)

31. A. Ino et al: Phys. Rev. B, 62:4137 (2000)
32. A. Ino et al: Phys. Rev. B, 65:094504 (2002)
33. T. Yoshida et al: Phys. Rev. Lett., 91:027001 (2003)
34. R. S. Markiewicz et al: Phys. Rev. B, 62:1252 (2000)
35. Z.-X. Shen et al: Phys. Rev. B, 44:12098 (1991)
36. N. Harima et al: Phys. Rev. B, 67:172501 (2003)
37. Y. Kohsaka et al: Journal of the Physical Society of Japan, 72:1018 (2003)
38. J. J. M. Pothuizen et al: Phys. Rev. Lett., 78:717 (1997)
39. N. P. Armitage et al: Phys. Rev. Lett., 88:257001 (2002)
40. P. G. Steeneken et al: Phys. Rev. Lett., 90:247005 (2003)
41. K. Tanaka et al: Phys. Rev. B, 70:092503 (2004)
42. T. Tohyama and S. Maekawa: Supercond. Sci. Technol., 13:R17 (2000)
43. A. Damascelli et al: Phys. Rev. Lett., 85:5194 (2000)
44. K. M. Shen et al: Phys. Rev. B, 64:180502(R)(2001)
45. N. J. C. Ingle et al: Phys. Rev. B, 72:205114 (2005)
46. E. Dagotto: Rev. Mod. Phys., 66:763 (1994)
47. X. J. Zhou et al: Phys. Rev. Lett., 92:187001 (2004)
48. G. A. Sawatzky: Nature, 342:480 (1989)
49. A. S. Alexandrov and J. Ranninger: Phys. Rev. B, 45:13109 (1992)
50. J. E. Hirsch: Phys. Lett. A, 282:392 (2001)
51. C. Kim et al: Phys. Rev. B, 65:174516 (2002)
52. D. W. Turner: *Molecular Photoelectron Spectroscopy*. Wiley, New York, 1970
53. G. D. Mahan: *Many-Particle Physics*. Kluwer Academic, 3rd edition, 2000
54. K. M. Shen: *Angle-Resolved Photoemission Spectroscopy Studies of the Mott Insulator to Superconductor Evolution in $Ca_{2-x}Na_xCuO_2Cl_2$*. PhD thesis, Stanford University, 2005
55. C. Kim et al: Phys. Rev. Lett., 80:4245 (1998)
56. A. S. Mishchenko and N. Nagaosa: Phys. Rev. Lett., 93:036402 (2004)
57. O. Rosch and O. Gunnarsson: Eur. Phys. J. B, 43:11 (2005)
58. P. H. Citrin et al: Phys. Rev. Lett., 33:965 (1974)
59. P. H. Citrin et al: Phys. Rev. B, 15:2923 (1977)
60. L. Perfetti et al: Phys. Rev. Lett., 87:216404 (2001)
61. L. Perfetti et al: Phys. Rev. B, 66:075107 (2002)
62. D. S. Dessau et al: Phys. Rev. Lett., 81:192 (1998)
63. V. Perebeinos and P. B. Allen: Phys. Rev. Lett., 85:5178 (2000)
64. N. Mannella et al: Nature, 438:474 (2005)
65. K. Okazaki et al: Phys. Rev. B, 69:140506(R)(2004)
66. K. Okazaki et al: Phys. Rev. B, 69:165104 (2004)
67. D. Schrupp et al: cond-mat/0405623 (2004)
68. T. Holstein: Ann. Phys. 8:325 (1959)
69. T. Holstein: Ann. Phys., 8:343 (1959)
70. K. M. Shen: Submitted to *Physical Review B*, (2006)
71. O. Rosch et al: Phys. Rev. Lett., 95:227002 (2005)
72. X. J. Zhou et al: Phys. Rev. Lett., 95:117001 (2005)
73. X. J. Zhou et al: Nature, 423:398 (2003)
74. R. B. Laughlin: cond-mat/0209269, (2002)
75. A. Paramekanti et al: Phys. Rev. B, 70:054504 (2004)
76. M. Randeria et al: Phys. Rev. B, 69:144509 (2004)
77. A. S. Mishchenko: *in preparation* 2005

78. H. Matsui et al: Phys. Rev. B, 67:060501 (2003)
79. A. Kaminski et al: Phys. Rev. Lett., 86:1070 (2001)
80. A. Gromko et al: Phys. Rev. B, 68:174520 (2003)
81. A. Koitzsch et al: Phys. Rev. B, 69:140507(R)(2004)
82. T. Cuk et al: Phys. Rev. Lett., 93:117003 (2004)
83. G. H. Gweon et al: Nature, 430:187 (2004)
84. G. Kotliar and J. Liu: Phys. Rev. Lett., 61:1784 (1988)
85. X.-G. Wen and P. A. Lee: Phys. Rev. Lett., 76:503 (1996)
86. N. Furukawa et al: Phys. Rev. Lett., 81:3195 (1998)
87. N. Furukawa and T. M. Rice: J. Phys.: Condens. Matt., 10:L381 (1998)
88. M. L. Salkola et al: Phys. Rev. Lett., 77:155 (1996)
89. W. E. Pickett: Rev. Mod. Phys., 61:433 (1989)
90. X. J. Zhou et al: Science, 286:268 (1999)
91. K. M. Shen et al: Science, 307:901 (2005)
92. V. Brouet et al: Phys. Rev. Lett., 93:126405 (2004)
93. J. Schafer et al: Phys. Rev. Lett., 87:196403 (2001)
94. T. E. Kidd et al: Phys. Rev. Lett., 88:226402 (2002)
95. M. Plate et al: Phys. Rev. Lett., 95:077001 (2005)
96. G. S. Boebinger et al: Phys. Rev. Lett., 77:5417 (1996)
97. C. M. Varma: Phys. Rev. Lett., 83:3538 (1999)
98. J. L. Tallon et al: Phys. Stat. Sol. B, 215:531 (1999)
99. S. Sachdev: Science, 288:475 (2000)

10

Many-Body Interaction in Hole- and Electron-Doped High-T_c Cuprate Superconductors

T. Takahashi, T. Sato, and H. Matsui

Department of Physics, Tohoku University, Sendai 980-8578, Japan
t.takahashi@arpes.phys.tohoku.ac.jp

Abstract. Recent progress in energy and momentum resolution in angle-resolved photoemission spectroscopy (ARPES) has made it possible to directly observe the interaction of electrons with bosonic modes responsible for the occurrence of superconductivity in high-T_c cuprates. This chapter explains results of recent high-resolution ARPES on hole- and electron-doped high-T_c superconductors and discusses the validity of the BCS-like framework of the superconducting mechanism, the origin of the "kink" structure in the band dispersion near the Fermi level produced through the electron–boson interaction, and the electron–hole symmetry in high-T_c cuprate superconductors.

10.1 Introduction

It is almost twenty years since the superconductivity with an unprecedentedly high transition temperature (T_c) was discovered in cupper oxides [1]. The resultant enthusiastic "high-T_c fever", which involved many researchers over a wide area of fields, has promoted intensive interaction not only between theorists and experimentalists, but also among many fields with different backgrounds. The discovery of high-T_c superconductor has also accelerated advancement of experimental techniques since verification of many proposed models requests accurate and high-quality experimental data. In fact, the energy resolution of photoemission spectroscopy has been improved by almost three orders in magnitude, now reaching the less-than 1-meV region. The recent remarkable improvement of the energy and momentum resolutions in photoemission spectroscopy has made it possible to directly observe the interaction between electrons and "modes" in high-T_c superconductors. Here we show and discuss the results of recent high-resolution angle-resolved photoemission spectroscopy (ARPES) on hole- and electron-doped high-T_c superconductors to study the many-body interaction responsible for and/or related to the occurrence of high-T_c superconductivity.

The first fundamental question in high-T_c superconductors is whether or not the superconductivity is understood/described in the framework of the BCS theory [2] in the "wide" meaning. To answer this question, we have performed ultra-high resolution ARPES measurements on a Bi-based hole-doped high-T_c superconductor, $Bi_2Sr_2Ca_2Cu_3O_{10}$ (Bi2223), to search for the Bogoliubov quasi-particle [3]. According to the BCS theory, the Bogoliubov quasi-particle is produced through the Cooper-pairing of two electrons and plays an essential role in characterizing the superconductivity. The experimental observation is regarded as the most direct evidence for the validity of the BCS-like mechanism of the superconductivity in high-T_c superconductors. Once the Bogoliubov quasi-particle picture, in other words, the BCS-like superconductivity in the wide meaning, is established, the second question is on the driving force of the superconductivity. In the BCS theory in the "narrow" meaning, it is phonon-mediated (lattice vibration). From the beginning of the discovery of high-T_c cuprates, the possibility of phonon-mediated superconductivity has been almost ignored or regarded less important, simply because the superconducting transition temperatures (T_c) extend far beyond the "BCS wall" (40–50 K) predicted by the BCS theory in the narrow meaning. However, recently several theoretical and experimental studies have revived the role of phonons in the high-T_c superconductivity. In order to get an insight into this essential problem, we have performed a systematic ARPES study on a series of Bi-based high-T_c superconductors, $Bi_2Sr_2Ca_{n-1}Cu_nO_{2n+4}$ ($n = 1-3$), where we comprehensively measured the momentum- and temperature-dependence of the "kink" in the energy dispersion near the Fermi level (E_F) (for similar investigations see also the contributions by Johnson and Valla and by Fink et al. in this volume). Since the interaction of electrons with a certain bosonic mode (phonon, magnon etc.) responsible for the occurrence of superconductivity produces a quasi-particle near E_F and appears as a "kink" in the energy dispersion near E_F. Each bosonic mode shows a characteristic momentum and temperature dependence of the kink, so that the systematic ARPES study would distinguish the origin of the kink (mode). The third fundamental question in the high-T_c superconductivity which we discuss here is the "electron–hole (a)symmetry". It is well established that an insulating mother cuprate compound such as La_2CuO_4 and Nd_2CuO_4 becomes a superconductor when excess electrons or holes are introduced (doped). But it is still not clear whether the electronic structure is just reversed (electron–hole symmetric) or not, and further whether the mechanism of superconductivity is the same (similar) or not. This fundamental problem has been left unresolved because the energy scale of electron-doped high-T_c superconductors is one order of magnitude smaller than that of hole-doped ones. We have performed precise ARPES measurements with ultrahigh-resolution on electron-doped high-T_c superconductors, $Nd_{2-x}Ce_xCuO_4$ (NCCO: $x = 0.13$) and $Pr_{1-x}LaCe_xCuO_4$ (PLCCO, $x = 0.11$), to study the electronic structure, the Fermi surface, and the superconducting gap symmetry, and compare them with those of hole-doped counterparts.

10.2 Experiments

High-quality hole-doped $Bi_2Sr_2Ca_{n-1}Cu_nO_{2n+4}$ ($n = 1$–3), electron-doped NCCO and PLCCO single crystals were grown by the traveling-solvent floating-zone (TSFZ) method [4–6]. ARPES measurements were performed using a Gammadata-Scienta SES-200 spectrometer with a high-flux discharge lamp and a toroidal grating monochromator at Tohoku University, and with a same-type spectrometer at the undulator 4m-NIM (normal incidence monochromator) beamline at the Synchrotron Radiation Center, Wisconsin. He Iα resonance line (21.218 eV) and 22 eV photons were used to excite photoelectrons. The energy and angular (momentum) resolutions were set at 9–15 meV and 0.2° (0.007 Å^{-1}), respectively. Samples were cleaved in situ in an ultra-high vacuum better than 5×10^{-11} Torr to obtain a clean surface. The Fermi level (E_F) of the samples was referred to that of a gold film evaporated onto the sample substrate. The samples are labeled by their doping levels (UD for underdoped, OP for optimally-doped, OD for overdoped) together with their onset T_c. For example, UD100K means an underdoped sample with the T_c of 100 K.

10.3 Results and Discussion

10.3.1 Fermi Surface and Superconducting Gap

In Fig. 10.1(a), we show ARPES-intensity maps of Bi2223 as a function of two-dimensional wave vector for an underdoped sample with $T_c = 100$ K (UD100K) and an overdoped sample with $T_c = 108$ K (OD108K). Two intensity maxima symmetric to M point ((π, 0) point) are clearly seen on M–Y(X) line ((π, 0)–(π, π) line), but not on Γ–M line ((0, 0)–(π, 0) line). This clearly defines a large hole-like Fermi surface centered at X(Y) for both samples. In Figs. 10.1(b) and 10.1(c), we plot the ARPES intensity for Bi2212 (UD70K and optimally-doped $T_c = 90$ K; OPT90K) and Bi2201 (UD19K and overdoped $T_c < 4$ K; OD< 4 K). One can clearly see that ARPES intensity distributions are essentially similar to that in Bi2223, showing that the hole-like Fermi surface is a generic feature of the CuO_2 plane in HTSCs.

In Fig. 10.1(d), we show the k-dependence of superconducting gap size Δ for Bi-based HTSCs [7,8]. The overall "V"-shape of the gap curves, as well as the node along Γ–Y line ($\phi = 45°$), strongly supports the $d_{x^2-y^2}$-wave nature of the gap in all the Bi-based HTSCs.

10.3.2 Bogoliubov Quasi-Particles

Figures 10.2(a) and 10.2(b) show ARPES spectra of overdoped $Bi_2Sr_2Ca_2Cu_3O_{10+\delta}$ (Bi2223, $T_c = 108$ K) at 140 K and 60 K, respectively, measured along

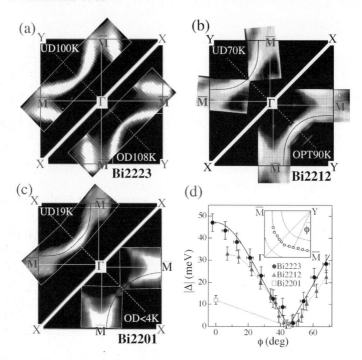

Fig. 10.1. Plots of ARPES intensity integrated within 50 meV centered at E_F for (**a**) Bi2223, (**b**) Bi2212, and (**c**) Bi2201 (Pb-substituted); (**d**) comparison of superconducting gap as a function of Fermi surface angle in Bi-based cuprates. Note that Figs. (**a**)–(**c**) each display the results for two samples, separated by the white line

the cut in the Brillouin zone shown in the inset. We clearly find that a superconducting gap of about 20 meV opens at the Fermi vector ($\boldsymbol{k}_F$) on lowering the temperature from 140 K to 60 K across T_c and at the same time a sharp coherent peak grows up in the vicinity of E_F in the spectra. As the wave vector ($\boldsymbol{k}$) is changed from (π, 0) to ($3\pi/2, -\pi/2$), the coherent peak gradually disperses toward E_F, showing a minimum energy gap at $\boldsymbol{k}_F$ (shown by thick line) which defines the superconducting gap, and then disperses back to the higher binding energy with rapidly reducing its intensity. This spectral change *below* E_F is consistent with a previous ARPES result on $Bi_2Sr_2CaCu_2O_{8+\delta}$ (Bi2212) [9]. More importantly, we find additional weak but discernible structures about 20 meV *above* E_F in the spectra, which are more clearly seen in Fig. 10.2(c) where the intensity scale is expanded. This new structure shows a clear momentum dependence with a stronger intensity in the region of $|\boldsymbol{k}| > |\boldsymbol{k}_F|$, opposite to the behavior of the band below E_F. In order to see more clearly the band dispersion above E_F, we have divided the ARPES spectra by the Fermi–Dirac (FD) function at 60 K convoluted with a Gaussian representing the instrumental resolution. The result is shown in

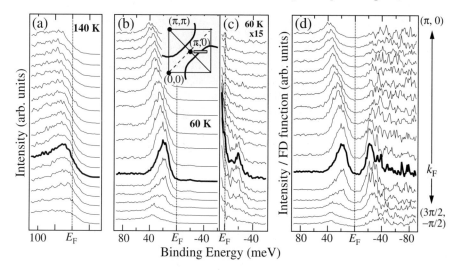

Fig. 10.2. (a) and (b) ARPES spectra of overdoped Bi2223 (OP108K) at 140 K and 60 K, respectively, measured along the cut in the Brillouin zone shown in the inset. Spectra shown by thick lines are measured at k_F; (c) same as (b) above E_F in an expanded intensity scale. (d) Renormalized ARPES spectra at 60 K divided by the Fermi–Dirac (FD) function at 60 K convoluted with a Gaussian representing the instrumental resolution

Fig. 10.2(d), where we find a dispersive structure above E_F with a comparable intensity to that below E_F, although the signal-to-noise ratio is relatively low because of the originally small ARPES intensity.

We plot in Fig. 10.3(a) the renormalized ARPES intensity (Fig. 10.2(d)) as a function of the momentum and the binding energy. Bright areas correspond to the strong intensity in the renormalized ARPES spectra. We observe several characteristic behaviors for the two branches of dispersive bands: (i) the dispersive feature is almost symmetric with respect to E_F while the intensity is not; (ii) the energy separation of the two bands is minimum at k_F; (iii) both bands show the bending-back effect at k_F; (iv) the spectral intensity of the two bands show the opposite evolution as a function of k in the vicinity of k_F. All these features qualitatively agree with the behavior of Bogoliubov quasi-particles (BQP) predicted from the BCS theory [2], indicating the basic validity of the BQP concept in high-T_c superconductors.

Next, we compare the experimental result with the theoretical prediction to examine the quantitative validity of the BCS theory. We show in Fig. 10.3(a) the theoretical band dispersion of BQP (broken lines denoted E_k and $-E_k$) for comparison. In the BCS theory, the band dispersion of BQP (E_k) is expressed as

$$E_k = (\epsilon_k^2 + |\Delta(k)|^2)^{1/2} , \tag{10.1}$$

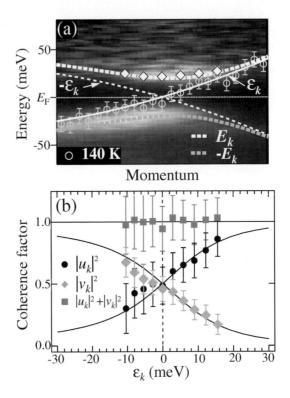

Fig. 10.3. (a) Plot of renormalized ARPES intensity (Fig. 10.2(d)) as a function of the momentum and the binding energy. Bright areas correspond to the strong intensity in the renormalized ARPES spectra, showing the experimentally determined band dispersion of the Bogoliubov quasi-particles. *White* and *gray broken lines* denoted by $E_{\bm{k}}$ and $-E_{\bm{k}}$ are the theoretical band dispersions below and above E_F, respectively, predicted from the BCS theory. *Open circles* show the experimental normal-state band dispersion, and white solid line ($\epsilon_{\bm{k}}$) is the fitting with a parabolic function; (b) comparison of the coherence factors above ($|u_{\bm{k}}|^2$) and below ($|v_{\bm{k}}|^2$) E_F between the ARPES experiment and the BCS theory

where $\epsilon_{\bm{k}}$ and $\Delta(\bm{k})$ are the normal-state dispersion and the superconducting gap, respectively. We have determined $\epsilon_{\bm{k}}$ (white solid line in Fig. 10.3(a)) from the ARPES spectra at 140 K. The superconducting gap $\Delta(\bm{k})$ is assumed to be the $d_{x^2-y^2}$-wave superconducting order parameter $\Delta(\bm{k}) = \Delta_0 |\cos(k_x) - \cos(k_y)|/2$, where Δ_0 is determined with the 60-K spectrum at $\bm{k}_F$. We find in Fig. 10.3(a) that the calculated dispersion well traces the strong intensity of ARPES spectra (bright areas), showing a good agreement in the band dispersion between the experiment and the theory.

To further study the validity of the BQP concept, we compare the coherence factors above/below E_F, $|\bm{k}|^2$ and $|v_{\bm{k}}|^2$, between the experiment and the theory. According to the BCS theory, the coherence factors are expressed as,

$$|u_{\bm{k}}|^2 = 1 - |v_{\bm{k}}|^2 = (1 - \epsilon_{\bm{k}}/E_{\bm{k}})/2 , \qquad (10.2)$$

where $\epsilon_{\bm{k}}$ and $E_{\bm{k}}$ are the energy dispersion of the normal quasi-particle and BQP bands, respectively. We show these "theoretical" coherence factors by smooth solid lines in Fig. 10.3(b). On the other hand, we experimentally deduced the coherence factors by fitting the original ARPES spectra with the following equation,

$$I(\bm{k}, \omega) = I_0(\bm{k})\{A_{\mathrm{BCS}}(\bm{k}, \omega) + A_{inc.}(\bm{k}, \omega)\} f(\omega, T) @ R(\omega) , \qquad (10.3)$$

where $I_0(\bm{k})$ is a prefactor which includes the kinematical factors and the dipole matrix element. $A_{\mathrm{BCS}}(\bm{k}, \omega)$ is the BCS spectral function expressed as,

$$A_{\mathrm{BCS}}(\bm{k}, \omega) = \frac{1}{\pi}\left[\frac{|u_{\bm{k}}|^2 \Gamma}{(\omega - E_{\bm{k}})^2 + \Gamma^2} + \frac{|v_{\bm{k}}|^2 \Gamma}{(\omega + E_{\bm{k}})^2 + \Gamma^2}\right] , \qquad (10.4)$$

where Γ is a linewidth broadening due to the finite lifetime of photoholes. $A_{inc.}(\bm{k}, \omega)$ in Eq.(3) is an empirical function to represent the incoherent background [10], $f(\omega, T)$ is the Fermi–Dirac function, and $@R(\omega)$ denotes the convolution with the resolution function $R(\omega)$. To remove the effect of $I_0(\bm{k})$, we have divided the spectral intensity of the superconducting state (60 K) by the integrated normal-state (140 K) spectral intensity at each k point. We determined the peak weights below and above E_F at each $\bm{k}$ point by decomposing the spectrum, and then divided them by the average value of the total peak weight at each $\bm{k}$ point [13]. We define these normalized weights as the experimental coherence factors $|v_{\bm{k}}|^2$ and $|u_{\bm{k}}|^2$ and show them in Fig. 10.3(b). As seen in Fig. 10.3(b), the coherence factors show a surprisingly good quantitative agreement between the experiment and the theory. It is also remarked that the sum of the experimental coherence factors is almost unity over the measured momentum region in good agreement with the prediction from the BCS theory, although the experimental coherence factors are determined totally independently of each other without using the sum rule. The present ARPES results thus indicate the basic validity of the BCS-like Bogoliubov quasi-particle picture in describing the superconducting state of high-T_c cuprates.

10.3.3 Many-Body Interaction in Hole-Doped HTSCs

Figure 10.4 shows ARPES-intensity plots for underdoped Bi2223 with $T_c = 100$ K (UD100K) at the normal and superconducting states, measured along five cuts in the Brillouin zone (cuts a–e) shown in Fig. 10.4 [14]. For cut a (nodal cut) at 40 K, the band rapidly approaches E_F from high binding energy and suddenly bends at 50-80 meV, showing a characteristic kink in the dispersion. The kink is weakened with increasing temperature, but still survives

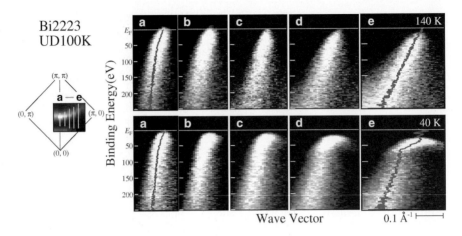

Fig. 10.4. ARPES-intensity plots of Bi2223 (UD100 K) along several cuts in the Brillouin zone, measured with 22-eV photons at the normal (140 K, upper panel) and the superconducting state (40 K, lower panel). Peak positions of momentum distribution curves (MDC) are indicated by *solid lines* for cuts a and e. Solid lines reach E_F even in the superconducting state because a small finite intensity due to the tail of the coherent peak remains near E_F. Left panel shows the Brillouin zone and the Fermi surface. *White bars* (a–e) indicate the cuts where the ARPES measurements (right panel) were done

even above T_c. In contrast to the relatively small temperature dependence in cut a, a drastic change is observed in the momentum region away from the nodal direction. In cut e just between the nodal cut and $(\pi, 0)$ point [15], a sharp kink at 50–80 meV together with opening of the superconducting gap is clearly seen at 40 K, while the dispersion is almost straight at 140 K. It is noted here that the kink in cut e at the superconducting state is not due to the opening of the superconducting gap, because, as shown in Fig. 10.3, the Bogoliubov quasi-particle band is smoothly rounded near E_F and does not make a kink like that in cut e [3,9].

We find in Fig. 10.4 that the kink appears only near the nodal direction in the normal state and gradually smears out on approaching $(\pi, 0)$, while the kink becomes more pronounced near $(\pi, 0)$ at the superconducting state (40 K). This totally opposite behavior of kinks as a function of momentum at two different temperatures above/below T_c suggests that the kink at the nodal cut in the normal state is different from the kink near $(\pi, 0)$ at the superconducting state (for similar results see contribution by Johnson and Valla and Fink et al. in their contribution).

Figure 10.5 shows ARPES-intensity plots of Bi2223 (UD100K) as a function of temperature across T_c measured along cuts a and e. For cut e, a strong renormalization of the dispersion is clearly seen at low temperatures below T_c, while the temperature-induced change is almost negligible at high

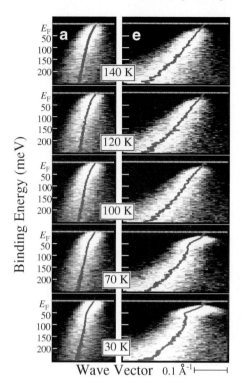

Fig. 10.5. ARPES-intensity plots of Bi2223 (UD100K) as a function of temperature across T_c measured with the He Iα resonance line along cuts a and e in Fig. 10.4

temperatures above T_c. This strongly suggests that the kink at cut e is closely related to the superconductivity. In contrast, the kink at cut a shows much smaller temperature dependence in the superconducting state although the kink is slightly enhanced at low temperatures.

In Fig. 10.6, we show the temperature dependence of MDC (Momentum Distribution Curve)-peak position for $n = 1$–3 measured along the nodal and off-nodal cuts. For $n = 2$ and 3, the dispersion along the off-nodal cut shows a strong bending behavior below T_c, while that above T_c is almost straight with very small temperature dependence. This indicates that the kink around $(\pi, 0)$ disappears around T_c for $n = 2$ and 3, showing a close correlation to the superconductivity. In sharp contrast to the remarkable temperature dependence of kink for $n = 2$ and 3, the dispersion of $n = 1$ shows almost no temperature dependence for both directions. This implies that the electron-mode coupling, which give rise to the strong temperature dependence in $n = 2$ and 3, is absent or very weak in $n = 1$.

In Fig. 10.7, we plot maximum value of the real part of self-energy $\text{Re}\Sigma(\omega)^{max}$, defined as an energy difference between the kink and the linear

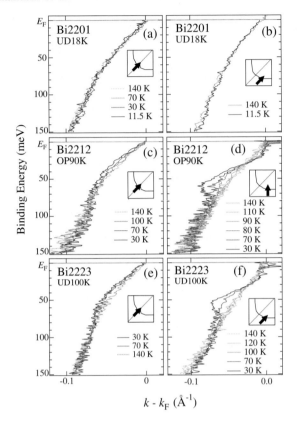

Fig. 10.6. Temperature dependence of MDC-peak position for Bi2201 [(**a**) and (**b**)], Bi2212 [(**c**) and (**d**)], and Bi2223 [(**e**) and (**f**)], measured along an arrow shown in each inset. Note that the direction of measurements (*arrow*) in (d) differs from that of (b) and (f)

bare band dispersion which passes the experimental dispersion at E_F and 250 meV (see the inset to Fig. 10.7) [17]. It is known that $\text{Re}\Sigma(\omega)^{max}$ serves as a good measure of the coupling strength when $\text{Re}\Sigma(\omega)^{max}$ comes dominantly from the interaction of electrons with collective excitation [17,18]. As seen in Fig. 10.7, $\text{Re}\Sigma(\omega)^{max}$ of $n = 2$ and 3 at the off-nodal direction gradually decreases with increasing temperature for all samples. However, $\text{Re}\Sigma(\omega)^{max}$ is almost vanished at T_c in optimally- and over-doped samples while it has a finite value even at T_c and still gradually decreases at higher temperatures in underdoped samples. This characteristic behavior of $\text{Re}\Sigma(\omega)^{max}$ for different dopings resembles the temperature dependence of the magnetic resonance peak reported by the inelastic neutron scattering (INS) experiment [19]. The peak intensity of the resonance peak of $\text{YBa}_2\text{Cu}_3\text{O}_{6+\delta}$ with similar T_c [19] is superposed in panels (b) and (c) of Fig. 10.7. We find a surprisingly good quantitative agreement between $\text{Re}\Sigma(\omega)^{max}$ and the resonance-peak intensity.

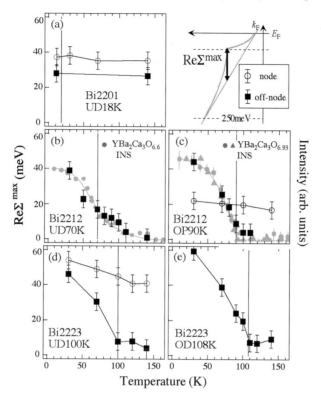

Fig. 10.7. Maximum value of the real part of the self-energy $\mathrm{Re}\Sigma(\omega)^{max}$ as a function of temperature measured along the nodal and the off-nodal cut for Bi2201, Bi2212, and Bi2223. Inset shows the definition of the experimentally determined $\mathrm{Re}\Sigma(\omega)^{max}$. A vertical *solid line* on each panel shows the T_c of sample. The peak intensity of resonance peak of $YBa_2Cu_3O_{6+\delta}$ with similar T_c [19] is superposed in panels (**b**) and (**c**)

In particular, it is remarked that the resonance peak has a finite intensity even at T_c and gradually decreases with increasing temperature in the underdoped sample, in a quite similar manner to the temperature dependence of $\mathrm{Re}\Sigma(\omega)^{max}$. This indicates that the kink around $(\pi, 0)$ in Bi2212 ($n = 2$) is of magnetic origin [17, 20, 21]. In contrast to the remarkable temperature dependence of $\mathrm{Re}\Sigma(\omega)^{max}$ in the off-nodal cut, that of the nodal cut shows much less temperature dependence as seen in Fig. 10.6. It is also remarked that $\mathrm{Re}\Sigma(\omega)^{max}$ in Bi2201 ($n = 1$) shows almost no temperature dependence in both directions in sharp contrast to the multi-layered compounds (Bi2212 and Bi2223).

Figures 10.8(a)–(c) show the energy dispersions near E_F in the superconducting state measured along the nodal direction for nearly-optimally-doped Bi-based HTSCs. We find that a kink has almost the same binding energy

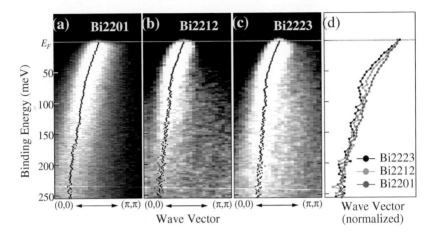

Fig. 10.8. ARPES intensity plots along $(0, 0)$-(π, π) direction in the superconducting state for (**a**) Bi2201 ($T_c = 19$ K), (**b**) Bi2212 ($T_c = 91$ K), and (**c**) Bi2223 ($T_c = 108$ K); (**d**) MDC peak dispersions for three samples after normalizing the wave vector

(50–80 meV) among these compounds. Remarkably, when we normalize the wave vector so as to align the peak energy at k_F and the highest binding energy (250 meV), the dispersions near E_F show a characteristic difference among the three compounds as shown in Fig. 10.8(d). It is evident that the kink becomes more pronounced as n increases.

From the ARPES results shown in Figs. 10.4–10.7, it is established that the kink around $(\pi, 0)$ has a close relation to the superconductivity. A possible interpretation of the kink is a coupling of electrons with the $Q = (\pi, \pi)$ magnetic mode because of the following reasons. First, it is clear from Figs. 10.4–10.6 that the kink at the superconducting state is stronger near $(\pi, 0)$ than around the nodal cut for $n = 2$ and 3. This indicates that electrons at $(\pi, 0)$ are easily scattered by the mode with a (π, π) vector [22, 23]. This is consistent with the $Q = (\pi, \pi)$ nature of the magnetic resonance mode. Second, as shown in Fig. 10.7, the temperature dependence of the magnetic resonance-peak intensity shows an excellent agreement with that of $\mathrm{Re}\Sigma(\omega)^{max}$. Third, the absence of a temperature-dependent kink around $(\pi, 0)$ for $n = 1$ is consistent with the magnetic-mode scenario because a resonance peak has not been observed in Bi2201 [24]. It is expected from the present ARPES results that Bi2223 shows a magnetic resonance peak similar to that of Bi2212 in the INS experiment, although the INS data are not available at present because of lack of a large Bi2223 single crystal. The stronger kink and the slightly larger value of $\mathrm{Re}\Sigma(\omega)^{max}$ at the off-nodal direction for $n = 3$ than that for $n = 2$ suggests that the resonance peak in Bi2223 would be much more intense than Bi2212. The observed dissimilarity in the behavior of kink between single- and multi-layered Bi-family compounds implies that the interlayer interaction is essential for the stronger coupling of electrons with magnetic modes as well as for higher T_c. This is supported by INS experiments [25], which show that the resonance-peak intensity of YBCO shows a modulation along Q_z, indicative of a strong coupling between adjacent CuO_2 layers [25].

Now we discuss the origin of the kink around the nodal direction at the normal state. We summarize the properties and behaviors of the kink in the following. (1) The kink around the nodal cut survives even above T_c, while the kink around $(\pi, 0)$ disappears around T_c. However, (2) the kink around the nodal cut does not show a clear temperature dependence above T_c. (3) A kink in the nodal cut is seen also in materials where a magnetic resonance peak is absent [26]. There are several candidates responsible for this kink. The first one is the marginal Fermi-liquid-like excitation [17]. The temperature-independent behavior of kink at the normal state seems consistent with this interpretation. The second is a contribution from the magnetic resonance mode which survives even above T_c [19]. However, the experimental fact that the kink appears strongest in the nodal cut is hardly explained in this framework, since the nodal cut is not connected by the $Q = (\pi, \pi)$ vector of magnetic mode. The third is the broad magnetic excitation observed in the Q-independent INS [27], which has been proposed to contribute to the kink at the normal state [21]. The last is the coupling of electrons with LO phonon [26]. The location in the Brillouin zone and the temperature-independent nature above T_c are consistent with the $Q = (\pi, 0)$ character of the LO phonon [28]. The phonon energy estimated from the INS experiment [28] is 80 meV, while the characteristic energy due to the magnetic mode is also 80 meV [40 meV (mode energy) + 40 meV (superconducting-gap energy) for optimally-doped Bi2212] [23]. This may be the reason for a similar energy scale for two kinks located near nodal cut and around $(\pi, 0)$. It is noted that the kink around the nodal cut shows a finite temperature dependence below T_c for Bi2212 and Bi2223 (see Fig. 10.6). This suggests that the magnetic mode, which is dominant around $(\pi, 0)$, has a finite influence even around the nodal cut in the superconducting state. It is noted that the stronger kink in the superconducting state along the nodal cut in $n = 3$ than in $n = 2$ as seen in Fig. 10.8 is consistent with the larger value of $\mathrm{Re}\Sigma(\omega)^{max}$ at off-nodal cut in $n = 3$.

10.3.4 Many-Body Interaction in Electron-Doped HTSCs

Figure 10.9 shows the plot of ARPES intensity near E_F at 30 K as a function of the two-dimensional wave vector to illustrate the Fermi surface of NCCO ($x = 0.13$). Bright areas correspond to the experimental Fermi surface. We normalized the intensity with respect to the highest binding energy of spectrum (400 meV) [29]. We find in Fig. 10.9 that the ARPES intensity at E_F shows a characteristic k-dependence while the experimental Fermi surface looks circle-like centered at (π, π) as predicted from the LDA band calculation [30]. On the experimental Fermi surface, the strongest ARPES intensity appears near $(\pi, 0)$, and a weak but observable intensity is seen around $(\pi/2,$

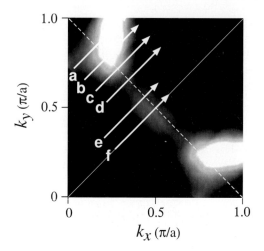

Fig. 10.9. Plot of near-E_F ARPES intensity for NCCO ($x = 0.13$) at 30 K integrated within 25 meV with respect to E_F and symmetrized with respect to the (0, 0)–(π, π) nodal line. Arrows denoted a-f show cuts where detailed ARPES measurements shown in Fig. 10.10 were done

$\pi/2$), while there is negligible or no intensity between these two momentum regions. It is noted that the area with negligible ARPES intensity on the Fermi surface coincides with the hot spot, namely the intersecting point of the LDA-like Fermi surface and the antiferromagnetic zone boundary. A similar ARPES-intensity modulation has been reported in the case of x = 0.15 with different photon energies [31], suggesting that the observed intensity modulation is not due to the matrix-element effect.

Figure 10.10 shows the ARPES spectra near E_F measured along several cuts across the Fermi surface (cuts a-f in Fig. 10.9) and the corresponding band dispersions derived from the spectra. Near the (π, 0) point (cuts a–c), we find two separated band dispersions; one is a very steep band dispersion located below ∼ 0.1 eV and another is a flat band very close to E_F. The strong ARPES intensity on the Fermi surface near the (π, 0) point as shown in Fig. 10.9 is due to this flat band located very close to E_F. As seen in Fig. 10.10, the presence of two separated bands in the same momentum region produces the characteristic "peak-dip-hump" structure in the ARPES spectrum measured near k_F. The band near E_F becomes flatter and the intensity is weakened on going from cut a to cut c, namely on approaching the hot spot. At the same time, the energy separation between the peak and the hump gradually increases. In cut d, which passes the hot spot, the peak near E_F almost disappears and as a result an energy gap of about 100 meV opens between E_F and the lower-lying steep band. The energy gap becomes gradually small on approaching the nodal line and finally the steep band appears to almost touch E_F in cut f, as evidenced by the recovery of the Fermi-edge-like structure in the spectrum. Here, it is

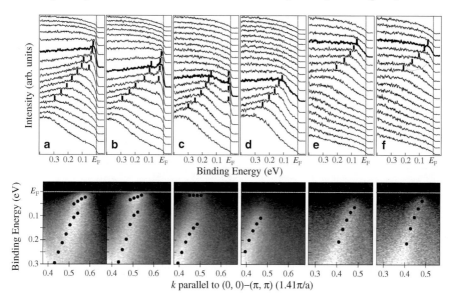

Fig. 10.10. *Upper panel:* ARPES spectra of NCCO ($x = 0.13$) measured at 30 K along several cuts parallel to the $(0, 0)$–(π, π) direction shown by arrows in Fig. 10.9. *Thick spectra* are at the Fermi surface. *Lower panel:* ARPES-intensity plot as a function of the wave vector and binding energy, showing the experimental band dispersion. Peak positions in ARPES spectra are shown by *bars* and *dots*

noted that the change of band dispersions among different cuts is continuous, indicating that the peak-dip-hump structure in the ARPES spectra in cuts a-c has a same origin as the energy gap in cuts d and e. The opening of a large energy gap at E_F in cut d is attributed to the antiferromagnetic spin-correlation, since it is located at the intersecting point of the "original" Fermi surface and the "shadow" Fermi surface produced by the antiferromagnetic interaction [31]. The present ARPES results in Fig. 10.10 clearly show that the gap at the hot spot is smoothly connected to the two separated bands near the $(\pi, 0)$ point, suggesting the effect of the antiferromagnetic correlation to modify the band dispersion.

We show in Fig. 10.11 a schematic diagram to explain how the quasi-particle dispersion is modified by the antiferromagnetic electron correlation. It is reminded that the intersecting point between the original band and the shadow band folded back into the magnetic Brillouin zone is always on the diagonal line $[(\pi, 0)$–$(0, \pi)]$, and more importantly, the intersecting point is *below* E_F at $(\pi, 0)$ and *above* E_F at $(\pi/2, \pi/2)$ in the presence of a nearly half-filled circle-like Fermi surface centered at (π, π) as shown in Fig. 10.9. The relative position of this intersecting point with respect to E_F plays an essential role in characterizing the band dispersion and the ARPES intensity on the Fermi surface. In the case I in Fig. 10.11, where the intersecting point is below

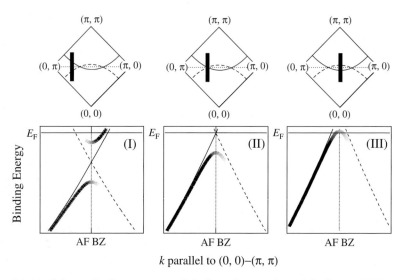

Fig. 10.11. Schematic diagram to explain how the quasi-particle dispersion is modified by the antiferromagnetic correlation for three different cases. In case I, the original quasi-particle band (*thin solid line*) and the shadow band (*thin broken line*) intersect each other below E_F. In cases II and III, the intersecting point is at and above E_F, respectively. *Thick solid lines* show the quasi-particle dispersions modified by the antiferromagnetic correlation

E_F, the strong antiferromagnetic scattering splits the original dispersion into two pieces above and below the intersecting point, respectively, producing an energy gap between the two separated bands. It is expected that the occupied band just below E_F is strongly bent and the quasi-particle effective mass is remarkably enhanced in this momentum region. This effect becomes stronger when one approaches the hot spot, because the intersecting point with the strongest antiferromagnetic scattering is gradually shifted to E_F. In the case II, where the intersecting point is just on E_F, the antiferromagnetic scattering eliminates the electronic states at E_F, producing a large energy gap at E_F.

Finally in the case III, where the intersecting point is above E_F, the antiferromagnetic interaction affects the band dispersion mainly in the unoccupied states, leaving the original band dispersion in the occupied states almost unaffected. We find that the gross feature of band dispersions in different cuts in Fig. 10.10 shows a good agreement with this simple picture. For example, cuts *a*, *d*, and *f* correspond to the cases I, II, and III, respectively. Thus the observed heavy-mass quasi-particle state around $(\pi, 0)$ and the *k*-dependence of quasi-particle dispersion are well explained in terms of the effect from the antiferromagnetic correlation. The continuous evolution of band dispersion along the Fermi surface shown in Fig. 10.10 strongly suggests the antiferromagnetic origin of the energy gap and the resulting mass-enhancement in NCCO.

We find in Fig. 10.10 that the energy separation between the peak and the hump (namely the energy separation between the upper and the lower bands separated by the antiferromagnetic correlation) gradually increases from cut a ($\sim$50 meV) to cut c ($\sim$120 meV). This k-dependence of the antiferromagnetic gap is not necessarily obvious in the simple picture in Fig. 10.11 and may suggest that the strength of the antiferromagnetic scattering has momentum dependence, with the stronger amplitude close to the hot spot. A recent tunneling spectroscopy reported a pseudogap comparable in the size to the superconducting gap, suggesting the second order parameter hidden within the supercondcting state in electron-doped HTSCs [32]. However, the one-order smaller energy scale compared to the antiferromagnetic gap suggests the different nature between these two gaps.

The mass-enhancement effect and the peak-dip-hump structure in Fig. 10.10 look similar to those in hole-doped HTSCs, which have been interpreted with some corrective modes such as the magnetic-resonance mode [20, 22, 33–35]. However, such arguments is not applicable to the electron-doped case, because the quasi-particle state is clearly observed even above T_c. As described above, the mass-enhancement in NCCO is due to the band folding caused by the antiferromagnetic order/fluctuation. In this case, the energy separation between the *peak* and the *dip* does not reflect the energy of collective mode, but the separation between the *peak* and the *hump* is related to the antiferromagnetic exchange interaction. It is also remarked here that the k-region where the heavy-mass quasi-particle state is observed coincides with the k-region where a large d-wave superconducting gap opens [36, 37]. This suggests that the superconductivity in electron-doped HTSCs occurs in the antiferromagnetically correlated quasi-particle state [38]. The present experimental result that the quasi-particle effective mass at E_F and the antiferromagnetic gap increase as moving away from $(\pi, 0)$ suggests a slight deviation in the superconducting order parameter from the simple $d_{x^2-y^2}$ symmetry, $\Delta(\boldsymbol{k}) \propto \cos(k_x a)$-$\cos(k_y a)$, in electron-doped HTSCs [39, 40]. This point will be clarified in the next section.

Next, we discuss how the heavy-mass quasi-particle state at cuts a-c in Fig. 10.10 changes as a function of temperature. Figure 10.12 shows the temperature dependence of ARPES spectrum measured at a point on the Fermi surface where the peak-dip-hump structure is observed (see the inset to Fig. 10.12). At low temperatures (50 and 90 K) below T_N (110 K), we clearly find a relatively sharp quasi-particle peak at E_F and a broad hump at 190 meV in the ARPES spectra, which are ascribed to the upper and lower pieces of the renormalized quasi-particle band, respectively. On increasing the temperature across T_N, the quasi-particle peak at E_F becomes substantially broadened and almost disappears at 130K. However, the suppression of the spectral weight near E_F ($E_F - 0.2$ eV), which defines "a large pseudogap", is still seen in the spectra at 130 K and 220 K. This suggests that while the long-range spin ordering disappears at T_N, the short-range spin-correlation survives even above T_N, affecting the electronic structure near E_F. It is remarked that the energy of the hump (190 meV) does not change with temperature. On further

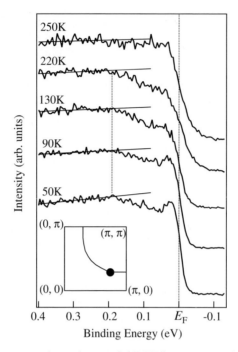

Fig. 10.12. Temperature dependence of ARPES spectrum of NCCO ($x = 0.13$) measured at a point on the Fermi surface shown by a *filled circle* in the inset, where the peak-dip-hump structure is clearly observed. *Solid straight lines* on the spectra show the linear fits to the high-energy region (0.2 – 0.5 eV).

increasing the temperature, the pseudogap is totally filled-in in the spectrum at 250 K, suggesting that the short range antiferromagnetic correlation disappears at around this temperature. The optical conductivity experiment has reported that a pseudogap-like suppression starts to develop in the energy-range lower than 0.18 eV at 190 K for $x = 0.125$ [41], consistent with the present study. Further, the optical conductivity and the Raman experiments have reported the simultaneous evolution of both the low-energy Drude-like response and the high-energy gap on decreasing the temperature [41,42]. This shows a good correspondence to the gradual development of the quasi-particle peak at E_F in ARPES spectrum at low temperatures. The reported sharpening of the low-energy optical response in NCCO [41,42] is well explained in terms of the quasi-particle mass-enhancement due to the antiferromagnetic correlation.

10.3.5 Superconducting Gap Symmetry in Electron-Doped HTSCs

The anisotropy of the superconducting (SC) gap is a direct clue for understanding the origin and mechanism of superconductivity. It is generally accepted that the superconducting-gap symmetry of hole-doped high-T_c superconductors (HTSCs) is $d_{x^2-y^2}$ wave and is described with the gap function of the monotonic $d_{x^2-y^2}$ form, $\Delta(\mathbf{k}) \propto \cos(k_x a) - \cos(k_y a)$ [7,43], where the maximum and zero superconducting gaps are located at the Brillouin-zone boundary and the diagonal, respectively. In electron-doped HTSCs, on the other hand, the superconducting-gap symmetry is still under hot debate. Although a general consensus for the overall $d_{x^2-y^2}$ wave in the optimally doped region seems to be established by microwave penetration depth measurements [44,45], scanning SQUID microscopy [46], and angle-resolved photoemission (ARPES) experiments [36,37], it has been proposed that the gap function in electron-doped HTSCs substantially deviates from the monotonic $d_{x^2-y^2}$ wave [39,40,47] and further may change into a different symmetry such as s wave in the over-doped region [47–51].

These arguments on the superconducting-gap anisotropy are related to the Fermi-surface geometry with respect to the magnetic Brillouin zone. If the antiferromagnetic spin fluctuation mediates the pairing in HTSCs, the superconducting gap is expected to have a large value at particular Fermi momenta (k_F) connected to each other by the antiferromagnetic scattering vector $Q = (\pi,\pi)$ [52]. This $\mathbf{k}_F$ point, so-called "hot spot", is defined as an intersection of the Fermi surface and the antiferromagnetic Brillouin-zone boundary as shown in Fig. 10.13. In the hole-doped case, the large circular Fermi surface centered at (π, π) cuts the antiferromagnetic Brillouin-zone boundary very close to $(\pi, 0)$, producing the hot spot near $(\pi, 0)$. This situation does not alter the characteristics of the original monotonic $d_{x^2-y^2}$ gap function with the maximum gap at $(\pi, 0)$. In contrast, in the electron-doped case the hot spot is moved toward the zone diagonal due to the shrinkage of hole-like Fermi surface, which may distort the monotonic $d_{x^2-y^2}$ gap function by displacing the maximum gap from $(\pi, 0)$ toward $(\pi/2, \pi/2)$ [39, 40, 47].

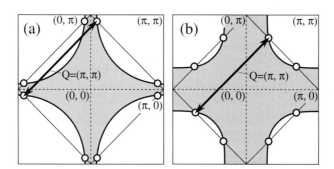

Fig. 10.13. Relation between the Fermi surface and the antiferromagnetic Brillouin zone for (*a*) hole- and (*b*) electron-doped HTSCs. *Thick solid curve* and *thin straight line* show the Fermi surface and the antiferromagnetic-Brillouin zone, respectively. *Arrow* and *open circle* show the antiferromagnetic scattering vector $Q = (\pi,\pi)$ and the hot spot, respectively

Furthermore, the proximity of the pairing potential with the opposite sign around the zone diagonal may suppress the $d_{x^2-y^2}$ gap symmetry itself [47]. Although the detailed momentum dependence of the superconducting gap has been well studied by ARPES for hole-doped HTSCs [7, 29, 43, 53–55], that of electron-doped HTSCs has been hardly measured because of the small (one order of magnitude smaller) superconducting gap compared with that of hole-doped ones. However, the experimental elucidation of the gap anisotropy in electron-doped HTSCs is highly desired to understand the origin and mechanism of the high-T_c superconductivity.

Figure 10.14 shows ARPES spectra in the close vicinity of E_F of PLCCO ($x = 0.11$; OP26K) measured at temperatures below and above T_c (8 K and 30 K, respectively) for three different k_F points as shown in the inset. Points A and C are on the $(\pi, 0)$–(π, π) and the diagonal cut, respectively, and point B corresponds to the hot spot [56]. We find in Fig. 10.14 that the leading-edge midpoint of the 8-K spectrum is shifted toward the high binding energy with respect to that of the 30-K spectrum by a few meV at points A and B, while that of point C does not show such a remarkable temperature-induced shift. This suggests that a $d_{x^2-y^2}$-like superconducting gap opens

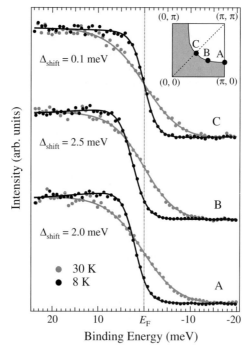

Fig. 10.14. Near-E_F ARPES spectra of PLCCO ($x = 0.11$; OP26K) measured below and above T_c at three k_F points on the Fermi surface shown in the inset. The 8-K and 30-K spectra are shown by *black* and *gray dots*, respectively. *Solid curves* show the fitting of the spectra

at low temperatures in PLCCO. However, it is remarked that the shift of midpoint at point B looks slightly larger than that at point A, exhibiting a striking contrast to the previous ARPES results on the hole-doped HTSCs [7, 29, 43, 53–55].

In order to quantitatively estimate the momentum dependence of the superconducting gap in PLCCO, we numerically fit the ARPES spectra by using the phenomenological Fermi–Dirac function with the onset as a free parameter, multiplied by a linear function and convoluted with a Gaussian resolution function [37]. Although the shift of leading-edge midpoint (Δ_{shift}) in the spectrum is not equal to the superconducting-gap size, it is empirically known that the Δ_{shift} is about a half of the superconducting gap and serves as a good measure for it [7, 29, 36, 37, 43, 54]. Estimated Δ_{shift}'s are 2.0, 2.5, and 0.1 meV with the accuracy of ± 0.2 meV at points A, B, and C, respectively. This clearly indicates that the gap function of PLCCO obviously deviates from the monotonic $d_{x^2-y^2}$ gap function. We have measured the near-E_{F} ARPES spectrum at 8 K for other several $\boldsymbol{k}_{\text{F}}$ points and estimated the Δ_{shift} value. Including these points, we plot the Δ_{shift}'s as a function of the Fermi surface angle (ϕ) in Fig. 10.14, together with the monotonic $d_{x^2-y^2}$ gap function for comparison. The deviation of the measured Δ_{shift} from the monotonic gap function is obvious. The Δ_{shift} is about 2 meV at around $\phi = 0°$, and gradually *increases* on increasing the Fermi surface angle, reaching the maximum value of about 2.5 meV at $\phi \sim 25°$, which corresponds to the hot spot. After passing the hot spot, the Δ_{shift} rapidly decreases and becomes almost zero at the diagonal ($\phi = 45°$). We have fit this experimental curve $\Delta_{\text{shift}}(\phi)$ with the non-monotonic $d_{x^2-y^2}$ gap function $\Delta(\phi) = \Delta_0 \left[B \cos(2\phi) + (1 - B) \cos(6\phi) \right]$ which includes the next higher harmonic ($\cos(6\phi)$) [39, 53, 55]. As shown in Fig. 10.15, the experimental curve is well fitted with the parameter set of $\Delta_0 = 1.9$ meV and $B = 1.43$, indicating the substantial contribution from the second harmonic to the gap function.

Finally we discuss the present observation in comparison with previous studies. The polarized Raman spectroscopy [39] observed that the 2Δ peak in the B_{2g} channel (67 cm^{-1}) is located at higher frequency than in the B_{1g} channel (50 cm^{-1}). The former and latter Raman channels probe mainly the $\boldsymbol{k}$ regions around $(0, 0)$–(π, π) and $(\pi, 0)$, respectively. Provided that the 2Δ peak in the B_{2g} channel is mainly contributed from the hot spot, the Raman experimental result indicates a 1.3-times larger superconducting gap at the hot spot than that around $(\pi, 0)$, in good agreement with the present ARPES result (2.5 meV / 2 meV = 1.25). It has been theoretically predicted [47] that the gap symmetry gradually changes from the $d_{x^2-y^2}$ wave to a different one such as s or p wave, when the hot spot is moved from $(\pi, 0)$ to $(\pi/2, \pi/2)$ with electron doping. The calculated gap function in the intermediate state exhibits the maximum gap around the hot spot, in good agreement with the present observation in PLCCO. The theory has predicted that the transition of gap symmetry occurs when the Fermi-surface angle of the hot spot (ϕ_{hs}) reaches the critical value of $\sim 23°$, which is similar to the ϕ_{hs} observed in this study,

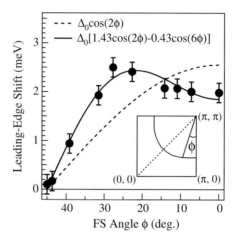

Fig. 10.15. The energy position of leading-edge midpoint (Δ_{shift}) plotted as a function of the Fermi-surface angle (ϕ). *Solid curve* is the result of fitting $\Delta_{\text{shift}}(\phi)$ with the non-monotonic $d_{x^2-y^2}$ gap function, $\Delta_0 \left[B \cos(2\phi) + (1-B) \cos(6\phi) \right]$, compared with the monotonic $d_{x^2-y^2}$ gap function (*broken line*)

suggesting that the present sample, $Pr_{1-x}LaCe_xCuO_4$ with $x = 0.11$, is on the boundary of the transition. ARPES on electron-doped HTSC samples with stronger doping is highly desired to study the transition of the gap symmetry.

References

1. J. G. Bednorz, and K. A. Müller: Z. Phys. B **64**, 189 (1986)
2. J. Bardeen et al: Phys. Rev. **108**, 1175 (1957)
3. N. N. Bogoliubov: Nuovo Cimento **7**, 794 (1958)
4. T. Fujii et al: J. Cryst. Growth **223**, 175 (2001)
5. I. Chong et al: Physica (Amsterdam) **290C**, 57 (1997)
6. T. Uefuji et al: Physica (Amsterdam) **357C-360C**, 208 (2001); *ibid.* **378C-381C**, 273 (2002)
7. H. Ding et al: Phys. Rev. B **54**, R9678 (1996)
8. T. Sato et al: Phys. Rev. B **63**, 132502 (2001)
9. J. C. Campuzano et al: Phys. Rev. B **53**, R14737 (1996)
10. We assumed a linear background with a cutoff with respect to E_F to represent the incoherent part of the spectra, as employed in previous ARPES studies on Bi2212 [11, 12]
11. D. L. Feng et al: Science **289**, 277 (2000)
12. H. Ding et al: Phys. Rev. Lett. **87**, 227001 (2001)
13. We did not use the normal-state quasi-particle weight for division, because the quasi-particle weight in ARPES spectrum shows temperature dependence in cuprates [11, 12]
14. T. Sato et al: Phys. Rev. Lett. **91**, 157003 (2003)

15. We do not use the ARPES data around $(\pi, 0)$ for estimating the peak dispersion to avoid possible complication from the flat band, the bilayer splitting [16], and the superlattice bands
16. A. A. Kordyuk et al: Phys. Rev. Lett. **89**, 077003 (2002)
17. P. D. Johnson et al: Phys. Rev. Lett. **87**, 177007 (2001)
18. A. Kaminski et al: Phys. Rev. Lett. **86**, 1070 (2001)
19. P. Dai et al: Science **284**, 1344 (1999)
20. J. C. Campuzano et al: Phys. Rev. Lett. **83**, 3709 (1999)
21. A. D. Gromko et al: Phys. Rev. B **68**, 174520 (2003)
22. Z. X. Shen and J. R. Schrieffer: Phys. Rev. Lett. **78**, 1771 (1997)
23. M. R. Norman et al: Phys. Rev. Lett. **79**, 3506 (1997)
24. K. Hirota: private communication.
25. H. F. Fong et al: Phys. Rev. Lett. **75**, 316 (1995); Nature (London) **398**, 588 (1999)
26. A. Lanzara et al: Nature (London) **412**, 510 (2001)
27. P. Bourges, in *The Gap Symmetry and Fluctuations in High Temperature Superconductors*, edited by J. Bok, G. Deutscher, D. Pavuna, and S. A. Wolf (Plenum Press, New York, 1998)
28. R. J. McQueeney et al: Phys. Rev. Lett., **82**, 628 (1999)
29. S. V. Borisenko et al: Phys. Rev. B **66**, 140509(R) (2002)
30. S. Massidda et al: Physica (Amsterdam) **157C**, 571 (1989)
31. N. P. Armitage et al: Phys. Rev. Lett. **87**, 147003 (2001), *ibid.* **88**, 257001 (2002)
32. L. Alff et al: Nature (London). **422**, 698 (2003)
33. T. Sato et al: Phys. Rev. Lett. **89**, 067005 (2002)
34. S. V. Borisenko et al: Phys. Rev. Lett. **90**, 207001 (2003)
35. T. K. Kim et al: Phys. Rev. Lett. **91**, 167002 (2003)
36. T. Sato et al: Science **291**, 1517 (2001)
37. N. P. Armitage et al: Phys. Rev. Lett. **86**, 1126 (2001)
38. G.-q. Zheng et al: Phys. Rev. Lett. **90**, 197005 (2003)
39. G. Blumberg et al: Phys. Rev. Lett. **88**, 107002 (2002)
40. H. Yoshimura and D. S. Hirashima: J. Phys. Soc. Jpn **73**, 2057 (2004)
41. Y. Onose et al: Phys. Rev. Lett. **87**, 217001 (2001)
42. A. Koitzsh et al: Phys. Rev. B **67**, 184522 (2003)
43. Z.-X. Shen et al: Phys. Rev. Lett. **70**, 1553 (1993)
44. J. D. Kokales et al: Phys. Rev. Lett. **85**, 3696 (2000)
45. R. Prozorov et al: Phys. Rev. Lett. **85**, 3700 (2000)
46. C. C. Tsuei, J. R. Kirtley: Phys. Rev. Lett. **85**, 182 (2000)
47. V. A. Khodel et al: Phys. Rev. B **69**, 144501 (2004)
48. J. A. Skinta et al: Phys. Rev. Lett. **88**, 207005 (2002)
49. A. Biswas et al: Phys. Rev. Lett. **88**, 207004 (2002)
50. A. V. Pronin et al: Phys. Rev. B **68**, 054511 (2003)
51. H. Balci and R. L. Greene: Phys. Rev. Lett. **93**, 067001 (2004)
52. D. J. Scalapino: Phys. Rep. **250**, 329 (1995)
53. J. Mesot et al: Phys. Rev. Lett. **83**, 840 (1999)
54. D. L. Feng et al: Phys. Rev. Lett. **88**, 107001 (2002)
55. H. Matsui et al: Phys. Rev. B **67**, 060501(R) (2003)
56. H. Matsui et al: Phys. Rev. Lett. **94**, 047005 (2005)

11

Dressing of the Charge Carriers in High-T_c Superconductors

J. Fink[1,2], S. Borisenko[1], A. Kordyuk[1,3], A. Koitzsch[1], J. Geck[1],
V. Zabolotnyy[1], M. Knupfer[1], B. Büchner[1], and H. Berger[4]

[1] Leibniz Institute for Solid State and Materials Research Dresden, P.O. Box 270016, 01171 Dresden, Germany
J.Fink@ifw-dresden.de
[2] Ames Laboratory, Iowa State University, Ames, Iowa 50011, USA
[3] Institute of Metal Physics of the National Academy of Sciences of Ukraine, 03142 Kyiv, Ukraine
[4] Institut de Physique de la Matière Complex, Ecole Politechnique Fédérale de Lausanne, 1015 Lausanne, Switzerland

Abstract. In this contribution we first present a short introduction into the lattice structure, the phase diagram and the electronic structure of high-T_c superconductors. Then we explain the principles of angle-resolved photoemission spectroscopy (ARPES) and the influence of the dressing of the charge carriers, which is normally described by the complex self-energy function, on the spectral function in the normal and the superconducting state. Finally we review our recent ARPES results on high-T_c superconductors at various k-points in the Brillouin zone near the Fermi surface. Information on the dressing of the charge carriers, i.e., on the effective mass and the scattering length, is obtained as a function of doping concentration, temperature, momentum and energy. The strong renormalization of the bandstructure due to the dressing can be explained in terms of a coupling to a continuum of spin fluctuations and in the superconducting state by an additional coupling to a triplet exciton excitation. Possibly, this dressing is related to the glue for the pair formation in cuprate superconductors.

11.1 Introduction

One hundred years ago, in the first of five famous papers [1] of his *annus mirabilis*, Albert Einstein postulated the dual nature of light, at once particle and wave, and thereby explained among other phenomena the photoelectric effect, originally discovered by H. Hertz [2]. This work of Einstein was also singled out by the Nobel committee in 1921. The photoelectric effect has since become the basis of one of the most important techniques in solid state research. In particular, angle-resolved photoemission spectroscopy (ARPES), first applied by Gobeli et al. [3], has developed to *the* technique to determine the bandstructure of solids. During the last decade, both the energy and the angular resolution of ARPES has increased by more than one order of magnitude. Thus it is possible to measure the dispersion very close to the Fermi level, where the spectral function, which is measured by ARPES, is renormalized by many-body effects such as electron–phonon, electron–electron, or electron–spin interactions. The mass enhancement due to such effects leads to a reduced dispersion and the finite life-time of the quasi-particles leads to a broadening of the spectral function. Thus the increase in resolution, achieved by new analyzers using two-dimensional detectors, together with new photon sources provided by undulators in 3rd generation synchrotron storage rings and new cryo-manipulators have opened a new field in ARPES: the determination of the low-energy many-body properties of solids which is termed very often the "dressing" of the charge carriers.

In high-T_c superconductors (HTSCs) discovered by Bednorz and Müller [4] the many-body effects are supposed to be particularly strong since these doped cuprates are close to a Mott–Hubbard insulator or, to be more precise, to a charge-transfer insulator [5]. Since in the normal and the superconducting state the renormalization effects are strong, the HTSCs are a paradigm for the new application of ARPES. Moreover, since in these compounds the mass enhancement and the superconducting gap are large, they can be measured using ARPES even without ultra-high resolution.

On the other hand, the understanding of the renormalization effects in the HTSCs is vital for the understanding of the mechanism of high-T_c superconductivity, since the dressing of the charge carriers may be related with the glue forming the Cooper pairs. Up to now there is no widely accepted microscopic theory, although the phenomenon has been discovered already 20 years ago. Similar to the conventional superconductors, before the development of a microscopic theory for the mechanism of superconductivity, first one has to understand the many-body effects in the normal state of these highly correlated systems. ARPES plays a major role in this process. Not only can it determine the momentum dependent gap. It is at present also the only method which can determine the momentum dependence of the renormalization effects due to the interactions of the charge carriers with other degrees of freedom.

In this contribution we review ARPES results on the dressing of the charge carriers in HTSCs obtained by our spectroscopy group. There are previous reviews on ARPES studies of HTSCs [6–8], which complement what is discussed here. Similar results are presented by Johnson and Valla and by Takahashi et al. in this volume.

11.2 High-T_c Superconductors

11.2.1 Structure and Phase Diagram

It is generally believed that superconductivity is associated with the two-dimensional CuO_2 planes shown in Fig. 11.1(a). In these planes Cu is divalent, i.e., Cu has one hole in the $3d$ shell. The CuO_2 planes are separated by block layers formed by other oxides (see Fig. 11.1(b)). Without doping, the interacting CuO_2 planes in the crystal form an antiferromagnetic lattice with a Néel temperature of about $T_N = 400\,\mathrm{K}$. By substitution of the ions in the block layers, it is possible to dope the CuO_2 planes, i.e., to add or to remove electrons from the CuO_2 planes. In this review we focus on hole doped systems. With increasing hole concentration and increasing temperature, the long-range antiferromagnetism disappears (see the phase diagram in Fig. 11.1(c)) but one knows from inelastic neutron scattering that spin fluctuations still exist at higher dopant concentrations and higher temperatures.

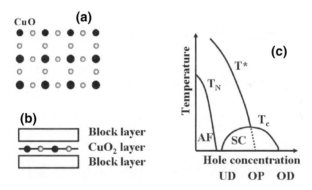

Fig. 11.1. (a) CuO_2 plane; (b) CuO_2 plane between block layers; (c) schematic phase diagram of hole-doped cuprates

With increasing dopant concentration the insulating properties transform into metallic ones and there is a high-T_c superconducting range. This range is normally divided into an underdoped (UD), an optimally doped (OP) and an overdoped (OD) region. Not only the superconducting state but also the normal state is unconventional. In the UD range there is a pseudogap between the T^* line and the T_c line. There are various explanations for the pseudogap [9]: preformed pairs which have no phase coherence, spin density waves, charge density waves, or the existence of a hidden order, caused, e.g. by circulating currents [10]. At low temperatures in the OP region the T^* line is very often related to a quantum critical point near the OP region. Possibly related to this quantum critical point, in the OP range, the normal state shows rather strange properties such as a linear temperature dependence of the resistivity over a very large temperature range or a temperature dependent Hall effect. Only in the OD range the system behaves like a normal correlated metal showing for example a quadratic temperature dependence of the resistivity.

11.2.2 Electronic Structure

In the following we give a short introduction into the electronic structure of cuprates. We start with a simple tight-binding bandstructure of a CuO_2 plane using for the beginning three hopping integrals, one between 2 neighboring Cu sites along the Cu-O bonding direction (t), one for a hopping to the second nearest Cu neighbor along the diagonal (t'), and one for the hopping to the third nearest neighbor (t''). The corresponding bandstructure is given by

$$E(\mathbf{k}) = \Delta\epsilon - 2t[\cos(k_x a) + \cos(k_y a)] + 4t' \cos(k_x a) \cos(k_y a)$$
$$- 2t''[\cos(2k_x a) + \cos(2k_y a)] \qquad (11.1)$$

where a is the length of the unit cell and $\Delta\epsilon$ fixes the Fermi level. This two-dimensional bandstructure is displayed in Fig. 11.2(a) for $t'/t = -0.3$, a value which is obtained from bandstructure calculations [11], and both t'' and $\Delta\epsilon$ equal to zero. It has a minimum in the center (Γ) and maxima at the corners of the Brillouin zone (e.g. at $(k_x, k_y) = (\pi, \pi)/a \equiv (\pi, \pi)$). Furthermore there are saddle points, e.g. at $(k_x, k_y) = (\pi, 0)$. In the undoped system there is one hole per Cu site and therefore this band should be half filled. This leads to a Fermi level just above the saddle points (see Fig. 11.2(a)). The Fermi surface consists of rounded squares around the corners of the Brillouin zone (see Fig. 11.2(b)). Upon hole doping the Fermi level moves towards the saddle point. It is interesting that for vanishing t' the Fermi surface would be quadratic and there would be no parallel sections (which could lead to a nesting) along x or y but along the diagonal. There are two special points on the Fermi surface (see Fig. 11.2(b)), which are also at the focus of most of the ARPES studies on HTSCs. There is the nodal point at the diagonal (N in Fig. 11.2(b)), where the superconducting order parameter is zero and the antinodal point (where the $(\pi, 0)$–(π, π) line cuts the Fermi surface), where the superconducting order parameter is believed to reach a maximum (AN in Fig. 11.2(b) [12, 13].

In many cuprates there is not just a single but several CuO_2 planes between the block layers. In these systems the CuO_2 planes are separated by additional ionic layers. This is illustrated for a bilayer system in Fig. 11.2(c). Such a bilayer system is for example $Bi_2Sr_2CaCu_2O_8$ which is the Drosophila for ARPES studies of HTSCs. In this compound the block layers are composed of BiO and SrO planes, while the ionic layer separating the two CuO_2 planes consist of Ca^{2+} layers. Doping is achieved in this compound by additional O atoms in the block layers. In those bilayer systems there is an interaction between the two adjacent CuO_2 planes which leads to a finite hopping integral $t_\perp$. This causes an additional term in the tight-binding calculations

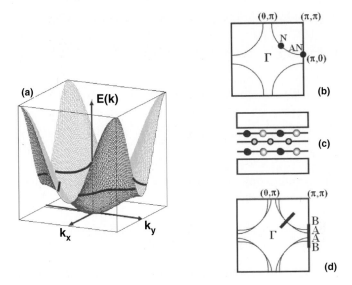

Fig. 11.2. (a) Tight-binding bandstructure of the CuO$_2$ plane; (b) Fermi surface of a CuO$_2$ plane. N: nodal point, AN: antinodal point; (c) bilayer system between block layers composed of two CuO$_2$ planes separated by one ionic layer; (d) Fermi surfaces of a bilayer system, B(A): (anti)bonding band. *Thick solid lines:* **k** values along which most of the present ARPES studies have been performed

$$E(\mathbf{k})_\perp = \pm t_\perp \left[\cos(k_x a) - \cos(k_y a)\right]^2 / 4 \quad (11.2)$$

leading to a splitting into a bonding and an antibonding band. This splitting is small at the nodal point [14] and it is largest at the antinodal point. In Fig. 11.2(d) we have illustrated this splitting of the Fermi surface caused by the interaction of the two CuO$_2$ planes.

The independent particle picture, describing just the interactions with the ion lattice and the potential of a homogeneous conduction electron distribution, is of minor use for the undoped systems since we know that those are not metallic but insulating. This comes from the Coulomb interaction U of two holes on the same Cu site which prohibits hopping of holes from one Cu site to the other. It causes the insulating behavior of undoped and slightly doped cuprates. The large on-site Coulomb repulsion of two holes on a Cu site is also responsible for the fact that the additional holes produced upon doping are formed on O sites [15]. The 2 eV energy gap is then a charge-transfer gap [5] between O 2p and Cu 3d states. Only when more and more holes are introduced into the CuO$_2$ planes is hopping of the holes possible and correlation effects get less important.

11.3 Angle-resolved Photoemission Spectroscopy

11.3.1 Principle

In photoemission spectroscopy monochromatic light with an energy $h\nu$ is shined onto a surface of a solid and the intensity as well as the kinetic energy, E_{kin}, of the outgoing photoelectrons is measured. Using the explanation of the photoelectric effect [1] one can obtain the binding energy of the electrons in the solid:

$$E_B = h\nu - \Phi - E_{kin} \equiv -E \ . \tag{11.3}$$

Here Φ is the workfunction. The charge carriers in HTSCs show a quasi-two-dimensional behavior. When the surface is parallel to the CuO_2 planes, the momentum $\hbar k_\parallel$ of the photoelectron is conserved when passing through the surface and thus this momentum is determined by the projection of the total momentum of the photoelectron to the surface:

$$\hbar k_\parallel = \sqrt{2mE_{kin}} \sin\theta \ . \tag{11.4}$$

Here θ is the angle between the direction of the photoelectron in the vacuum and the surface normal.

There are numerous treatises of the photoelectron process in the literature where the limitations of the models describing it are discussed [16]. They are not repeated in this contribution. Rather the essential points for the analysis of ARPES studies on the dressing of the charge carriers in HTSCs are restated. It is assumed that the energy and momentum dependence of the photocurrent in ARPES studies can be described by

$$I(E, \boldsymbol{k}) \propto M^2 A(E, \boldsymbol{k}) f(E) + B(E, \boldsymbol{k}) \tag{11.5}$$

where $M = \langle \psi_f | H' | \psi_i \rangle$ is a matrix element between the initial and the final state and H' is a dipole operator. $A(E, k)$ is the spectral function which is the essential result in ARPES studies. $f(E) = 1/[\exp(E/k_BT) + 1]$ is the Fermi function which takes into account that only occupied states are measured and $B(E, \boldsymbol{k})$ is an extrinsic background coming from secondary electrons. For a comparison of calculated data with experimental data, the former have to be convoluted with the energy and momentum resolution.

The dynamics of an electron in an interacting system can be described by a Green's function [17]

$$G(E, \boldsymbol{k}) = \frac{1}{E - \epsilon_{\boldsymbol{k}} - \Sigma(E, \boldsymbol{k})} \ . \tag{11.6}$$

$\Sigma(E, \boldsymbol{k}) = \Sigma'(E, \boldsymbol{k}) + i\Sigma''(E, \boldsymbol{k})$ is the complex self-energy function which contains the information on the dressing, i.e., on what goes beyond the independent-particle model. $\epsilon_{\boldsymbol{k}}$ gives the dispersion of the bare particles without many-body interactions. The spectral function can be expressed [18, 19] by

$$A(E, \boldsymbol{k}) = -\frac{1}{\pi}\mathrm{Im}G(E,\boldsymbol{k}) = -\frac{1}{\pi}\frac{\Sigma''(E,\boldsymbol{k})}{[E - \epsilon_{\boldsymbol{k}} - \Sigma'(E,\boldsymbol{k})]^2 + [\Sigma''(E,\boldsymbol{k})]^2} \quad (11.7)$$

For $\Sigma = 0$, i. e., for the non-interacting case, the Green's function and thus the spectral function is a delta function at the bare-particle energy $\epsilon_{\boldsymbol{k}}$. Taking interactions into account, the spectral function given in Eq. (11.7) is a rather complicated function. On the other hand, in many cases only local interactions are important which leads to a $\boldsymbol{k}$-independent or weakly $\boldsymbol{k}$-dependent self-energy function. Furthermore, in the case of not too strong interactions, often quasi-particles with properties still very close to the bare particles, can be projected out from the spectral function. To perform this extraction one expands the complex self-energy function around the bare particle energy $\epsilon_{\boldsymbol{k}}$: $\Sigma(E) \approx \Sigma(\epsilon_{\boldsymbol{k}}) + \partial\Sigma(E)/\partial E|_{E=\epsilon_{\boldsymbol{k}}}(E - \epsilon_{\boldsymbol{k}})$. Very often one introduces the coupling function $\lambda(\epsilon_{\boldsymbol{k}}) = -\partial\Sigma'(E)/\partial E|_{E=\epsilon_{\boldsymbol{k}}}$ and the renormalization function $Z(\epsilon_{\boldsymbol{k}}) = 1 + \lambda$. Note that most unfortunately the *renormalization function* as used here is defined in the literature as $Z(\epsilon_{\boldsymbol{k}}) = 1+\lambda$ whereas the *renormalization constant* $Z(\epsilon_{\boldsymbol{k}} = 0)$ is given by $Z(\epsilon_{\boldsymbol{k}} = 0) = [1 + \lambda(\epsilon_{\boldsymbol{k}} = 0)]^{-1}$. Neglecting the partial derivative of $\Sigma''(E)$ one obtains for the spectral function

$$A(E, \boldsymbol{k})_{coh} = -\frac{1}{\pi}Z(\epsilon_{\boldsymbol{k}})^{-1}\frac{Z(\epsilon_{\boldsymbol{k}})^{-1}\Sigma''(\epsilon_{\boldsymbol{k}})}{[E - \epsilon_{\boldsymbol{k}} - Z(\epsilon_{\boldsymbol{k}})^{-1}\Sigma'(\epsilon_{\boldsymbol{k}})]^2 + [Z(\epsilon_{\boldsymbol{k}})^{-1}\Sigma''(\epsilon_{\boldsymbol{k}})]^2} \quad (11.8)$$

This is the coherent fraction of the spectral function and its spectral weight is given by Z^{-1}. It is called coherent because it describes a (quasi-)particle which is very similar to the bare particle. Instead of a delta-function we have now a Lorentzian. The energy of the quasi-particle is determined by the new maximum of the spectral function which occurs at $E = \epsilon_{\boldsymbol{k}} - Z^{-1}\Sigma'(\epsilon_{\boldsymbol{k}})$. The life-time of the quasi-particle is determined in a cut at constant k by the FWHM of the Lorentzian which is given by $\Gamma = 2Z^{-1}\Sigma''(\epsilon_{\boldsymbol{k}})$.

Close to the Fermi we can assume that the real part of the self-energy is linear in energy, i.e., $\Sigma'(\epsilon_{\boldsymbol{k}}) = -\lambda\epsilon_{\boldsymbol{k}}$, with λ being a constant. Then also $Z(0)$ is a constant. We reiterate that in many contributions in the literature (also in all other contributions of this volume where the renormalization constant Z is used!) this renormalization constant $Z(0)$ is termed $1/Z$. For the renormalized energy of the quasi-particle, we now obtain $E_k = \epsilon_{\boldsymbol{k}}/(1 + \lambda)$. Thus close to the Fermi level we have in the case of a linear real part of the self-energy a renormalization by a factor of $1 + \lambda$ or in other words, due to the interactions we have for the coherent quasi-particles a mass enhancement $m^* = (1+\lambda)m$.

The incoherent part of the spectral function, the spectral weight of which is given by $1 - Z^{-1}$, contains all the spectral weight that cannot be described by the Lorentzian close to the bare particle energy, e.g., satellites. Z^{-1} also determines the size of the jump at k_{F} of the momentum distribution $n(k)$, which can be calculated from the energy integral of the spectral function $A(E, \boldsymbol{k})$. Thus if the jump in $n(k)$ comes to zero, at this very point the quasi-particle weight Z^{-1} vanishes logarithmically as one approaches the Fermi level. For such an electron liquid the term "marginal" Fermi liquid [20] has been introduced. This is related to another condition for the existence of (coherent) quasi-particles [21, 22]. The finite lifetime implies an uncertainty in energy. Only if this uncertainty is much smaller than the binding energy ($\Sigma''/E \to 0$) the particles can propagate coherently and the concept of quasi-particles has a physical meaning.

In principle, performing constant-k scans, commonly called energy distribution curves (EDCs), one can extract the spectral function along the energy axis and using Eq. (11.7) one can derive the complex self-energy function. In reality there is a background, the exact energy dependence of which is not known. In addition, close to the Fermi energy there is the energy-dependent Fermi function. These problems are strongly reduced when performing constant-energy scans, usually called momentum distribution curves (MDCs) [23]. Close to the Fermi level the bare particle bandstructure can be expanded as $\epsilon_k = v_F \hbar (k - k_F)$. Assuming again a weakly k-dependent $\Sigma(E, k)$, the spectral function along the particular k-direction is a Lorentzian (see Eq. (11.7)). The width is given by $\Sigma''/v_F \hbar$ and from the shift relative to the bare particle dispersion one can obtain the real part Σ'. This evaluation is much less dependent on a weakly k-dependent background and on the Fermi function.

11.3.2 Spectral Function in the Normal State

In a real solid there are several contributions to the self-energy. The important ones, related to inelastic scattering processes can be reduced to contributions which are related to bosonic excitations (see Fig. 11.3). In the case where the boson is a particle-hole excitation, which is depicted in Fig. 11.3(a), a photoelectron hole is filled by a transition from a higher energy level and the energy is used to excite an Auger electron to a state above the Fermi level. The final state is thus a photoelectron hole scattered into a higher state plus an electron–hole pair. For a normal Fermi liquid of a three-dimensional solid at $T = 0$ phase space arguments and the Pauli principle lead to the complex self-energy function $\Sigma = \alpha E - i\beta E^2$. In a two-dimensional solid the imaginary

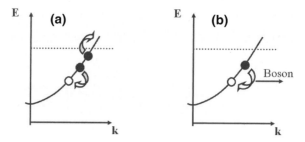

Fig. 11.3. Bosonic excitations contributing to the finite lifetime of a photohole in metallic solids. (**a**) electron–hole excitations, (**b**) discrete bosonic mode. The *dashed line* corresponds to the Fermi level.

part of Σ changes from a quadratic energy dependence to $\beta' E^2 \ln|E/E_F|$ [24] which is similar to the 3D case only as long as E is much smaller than the bandwidth. Increasing the interactions more and more, associated with a reduction of Z^{-1}, changes the self-energy function. For $Z^{-1} = 0$ where the spectral weight of the quasi-particles disappears one reaches the above mentioned marginal Fermi liquid [20]. In this case the self-energy is given by $\Sigma = \lambda_{MFL}[E \ln|x/E_c| + i(\pi/2)x]$ where $x = \max(E, k_B T)$ and E_c is a cutoff energy taking into account the finite width of the conduction band. This self-energy function is a phenomenological explanation, among others, of the linear temperature dependence of the resistivity observed in optimally doped HTSCs, since the imaginary part of the self-energy and thus the inverse scattering rate is linear in T.

Besides the particle-hole excitations described above, the photohole may be scattered to higher (lower) energies by the emission (absorption) of a discrete boson. This is illustrated in Fig. 11.3(b) for the emission of a bosonic excitation. Such discrete bosonic excitations may be phonons, spin excitations, plasmons, excitons, etc. The relevant excitations are listed in Table 11.1 together with their characteristic energies in optimally doped HTSCs.

Table 11.1. Bosonic excitations which couple to the charge carriers together with their characteristic energies in HTSCs

System	Excitations	Characteristic energy(meV)
ion lattice	phonons	90
spin lattice/liquid	magnons	180
e-liquid	plasmons	1000

The self-energy function for a coupling of the charge carriers to a bosonic mode for the case that the energy of the mode is much smaller than the band width has been treated by Engelsberg and Schrieffer [25]. The assumption of a strong screening of the bosonic excitations is probably adequate for the doped HTSCs but probably not for the undoped or slightly doped parent compounds [26,27]. In the well-screened case, Σ'' is zero up to the mode energy Ω_0. This is immediately clear from Fig. 11.3(b) since the photohole can only be filled when the binding energy is larger than Ω_0. Σ'' is constant above the mode energy (see Fig. 11.4(b)). Performing the Kramers–Kronig transformation one obtains Σ', which is given by $\Sigma' = (\lambda \Omega_0/2) \ln|(E+\Omega_0)/(E-\Omega_0)|$ (see Fig. 11.4(a)). It shows a logarithmic singularity at the mode energy, Ω_0. At low energies there is a linear energy dependence of Σ' and the slope determines the coupling constant λ. In this model it is related to the imaginary part of the self-energy function $\Sigma''(|E| > \Omega_0) \equiv \Sigma''(-\infty)$, by $\lambda = -\Sigma''(-\infty)/(\pi\Omega_0/2)$. From this it is clear that for a given Ω_0 both λ and $\Sigma''(-\infty)$ are a measure of the coupling strength to the bosonic mode.

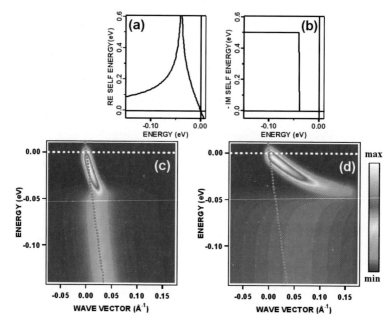

Fig. 11.4. Real part (**a**) and imaginary part (**b**) of the self-energy function for a coupling to a mode $\Omega_0 = 40$ meV and a coupling constant $\lambda = 8$. Calculated spectral function $A(E, \mathbf{k})$ for $\lambda = 1$ (**c**) and $\lambda = 8$ (**d**) in the normal state

In Fig. 11.4(c) and (d) we have displayed the calculated spectral function for $\lambda = 1$ and $\lambda = 8$, respectively. Compared to the bare particle dispersion, given by the red dashed line, for $|E| < \Omega_0$ there is a mass renormalization, i.e., a reduced dispersion and no broadening, except the energy and momentum resolution broadening, which was taken to be 5 meV and 0.005 Å^{-1}, respectively. For $|E| > \Omega_0$, there is a back-dispersion to the bare particle energy. Moreover, there is a broadening due to a finite Σ'', increasing with increasing λ. For large λ, the width for constant E scans is, at least up to some energy, larger than the energy of the charge carriers and therefore they can be called incoherent (see Subsect. 11.3.2) in contrast to the energy range $|E| < \Omega_0$ or at very high binding energies, where the width is much smaller than the binding energy and where they are coherent [25]. The change in the dispersion is very often termed a "kink" but looking closer at the spectral function, in particular for high λ, it is a branching of two dispersion arms.

11.3.3 Spectral Function of Solids in the Superconducting State

For the description of the spectral function in the superconducting state, two excitations have to be taken into account: the electron–hole and the pair excitations. This leads to a (2×2) Green's function [28]. Usually the complex renormalization function

$$Z(E,\bm{k}) = 1 - \Sigma(E,\bm{k})/E\,, \tag{11.9}$$

is introduced. For the one-mode model, the self-energy of the superconducting state corresponds to the self-energy of the normal state in which Ω_0 is replaced by $\Omega_0 + \Delta$. This can be easily seen from Fig. 11.3(b) and assuming a gap opening with the energy Δ. The coupling constant in the superconducting state, λ_{sc}, is related to the renormalization function by $\lambda_{sc} = Z(0) - 1$. For the Auger process shown in Fig. 11.3(a) the onset of the scattering rate is at 3Δ. The reason for this is that the bosonic (e-h) excitations have in this case a lower limit of 2Δ. The complex spectral function is given by [29]

$$A(E,\bm{k}) = -\frac{1}{\pi} Im \frac{Z(E,\bm{k})E + \epsilon_{\bm{k}}}{Z(E,\bm{k})^2(E^2 - \Delta(E,\bm{k})^2) - \epsilon_{\bm{k}}^2}\,. \tag{11.10}$$

In general, $\Delta(E,\bm{k})$ is also a complex function. In Fig. 11.5 we show for the one-mode model the calculated spectral function in the superconducting state using the same energy and momentum resolutions and the same mode energy as before. The imaginary part of Δ was neglected and the real part was set to 30 meV. One clearly realizes the BCS-Bogoliubov-like back-dispersion at the gap energy Δ and besides this, a total shift of the dispersive arms by the gap energy. Thus the branching energy occurs at $\Omega_0 + \Delta$.

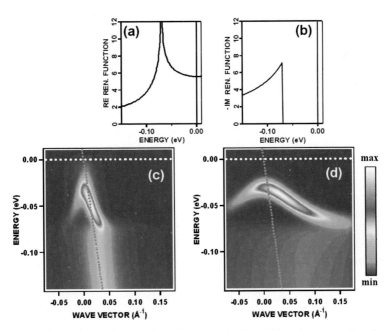

Fig. 11.5. The real part (**a**) and imaginary part (**b**) of the renormalization function Z(E) for $\Omega_0 = 40$ meV, $\Delta = 30$ meV, and $\lambda_{sc} = 5$. Spectral function $A(E,k)$ for a coupling constant $\lambda_{sc} = 1$ (**c**) and $\lambda_{sc} = 5$ (**d**) in the superconducting state

Looking at the phase diagram in Fig. 11.1(c) it is clear that the HTSCs are very close to a transition into a Mott-insulating state and therefore we expect a large fraction of incoherent spectral weight in the normal state. This, however changes when going into the superconducting state where for $\min(\Delta + \Omega_0, 3\Delta) > |E| > \Delta$ the incoherent states are transformed into coherent ones. The reason for this is that in the superconducting state a gap opens for Σ'' for $|E| < 3\Delta$ (e-h scattering rate) or for $|E| < \Omega_0 + \Delta$ (bosonic scattering rate).

The dispersion is given by [30]:

$$[ReZ(E, \mathbf{k})]^2[E^2 - \Delta(E, \mathbf{k})^2] - \epsilon_{\mathbf{k}}^2 = 0. \qquad (11.11)$$

In the conventional superconductors the mode energy is much larger than the gap and therefore for $|E|$ slightly larger than Δ, Z, and thus λ_{sc}, are constant. In this case Eq. (11.11) yields for the maxima of the spectral function

$$E = \sqrt{\Delta^2 + \epsilon_{\mathbf{k}}^2/(1+\lambda_{sc})^2}. \qquad (11.12)$$

For HTSC the gap is comparable to the mode energy and therefore Eq. (11.12) is no longer valid and the full Eq. (11.11) should be used to fit the dispersion. Then λ_{sc} is related to the normal state λ_n (from $\lambda_n = (Z(0)-1)|_{\Delta=0}$) by $\lambda_n = \lambda_{sc}(\Omega_0 + \Delta)/\Omega_0$. It is this λ_n which should be considered when comparing the coupling strength of the charge carriers to a bosonic mode of HTSCs and conventional superconductors.

When one measures an EDC at k_F a peak is observed followed by a dip and a hump. Such an energy distribution is well known from tunnelling spectroscopy in conventional superconductors which was explained in terms of a coupling of the electrons to phonons. A closer inspection indicates for the one-mode-model that at k_F the peak is followed by a region of low spectral weight and a threshold, which appears at $\Omega_0 + \Delta$. Far away from k_F this threshold is not contaminated by the tails of the peak.

11.3.4 Experimental

During the last decade ARPES has experienced an explosive period of qualitative and quantitative improvements. Previously ARPES was performed by rotating the analyzer step by step. In this way an enormous amount of information was lost because only one angle of the emitted photoelectrons was recorded. The development of the so-called "angle mode" [31], applied in the new generation of Gammadata-Scienta analyzers, allows the simultaneous recording of both an energy and an angle range. This was achieved by a multielement electrostatic lens system, by which each photoemission angle was imaged to a different spot of the entrance slit of the a hemispherical, electrostatic deflection analyzer. This angular information is then transferred to the exit of the analyzer and the energy and angle dispersion is recorded by a two-dimensional detector consisting of a microchannel plate, a phosphor plate, and a charge coupled device detector. This caused an improvement both of the energy and the momentum resolution by more than one order of magnitude and an enormous improvement of the detection efficiency, leading to a very strong reduction of measuring time. But not only new analyzers and detectors lead to a huge progress of the ARPES technique. Also new photon sources such as undulators in synchrotron storage rings [32], new microwave driven He discharge lamps, and new cryo-manipulators contributed to the rapid development of the method.

The measurements presented in this contribution were performed with Gammadata-Scienta SES 200 and 100 analyzers using the above-mentioned angle mode. The photon sources used were a high-intensity He resonance Gammadata VUV 5000 lamp or various beamlines, delivering linearly or circularly polarized light in a wide energy range between 15 and 100 eV: the U125/1 PGM beamline at BESSY [33], the 4.2R beamline "Circular Polarization" at ELETTRA, or the beamline SIS at the SLS. The angular rotation of the sample was achieved by a purpose built high-precision cryo-manipulator which allows the sample to be cooled to 25 K and a computer-controlled angular scanning around three perpendicular axes in a wide range of angles with a precision of $0.1°$. The energy and the angle/momentum resolutions were set in most cases in the ranges 8-25 meV and $0.2°/0.01 - 0.02\,\text{Å}^{-1}$, respectively, which is a compromise between energy and momentum resolution and intensity.

Almost all results presented in this review were obtained from high-quality and well characterized single crystals of $(Bi,Pb)_2Sr_2CaCu_2O_{8+\delta}$ (Bi2212). The reason for this is the following. There is only van der Waals bonding between two adjacent BiO planes and therefore it is easy to cleave the crystals. Upon cleaving, no ionic or covalent bonds are broken which would lead to polar surfaces and to a redistribution of charges at the surface. Moreover, we know from bandstructure calculations that among all HTSCs, the Bi-compounds have the lowest k_z dispersion, i.e., they are very close to a two-dimensional electronic system. This is very important for the evaluation of the ARPES data. Probably on all other HTSCs, upon cleaving there is a redistribution of charges and possibly a suppressed superconductivity at the surface. The bilayer system of the Bi-HTSC family is complicated by the existence of two bands at the Fermi surface. On the other hand, it is that system where the whole superconducting range from the UD to the OD range can be studied. The system without Pb has a further complication. It has a superstructure along the b-axis leading in ARPES to diffraction replicas which complicate the evaluation of the data [34,35]. In order to avoid this, about 20% of the Bi ions were replaced by Pb which leads to superstructure-free samples.

The potential of the new generation ARPES technique is illustrated in Fig. 11.6 where we show room temperature data of OP Bi2212 in the three-dimensional (E, k_x, k_y) space. The fourth dimension is symbolized by the color scale, representing the photoelectron intensity. The right front plane of the

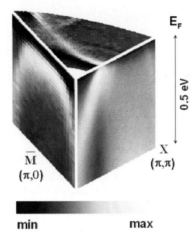

Fig. 11.6. Photoelectron intensity of a $(Bi,Pb)_2Sr_2CaCu_2O_{8+\delta}$ single crystal in the three-dimensional (E, k_x, k_y) space measured at room temperature by ARPES

section shown in Fig. 11.6 was taken simultaneously by setting the **k**-vector parallel to the $\Gamma - (\pi, \pi)$ direction. Then the sample was turned step by step until the k-vector was parallel to the $\Gamma - (\pi, 0)$ direction thus sampling 100000 data points of the whole section.

Such a "piece of cake" can be cut along different directions. A horizontal cut at the Fermi level yields the Fermi surface. A vertical cut along a certain **k**-direction yields the "bandstructure" (the bare particle dispersion plus the renormalization) along this direction. In these data, the essential points of the bandstructure shown in Fig. 11.2 are reproduced. Along the $\Gamma - (\pi, \pi)$ direction there is a crossing of the Fermi level at the nodal point (close to $(\pi/2, \pi/2)$). Along the $\Gamma - (\pi, 0)$ direction there is no crossing of the Fermi level but the saddle point is realized just below E_F.

11.4 The Bare-particle Dispersion

In order to extract the dressing of the charge carriers due to the many-body effects from the ARPES data, one has to know the bare-particle dispersion, i.e., the dispersion which is determined only by the interaction with the ions and the potential due to a homogeneous conduction electron distribution. We have suggested three different ways to obtain the bare-particle band structure.

The first one starts with the Fermi surface measured by ARPES. How to measure those has been already described in Subsect. 11.3.4. In Fig. 11.7 we show ARPES measurements of the Fermi surface of Bi2212 for various dopant concentrations [36]. Using a commonly employed empirical relation [37] between T_c and the hole concentration, x, determined from chemical analysis,

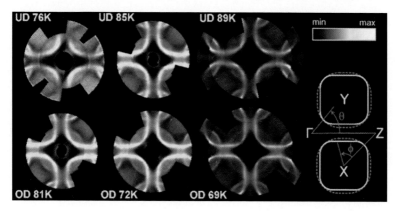

Fig. 11.7. Fermi surfaces of $(Bi,Pb)_2Sr_2CaCu_2O_{8+\delta}$ having various dopant concentrations and T_c (indicated in the panels in units of K) measured by ARPES at room temperature. *Upper row:* underdoped (UD) samples, *lower row:* overdoped (OD) samples

the measured samples cover a doping range of $x = 0.12$ to 0.22. The measurements were performed at room temperature.

Before we come to the evaluation of the bare-particle dispersion, we make some remarks on the character of the measured Fermi surfaces. Firstly, the topology does not change, which means that within the studied doping range there is no transition from a hole-like to an electron-like surface. Secondly, the shape of the Fermi surface around (π, π) changes from being quite rounded at low doping to taking on the form of a square with well rounded corners at higher doping. This is exactly what is expected within a rigid-band approximation and looking at Fig. 11.2. At low doping we are far away from the saddle point and we expect a more rounded Fermi surface. At higher doping we move E_F closer to the saddle point leading to a more quadratic Fermi surface. Thirdly, in underdoped samples, there is an intensity reduction close to $(\pi, 0)$ although the intensities are normalized to the total intensity along the particular $\boldsymbol{k}$-direction to reduce effects due to the $\boldsymbol{k}$-dependence of the matrix element in Eq. (11.5). This reduction in spectral weight is related to the formation of the pseudo-gap below T^*, which is above room temperature in the underdoped samples. This can be treated as a formation of arcs around the nodal points for low dopant concentrations.

It is possible to fit the measured Fermi surface using Eq. (11.1). Such a fit is shown for an optimally doped sample on the right hand side of Fig. 11.7 by a yellow line. Only recently, due to the improved resolution, the bilayer splitting in HTSCs has been resolved [38, 39], while in low resolution data the non-detection of this splitting was ascribed to a strong incoherence of the electronic states close to $(\pi, 0)$. From calculations of the energy dependence of the matrix element in Eq. (11.5) [40, 41] and from systematic photon-energy-dependent measurements (see below) we know that for the photon energy

$h\nu = 21.2$ eV the matrix element for the bonding band is more than a factor 2 larger than for the antibonding band. Therefore, we see in Fig. 11.7 mainly the Fermi surface of the bonding band. Utilizing other photon energies, the bilayer splitting can be clearly resolved, even for UD samples [36]. The red rounded squares in Fig. 11.7 illustrates the Fermi surface of the antibonding band. From the evaluation of the area of the Fermi surface and taking into account the bilayer splitting it is possible to derive the hole concentration which nicely agrees with those values derived from T_c using the universal relation, mentioned above. This is an important result supporting the validity of Luttinger's theorem (the volume of the Fermi surface should be conserved upon switching on the interactions) within the studied concentration range. Finally, we mention the existence of a shadow Fermi surface which corresponds to a (π, π) shifted (normal) Fermi surface in the Fermi surface data, shown in Fig. 11.7. After its first observation [42], it was believed to occur due to the emission of spin fluctuations. More recent measurements indicate that its origin is related to structural effects [43].

Now we come back to the determination of the bare-particle band structure. Assuming that the self-energy effects at E_F are negligible (which is supported by the experimental result that the Luttinger theorem is not violated in the concentration range under consideration), it is possible to obtain information on the unrenormalized bandstructure from the Fermi surface. By fitting the Fermi surface with a tight-binding bandstructure, one obtains relative values of the hopping integrals, i.e., the hopping integrals t', t'', and $t_\perp$ normalized to t. To obtain the absolute values we have measured the spectral function along the nodal direction. From the measured widths at constant energies one can derive the imaginary part of the self-energy function. Performing a Kramers–Kronig transformation, it is possible to derive the real part of Σ and using Eq. (11.7) it is possible to calculate the bare-particle dispersion from $\epsilon_k = E_M - \Sigma'$ where E_M is the measured dispersion (see Sect. 11.5). In this way [44] the absolute values of the hopping integrals for an UD and an OD sample has been obtained (see Table 11.2).

Table 11.2. Tight-binding parameters for an underdoped and overdoped $(Bi,Pb)_2Sr_2CaCu_2O_{8+\delta}$ sample

Sample	t(eV)	t'(eV)	t''(eV)	$t_\perp$(eV)	$\Delta\epsilon$
UD 77 K	0.39	0.078	0.039	0.082	0.29
OD 69 K	0.40	0.090	0.045	0.082	0.43

A second way to determine the bare-particle bandstructure is to evaluate the anisotropic plasmon dispersion which was measured by electron energy-loss spectroscopy for momentum transfers parallel to the CuO_2 planes [45,46]. This plasmon dispersion is determined by the projection of the Fermi velocity on the plasmon propagation directions, which could be varied in the experiment. Since the (unscreened) plasmon energy is at about 2 eV, these excitations are considerably higher than the renormalization energies (see Table 11.1) and therefore the plasmon dispersion is determined by the unrenormalized, averaged Fermi velocity. It is thus possible to fit the momentum dependence of the averaged Fermi velocity by a tight-binding bandstructure. Similar hopping integrals as those shown in Table 11.2 were obtained for an optimally doped sample. Of course no information on the bilayer splitting could be obtained from those measurements.

Finally, a third way to obtain the bare-particle bandstructure is to look at the LDA bandstructure calculations [11]. It is remarkable that the tight-binding parameters, obtained from a tight-binding fit of the LDA bandstructure, are very similar to those given in Table 11.2.

11.5 The Dressing of the Charge Carriers at the Nodal Point

The dynamics of the charge carriers with momentum close to the nodal point determine the transport properties in the normal state. This is particularly the case in the UD region, where a pseudogap opens along the other directions. In order to obtain information on the dressing of the charge carriers at the nodal point, we performed measurements with k parallel to the $(\Gamma - (\pi, \pi))$ direction (see Fig. 11.2(d)). In Fig. 11.8(a) we show the spectral function $A(E, k)$ in a false-color scale together with the bare-particle dispersion ϵ_k [47]. Already without a quantitative analysis, one can learn important facts from a simple visual inspection of Fig. 11.8(a). We clearly see that there is a strong mass renormalization over an energy range which extends up to at least 0.4 eV which is much larger than the energy of the highest phonon modes $E_{ph} = 90$ meV in these compounds [48] (for similar investigations see the contributions by Johnson and Valla and Takahashi et al. in this volume). In these normal state data the measured dispersion (red line) indicates a "soft" kink at about 70 meV but comparing the measured spectral function with that calculated for a single Einstein mode (see Fig. 11.4(c)) one realizes a clear difference. While in the one-mode model there is a sudden change of the k-dependent width from a resolution broadened delta-function to a larger width determined by the constant Σ'', in the experimental data there is a continuous increase of the width (at constant energy) with increasing binding energy. This clearly excludes the interpretation in terms of a coupling to a single phonon line and indicates that the dominant part of the renormalization must be due to a coupling to an electronic continuum. More information can be obtained by a quantitative analysis of the data, namely the extraction of the self-energy function. As described in Subsect. 11.3.2, Σ' can be derived from the difference between the bare-particle dispersion and the measured dispersion, as determined from a fit of the data by a Lorentzian at constant

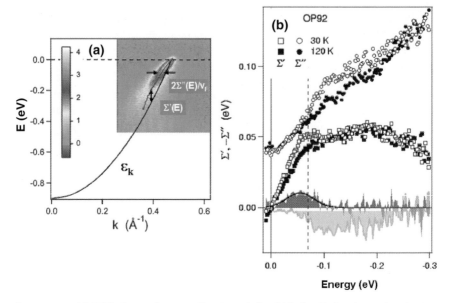

Fig. 11.8. ARPES data of optimally doped $(Bi,Pb)_2Sr_2CaCu_2O_{8+\delta}$ for $\boldsymbol{k}$ along to the nodal direction. (**a**) spectral function (in false-color scale) at $T = 130\,\mathrm{K}$ together with the bare particle dispersion $\varepsilon_{\boldsymbol{k}}$ (*black line*). The *red line* gives the dispersion derived from constant E cuts; (**b**) Real (*squares*) and negative imaginary (*circles*) part of the self-energy function at $T = 30\,\mathrm{K}$ (*open symbols*) and $T = 120\,\mathrm{K}$ (*closed symbols*). *Dark shaded area*: difference of the real part between the two temperatures. *Light shaded area*: negative difference of the imaginary part between the two temperatures

energy and taking the maximum. From the same fit the width (FWHM) of the Lorentzian, Γ_k, yields $\Sigma'' = \hbar\Gamma_k v_F/2$.

In Fig. 11.8(b) we show Σ' and Σ'' of an optimally doped BiPb2212 crystal measured in the superconducting state at $T = 30\,\mathrm{K}$ and in the normal state at $T = 130\,\mathrm{K}$ [49]. The data can be analyzed in terms of 3 different scattering channels. The first channel related to elastic scattering from the potential of the dopant atoms and possibly also from defects at the surface can explain about 20% the offset of Σ'' at zero energy. The other 80% of the offset are due to the finite momentum resolution. The second scattering channel in the normal state can be related to a coupling to a continuum of excitations extending up to about 350 meV. This leads in the normal state to a marginal Fermi-liquid behavior (see Subsect. 11.3.2): an almost linear energy dependence of the scattering rate and at low temperatures an energy dependence of Σ' close to $E \ln E$. The continuum to which the charge carriers couple has a cut-off energy for Σ' of about 350 meV. It is remarkable that this energy is close to the energy of twice the exchange integral, $J = 180$ meV. Assuming a coupling of the charge carriers to magnetic excitations [50] in a simple approximation [30] the self-energy function can be calculated by a convolution of the bare particle Greens function G_0 and the energy and momentum dependent magnetic susceptibility χ. This means $\Sigma = g^2(G_0 \otimes \chi)$ where g is a coupling constant. For a two-dimensional magnet it is expected that χ extends up to an energy of $2J$ and therefore if the self-energy is determined by magnetic excitations also Σ' should have a cutoff at that energy. This would support the interpretation of the continuum in terms of magnetic excitations. In this context one should mention recent ARPES measurements of an Fe film on W, where also a strong renormalization well above the phonon energies has been detected which was interpreted in terms of a coupling to magnetic excitations [51]. On the other hand the cutoff energy in Σ' may be also related to the finite width of the Cu-O band.

The third scattering channel exists mainly below T_c and its intensity is getting rather weak at higher temperatures. It causes a peak in Σ' near 70 meV and an edge in Σ'' at about the same energy. This leads to a pronounced change of the dispersion at the nodal point at $\sim$70 meV which was previously termed the "kink" [52]. The differences between the self-energy functions $\Delta\Sigma'$ and $\Delta\Sigma''$ when going from 30 K to 130 K are plotted in Fig. 11.8(b) by shaded areas. Both are typical of a self-energy function determined by a single bosonic mode. The energy of the mode may be either $\sim$70 meV, when the nodal point is coupled to gap-less other nodal states or $\sim$40 meV when they are coupled to states close to the antinodal point which in the superconducting state have a gap of 30 meV. A bosonic mode near 40 meV can be related to the magnetic resonance mode, first detected by inelastic neutron scattering experiments [53], a collective mode (spin exciton) which is formed inside the spin gap of 2Δ and which decays into single-particle excitations above T_c because of the closing of the gap. The mode energy $\Omega_0 = 40$ meV together with a gap energy $\Delta = 30$ meV yields a kink energy of 70 meV thus explaining the kink by a coupling of the antinodal point to the nodal point. Previous ARPES, optical, and theoretical studies [54–56] have been interpreted in terms of this magnetic resonance mode. On the other hand, theoretical work [57] has pointed out that because of kinematic constraints a coupling of the antinodal point to the nodal point via the 40 meV magnetic resonance mode should not be possible. Recently a new magnetic resonance mode (the Q^* mode) near 60 meV has been detected [58, 59] which may explain the above mentioned coupling between nodal points.

In principle the appearance of a sharper kink in the superconducting state and a *decrease* of the scattering rate in the superconducting state [60] has been also explained by the opening of a superconducting gap in the continuum [30]. On the other hand, the data shown in Fig. 11.8(b) could indicate that in the superconducting state when compared with the normal state, there is an *additional* scattering channel and not a *reduction* of the scattering rate.

In the following we discuss the doping and temperature dependence of the renormalization effects at the nodal point. In Fig. 11.9(a) we show the doping dependence of the real part of the self-energy function above T_c [47].

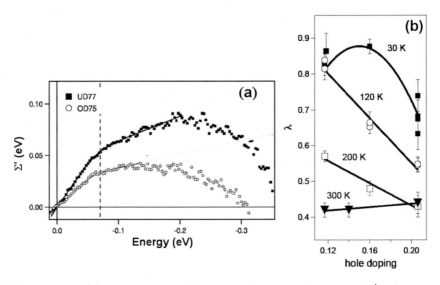

Fig. 11.9. (a) Real part of the self-energy function, Σ', for two $(Bi,Pb)_2Sr_2CaCu_2O_{8+\delta}$ samples at $T = 130\,K$ at the nodal point. UD77: underdoped with $T_c = 77\,K$, OD75: overdoped with $T_c = 75\,K$; (b) coupling constant λ at the nodal point as a function of hole concentration for various temperatures

Here the contributions from the third scattering channel, the coupling to a single bosonic mode, have almost disappeared and mainly a coupling to the continuum is observed. A rather strong doping dependence is realized. In the UD sample Σ' is much larger and extends to much higher energies compared to the OD sample. This could support the assumption that the continuum is related to magnetic excitations, which increase when approaching the Mott–Hubbard insulator. From the slope at zero energy (see Subsect. 11.3.2), λ values could be derived which are summarized in Fig. 11.9(b). The strong doping dependence of λ in the normal state questions the postulation that independent of the dopant concentration there is a universal Fermi velocity [61].

In the normal state λ decreases with increasing hole concentration and increasing temperature. This is expected in the scenario of a coupling to a continuum of overdamped spin excitations since for the susceptibility of these excitations a similar doping and temperature dependence is expected. At 300 K, λ is almost independent of the hole concentration. Possibly there the contribution from the coupling to a continuum of magnetic excitations has become smaller than the contributions from electron–hole excitations without spin reversal. The temperature dependence of the coupling constant at lower hole concentrations is consistent with the marginal Fermi-liquid model, since there at high temperatures the low-energy properties are no more determined by the energy dependence and therefore λ should decrease with increasing temperature. This is in stark contrast to the normal Fermi-liquid behavior which is observed in the OD sample (see below).

In the superconducting state there is an additional increase of λ, the concentration dependence of which is quite different from that in the normal state. This clearly indicates once more the existence of a new additional scattering channel below T_c.

The scattering rate being linear in energy for the OP doped sample at 130 K transforms *continuously* into a more quadratic one both in the normal and the superconducting state [62, 63]. This indicates that both the second and the third scattering channel decrease with increasing hole doping, which is expected in the magnetic scenario. The doping dependence shows in the normal state a transition from a marginal Fermi-liquid behavior to a more normal Fermi-liquid behavior at high hole concentrations. The quadratic increase in energy (see Subsect. 11.3.2) of Σ' is determined by the coefficient $\beta = 1.8$ (eV)$^{-1}$. This coefficient is much larger than the value 0.14 derived for electrons forming the Mo(110) surface states [64]. This indicates that even in OD HTSCs correlation effects are still important and electron–electron interactions and possibly still the coupling to spin fluctuations are strong.

The strong doping and temperature dependence of the additional (bosonic) channel is difficult to explain in terms of phonon excitations. We therefore offered for the additional third scattering channel an explanation in terms of a coupling to a magnetic neutron resonance mode, which only occurs below T_c. Finally we mention that an explanation of the extension of the renormalization to high energies in terms of a multi-bosonic excitation is very unlikely. A λ value below 1, which corresponds to a quasi-particle spectral weight Z^{-1} larger than 0.5 would not match with a coupling to polaronic multi-bosonic excitations.

11.6 The Dressing of the Charge Carriers at the Antinodal Point

Most of the ARPES studies in the past were focused on the nodal point, where narrow features in $(E, \mathbf{k})$ space have been detected, indicating the existence of quasi-particles far down in the underdoped or even slightly doped region. On the other hand the antinodal point is of particular interest concerning the superconducting properties, since in the d-wave superconductors the superconducting order parameter has a maximum at the antinodal point [12]. The region near the $(\pi, 0)$ point has been always much more difficult to investigate due to complications of the bilayer splitting, which could not be resolved by ARPES for 15 years. On the other hand, as mentioned above, only with bilayer systems of the Bi-HTSC family the entire superconducting range from the UD to the OD region can be studied. Thus due to the existence of two Fermi surfaces and two bands close to the Fermi level near $(\pi, 0)$, with a reduced resolution only a broad distribution of spectral weight could be observed, leading to the conclusion that in this (E, k) range very strong interactions appear causing a complete incoherence of the dynamics of charge carriers [65]. Moreover, in the superconducting state, very early a peak-dip-hump structure has been observed for all dopant concentrations which in analogy to the tunnelling spectra in conventional superconductors, was interpreted as a strong coupling to a bosonic excitation [65]. This picture partially changed with the advent of the improved experimental situation.

First of all it has been shown by photon-energy dependent measurements in the range $h\nu = 20$–60 eV using synchrotron radiation [66,67] that the peak-dip-hump structures strongly change as a function of the photon energy. This indicated that the matrix element in Eq. (11.5) has a different photon energy dependence for the bonding and the antibonding band at $(\pi, 0)$. This experimental observation was confirmed by calculations of the matrix element using LDA bandstructure calculations [40,41]. It turned out that the peak-dip-hump structure in the OD sample was dominated by the bilayer splitting, i.e, the peak is caused by the antibonding band and a hump is caused by the bonding band. In the UD range the complicated spectral shape could be traced back to a superposition of the bilayer effects and strong renormalization effects in the superconducting state. In this situation, only momentum-dependent measurements [68–70] along the $(\pi, \pi) - (\pi, -\pi)$ line could separate the two effects. In Fig. 11.10 a collection of our ARPES data along this direction, centered around the $(\pi, 0)$ point, is shown as a function of the dopant concentration in the superconducting state $(T = 30\,\mathrm{K})$. In the lowest row, normal state data $(T = 120\,\mathrm{K})$ are also shown for the UD sample. In the upper left corner the data for an OD sample clearly show the splitting into a bonding and an antibonding band related to four Fermi-surface crossings and two saddle points as expected from the tight-binding bandstructure calculations, shown in Fig. 11.2(a) taking into account the bilayer splitting visualized in Fig. 11.2(d). Looking in the same column at the low temperature data of the OP and UD sample the two bands are no more resolved. As mentioned before the matrix element for the excitation of the 2 bands is strongly photon energy dependent and it was shown [40, 41, 66, 67] that the spectra in the first column which were taken at $h\nu = 38$ eV represent mainly the bonding band with some contributions from the antibonding band. The data in the second column were taken with $h\nu = 50$ (or 55) eV and have almost pure antibonding character. Subtracting the second column from the first column yields almost the pure spectral weight from the bonding band. Using this procedure one clearly recognizes that even in the UD samples the bonding and the antibonding band can be well separated. In the superconducting state (first 3 rows) these data show strong changes upon reducing the dopant concentration. The bonding, and most clearly seen, the antibonding band move further and further below the Fermi level, indicating the reduction of holes. In the bonding band of the OD crystal almost no kink is observed but in the OP sample a very strong kink is realized, disclosed by the appearance of a flat dispersion between the gap energy at about $\sim$30 meV and the branching energy of $\sim$70 meV followed

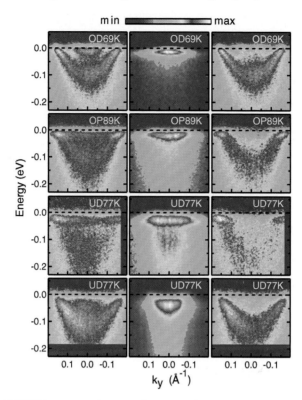

Fig. 11.10. ARPES intensity plots as a function of energy and wave vector along the $(\pi,\pi)-(\pi,-\pi)$ direction of overdoped (OD), optimally doped (OP) and underdoped (UD) $(Bi,Pb)_2Sr_2CaCu_2O_{8+\delta}$ superconductors taken at $T = 30\,K$ (upper 3 rows). Zero corresponds to the $(\pi,0)$ point. Fourth row: data for an UD sample taken at $T = 120\,K$. *Left column:* data taken with a photon energy $h\nu = 38\,eV$, at which the signal from the bonding band is maximal. *Middle column:* data taken at $h\nu = 50\,eV$ (or $55\,eV$), where the signal from the antibonding band is dominant. *Right column:* subtraction of the latter from the former yielding the spectral weight of the bonding band

by a steeper dispersion and a strong broadening. The strong renormalization effects increase even further when going from the OP doped sample to the UD sample. Remarkably, the renormalization effects (with the exception of the pseudogap) described above, completely disappear in the normal state as can be seen in the fourth row where data from an UD sample, taken at 120 K, are shown. As in the OD sample, a normal dispersion without a kink is now detected for the bonding band. Also for the antibonding band there is a transition from a flat band at low temperatures to a dispersive band above T_c. A comparison with the bare-particle band structure (not shown) indicates that there is reduction of the bandwidth by a factor of about 2 which means that there is a λ^w of about 1 in the normal state. A renormalization corresponding to a λ^w of about 1 is also detected above the branching energy near $(\pi, 0)$ in the superconducting state. This bandwidth renormalization in the range of the antinodal point is similar to that one at the nodal point and is probably also related to a coupling to a continuum of magnetic excitations.

In Fig. 11.11(a) we show an ARPES intensity distribution of the antibonding band near k_F measured with a photon energy $h\nu = 50\,\text{eV}$ along the $(1.4\pi, \pi) - (1.4\pi, -\pi)$ line for OP Pb-Bi2212 at 30 K. At this place in the second Brillouin zone the bare particle dispersion of the antibonding band reaches well below the branching energy $E_B = 70\,\text{meV}$. Therefore, contrary to the data shown in Fig. 11.10 (second row, second column), which were taken along the $(\pi, \pi) - (\pi, -\pi)$ line, the branching into two dispersive arms can be clearly realized. The data in Fig. 11.11(a) together with those shown in Fig. 11.10 (second row, third column) for the bonding band, when compared with the model calculations shown in Fig. 11.5, clearly show that the dominant renormalization effect in the superconducting state is a coupling to a bosonic mode [60, 71].

To obtain more information about the renormalization and the character of the mode, the spectral function was analyzed quantitatively [72]. Cutting the measured intensity distribution of the bonding band (see Fig. 11.10) at k_F yields the peak-dip-hump structure shown in Fig. 11.11(b). From the peak

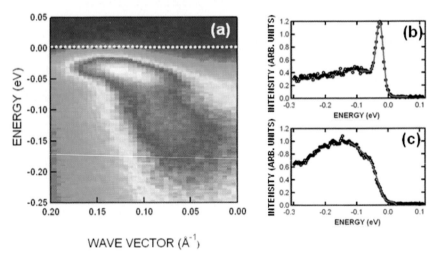

Fig. 11.11. (a) Spectral function for k-values near the $(1.4\pi, \pi) - (1.4\pi, -\pi)$ direction of the OP $(\text{Bi,Pb})_2\text{Sr}_2\text{CaCu}_2\text{O}_{8+\delta}$ superconductor taken at $T = 30\,\text{K}$. Zero corresponds to the $(1.4\pi, 0)$ point. The data were taken with a photon energy $h\nu = 50\,\text{eV}$ in order to maximize the intensity of the antibonding band; (b) constant-k cut of the spectral weight of the bonding band (see Fig. 11.10, optimally doped sample, $T = 30\,\text{K}$) at k_F; (c) cut of the data at about one third of k_F (starting from $(\pi, 0)$)

energy one can derive a superconducting gap energy of $\Delta = 30$ meV. Cutting the data at about $1/3$ of $\boldsymbol{k}_F$ (starting from $(\pi,0)$) yields the spectrum shown in Fig. 11.11(c). There the coherent peak is strongly reduced and in the framework of a one-mode model, the threshold after the coherent peak, derived from a fit of the spectrum, yields the branching energy $\Omega_0 + \Delta = 70$ meV. Another way to obtain the branching energy is to determine the threshold of Σ'' which can be obtained by fitting the spectral weight of constant energy cuts using Eq. (11.10). From this, the branching energy $\Omega_0 + \Delta = 70$ meV can be derived. From fits of constant energy cuts just below the branching energy the parameter $\Sigma''(-\infty) \sim 130$ meV can be obtained which is also a measurement of the coupling of the charge carriers to a bosonic mode (see Subsect. 11.3.2).

Important information comes from the dispersion between the gap energy and the branching energy. Originally [68] the data were fitted using Eq. (11.12) yielding λ values as a function of the dopant concentration shown in Fig. 11.12. However, as pointed out in Subsect. 11.3.3, the situation in HTSCs is quite different from conventional superconductors. In the former Ω_0 is not much larger than Δ and therefore the function $Z(E)$, from which λ is derived, is energy dependent. Furthermore, as shown in Subsect. 11.3.3 the λ values, evaluated in this way, depend on the gap energy. In the one-mode model the gap energy dependence of λ is determined by the factor $(\Omega_0 + \Delta)/\Omega_0$. Therefore it is questionable whether those λ values are a good measure of the coupling strength to a bosonic mode. More recently [72], we have fitted the dispersion of the coherent peak of an OP doped sample using the full Eq. (11.11) taking into account the above mentioned band renormalization by a factor of 2 using a $\lambda^w \sim 1$. From the derived $Z(E)$ in the superconducting state, $Z(E)$ in the normal state could be calculated by setting Δ to zero and then a total coupling constant $\lambda_n^t = 3.9$ could be obtained which is composed of a $\lambda_n^b = 2.6$ due to the coupling to the bosonic mode and a $\lambda_n^w = 1.3$ from

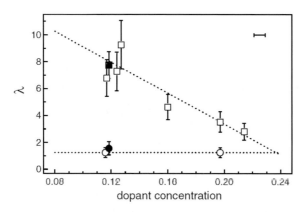

Fig. 11.12. The coupling strength parameter λ at the antinodal point as a function of doping concentration. *Squares*: superconducting state; *circles*: normal state; *open (solid) symbols*: bonding (antibonding) band

the band renormalization. It is interesting that this value is close to the value derived using Eq. (11.12). The reason for this is that the reduction due to the energy dependence of Z is partially compensated by the transformation into the normal state. A first estimate shows that the values at other dopant concentrations are also not drastically changed.

One may argue that those very large λ-values are unphysical and meaningless because, in the case of electron–phonon coupling, lattice instabilities may be expected. On the other hand also another measure of the coupling strength, the imaginary part of the self-energy function above the branching energy is very large. From $\Sigma''(-\infty) = 130\,\mathrm{meV}$ and $\Omega_0 = 40\,\mathrm{meV}$ one obtains (see Sect. 11.3.2) $\lambda = \Sigma''(-\infty)/(\pi\Omega_0/2) = 2.1$ which is not far from the above given value $\lambda_n^b = 2.6$ for the coupling to the bosonic mode derived from the dispersion.

It is interesting to compare the present values of $\lambda_n^b = 2.6$ and $\Sigma''(-\infty) = 130\,\mathrm{meV}$ derived for an OP HTSC in the superconducting state with those obtained for the electron–phonon coupling of surface electrons on a Mo(110) surface ($\lambda_n^b = 0.42$ and $\Sigma''(-\infty)_{el-ph} = 30\,\mathrm{meV}$) [23]. So both parameters are for the HTSC a factor 4–6 larger than for the Mo(110) surface. This indicates that in the HTSCs in the superconducting state there is really an anomalous strong coupling to a bosonic mode, which manifests itself both in the high coupling constant and in the high scattering rates above the branching energy. Finally it is remarkable that there is no indication of a multibosonic excitation comparable to that in the undoped cuprates [26, 27] since in that case, taking the above derived λ^b values, the intensity of the coherent state relative to the incoherent states should be strongly reduced in disagreement with the data shown in Figs. 11.10 and 11.11.

In Fig. 11.13 we show the renormalization of the antibonding band in the superconducting state for an OP sample when going from the antinodal point to the nodal point [73]. By looking at the dispersion close to E_F the

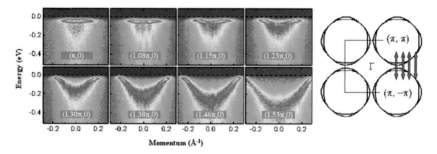

Fig. 11.13. Intensity distribution for cuts in the Brillouin zone indicated in the right-hand sketch of the optimally doped $(\mathrm{Bi,Pb})_2\mathrm{Sr}_2\mathrm{CaCu}_2\mathrm{O}_{8+\delta}$ superconductor taken at $T = 30\,\mathrm{K}$. *Upper left panel:* antinodal point. *Lower right panel:* nodal point. The data were taken with a photon energy $h\nu = 50\,\mathrm{eV}$ in order to maximize the intensity of the antibonding band

renormalization is large at the antinodal point and it is much weaker at the nodal point. This is in line with a strong coupling to a mode which is related to a high susceptibility for a wave vector (π,π) which leads to a coupling between antinodal points.

This leads to the above mentioned spin fluctuation scenario, in which below T_c, the opening of the gap causes via a feed-back process the appearance of a magnetic resonance mode, detected by inelastic neutron scattering [53]. This mode has a high spin susceptibility at the wave vector (π,π), the energy is ~ 40 meV, and as mentioned above it exists only below T_c. Thus from the measurements of the spectral function around $(\pi,0)$, in particular from the energy, the momentum, and the temperature dependence we conclude that the mode to which the charge carriers at the antinodal point so strongly couple, is the magnetic resonance mode. In a recent theoretical work [56] it was pointed out that according to magnetic susceptibility measurements using inelastic neutron scattering the magnetic resonance mode couples the antibonding band predominantly to the bonding band and vice versa. This means only the odd susceptibilities χ_{AB} and χ_{BA} are large and the even susceptibilities χ_{AA} and χ_{BB} are small. There is no coupling via the resonance mode within a band. It is remarkable that the coupling of the bonding band to the resonance mode starts in the OD region near 22% doping when the saddle point of the antibonding band just crosses the Fermi level (see Fig. 11.12). The result that in the UD region λ is similar for both bands is understandable, since the Fermi velocities and therefore the density of states and the odd susceptibilities χ_{AB} and χ_{BA} should be comparable. This scenario is supported by recent measurements of the energy dependence of the different scattering rates of the bonding and the antibonding band close to the nodal point [74]. Similar data have been recently presented for the system $YBa_2Cu_3O_7$ [75]. Furthermore, recently our group has observed large changes of the renormalization effects at the nodal and the antinodal point upon substituting 1 or 2% of the Cu ions by nonmagnetic Zn ($S=0$) or magnetic Ni ($S=1$), respectively [76]. These strong changes also strongly support the magnetic scenario since this substitution of a very small amount of the Cu ions should not change the coupling of the charge carriers to phonons. On the other hand we do not want to conceal that there are also interpretations of the above discussed bosonic mode in terms of phonon excitations [77].

At the end of this Section we would like to mention some ARPES results on the spectral function in the pseudogap region [78]. The pseudogap is one of the most remarkable properties of HTSCs in the UD region above T_c. In Fig. 11.14 we compare the dispersion along the $(\pi,0)-(\pi,-\pi)$ direction close to the antinodal point of an UD sample in the superconducting and in the pseudogap state. In the superconducting state one realizes the characteristic BCS-Bogoliubov-like back-dispersion at k_F. In the pseudogap state no more a bending back of the dispersion is observed. Instead the spectral weight fades when the binding energy approaches the gap energy. This is in line with the observed disappearance of the coherence peak in tunnelling spectra of HTSCs

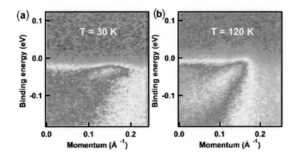

Fig. 11.14. Intensity distribution near the antinodal point along the $(\pi,0)-(\pi,-\pi)$ direction of an underdoped $(Bi,Pb)_2Sr_2CaCu_2O_{8+\delta}$ crystal with $T_c = 77$ K. The photon energy has been chosen to be 38 eV in order to suppress the antibonding band. (**a**) superconducting state; (**b**) pseudogap state

in the pseudogap phase [79]. The experimental observation of this behavior was explained in terms of phase fluctuations of the superconducting order parameter. The BCS wave function is a coherent superposition of wave functions with different number of electrons, N, and usually also a phase factor $e^{i\phi}$ is introduced. This leads to a general uncertainty relation for the particle number and the phase: $\Delta N \Delta \phi \geq 1$ [80]. In the superconducting state at $T = 0$ where there is a large phase stiffness, i.e., $\Delta \phi = 0$, the particle number is completely uncertain leading to a large particle-hole mixing and thus to a large back-dispersion. With increasing temperature, the phases get completely uncorrelated and one obtains $\Delta N = 0$. Then the back-dispersion must disappear. In this way a crossover from a BCS-like phase-ordered bandstructure to a completely new phase-disordered pseudogap bandstructure is obtained.

11.7 Conclusions

In this chapter we presented part of our ARPES results on HTSCs. They were obtained with an energy and a momentum resolution of 8-25 meV and 0.015 Å^{-1}. This is at present the state-of-the-art ARPES when reasonable intensities are used during the measurements. Using this resolution, a lot of information on the dressing of the charge carriers in HTSCs at different k-points in the Brillouin zone has been obtained during the last 10 years. It will be really one of the big challenges of experimental solid state physics to enter in the range of sub-1meV-resolution in *angle-resolved* photoemission spectroscopy. In the next 10 years it is predictable that there will be a further improvement of the energy resolution by one order of magnitude for angle-resolved measurements. It can be anticipated that further interesting results on the dressing and possibly also on the pairing mechanism in HTSCs can be realized.

Acknowledgements

We acknowledge financial support by the DFG Forschergruppe under Grant No. FOR 538. One of the authors (J.F.) appreciates the hospitality during his stay at the Ames Lab and thanks for the critical reading of the manuscript by D. Lynch. We thank R. Follath, T.K. Kim, S. Legner, and K.A. Nenkov for fruitful collaboration. In particular we thank Mark Golden for his contributions in the early stage of the project. Finally we acknowledge helpful discussions and collaboration with colleagues from theory: A.V. Chubukov, T. Eckl, M. Eschrig, and W. Hanke.

References

1. A. Einstein: Ann. d. Phys. **31**, 132 (1905)
2. H. Hertz: Ann. d. Phys. **17**, 983 (1887)
3. G. W. Gobeli et al.: Phys. Rev. Lett. **12**, 94 (1964)
4. G. Bednorz and K. A. Müller: Z. Phys. B **64**, 189 (1986)
5. J. Zaanen et al.: Phys. Rev. Lett. **64**, 189 (1986)
6. D. W. Lynch and C. G. Olson: *Photoemission Studies of High-Temperature Superconductors* (Cambridge University Press 1999), pp 1–432
7. A. Damascelli et al.: Rev. Mod. Phys. **75**, 473 (2003)
8. J. C. Campuzano et al.: Photoemission in the High-T_c Superconductors. In: *Physics of Superconductors*, vol II, ed by K. H. Bennemann and J. B. Ketterson (Springer, Berlin Heidelberg New York 2004), pp 167–273
9. T. Timusk: Rep. Progr. Phys. **62**, 61 (1999) **75**, 473 (2003)
10. C. M. Varma: Phys. Rev. B **55**, 14554 (1997)
11. O. K. Andersen et al.: J. Phys. Chem. Solids **56**, 1573 (1995)
12. H. Ding et al.: Phys. Rev. B **54**, R9678 (1996)
13. S. V. Borisenko et al.: Phys. Rev. B **66**, 140509(R) (2002)
14. A. A. Kordyuk et al.: Phys. Rev. B **70**, 214525 (2004)
15. N. Nücker et al.: Phys. Rev. B **37**, 6827 (1988)
16. S. Hüfner: *Photoelectron Spectroscopy* (Springer, Berlin Heidelberg New York 1996) and references therein.
17. G. D. Mahan: *Many-Particle Physics* (Plenum Press, New York 1990)
18. L. Hedin and S. Lundqvist: Solid State Physics **23**, 1963 (1969)
19. C. O. Almbladh and L. Hedin: Beyond the one-electron model in *Handbook of Synchrotron Radiation*, vol 1b, ed by E. E. Koch (North Holland, Amsterdam 1983), pp 607–904
20. C. M. Varma et al.: Phys. Rev. Lett. **63**, 1996 (1989)
21. D. Pines and P. Nozières: *The Theory of Quantum Liquids*,vol 1, (W. A. Benjamin, New York 1966), p. 64
22. M. Imada et al.: Rev. Mod. Phys. **70**, 1039 (1998)
23. T. Valla et al.: Science **285**, 2110 (1999)
24. C. Hodges et al.: Phys. Rev. B **4**, 302 (1971)
25. S. Engelsberg and J. R. Schrieffer: Phys. Rev. **131**, 993 (1963)
26. K. M. Shen et al.: Phys. Rev. Lett. **93**, 267002 (2004)
27. O. Rösch and O. Gunnarsson: Phys. Rev. Lett. **92**, 146403 (2004)

28. Y. Nambu: Phys. Rev. **117**, 648 (1960)
29. D. J. Scalapino, in *Superconductivity*, vol. 1, ed. R. D. Parks (Marcel Decker, New York 1969), p. 449
30. A. V. Chubukov and M. R. Norman: Phys. Rev. B **70**, 174505 (2004)
31. N. Martensson et al.: Journal of Electron Spectroscopy and Related Phenomena **70**, 117 (1994)
32. *Handbook on Synchrotron Radiation*, vol 1-4, ed by E. E. Koch et al (North Holland, Amsterdam 1983–1991)
33. R. Follath: Nucl. Instrum. Meth, Phys. Res. A **467–468**, 418 (2001)
34. S. V. Borisenko et al.: Phys. Rev. Lett. **84**, 4453 (2000)
35. S. Legner et al.: Phys. Rev. B **62**, 154 (2000)
36. A. A. Kordyuk et al.: Phys. Rev. B **66**, 014502 (2002)
37. J. L. Tallon et al.: Phys. Rev. B **51**, 12911 (1995)
38. D. L. Feng et al.: Phys. Rev. Lett. **86**, 5550 (2001)
39. Y.-D. Chuang et al.: Phys. Rev. Lett. **87**, 117002 (2001)
40. A. Bansil and M. Lindroos, Phys. Rev. Lett. **83**, 5154 (1999)
41. J. D. Lee and A. Fujimori, Phys. Rev. Lett. **87**, 167008 (2001)
42. P. Aebi et al.: Phys. Rev. Lett. **72**, 2757 (1994)
43. A. Koitzsch et al.: Phys. Rev. B **69**, 220505(R) (2004)
44. A. A. Kordyuk et al.: Phys. Rev. B **67**, 064504 (2003)
45. N. Nücker et al.: Phys. Rev. B **44**, 7155 (1991)
46. V. G. Grigorian et al.: Phys. Rev. B **60**, 1340 (1999)
47. A. A. Kordyuk et al.: Phys. Rev. B **71**, 214513 (2005)
48. L. Pintschovius and W. Reichardt: Neutron Scattering in Layered Copper-Oxide Superconductors. In: *Physics and Chenistry of Materials with Low Dimensional Structures*, vol 20, ed. by A. Furrer (Kluwer Academic, Dordrecht, 1998), p. 165
49. A. A. Kordyuk et al.: Phys. Rev. Lett. **97**, 017002 (2006)
50. A. Abanov et al.: Adv. Phys. **52**, 119 (2003).
51. J. Schäfer et al.: Phys. Rev. Lett. **02**, 097205 (2005)
52. P. V. Bogdanov et al.: Phys. Rev. Lett. **85**, 2581 (2000)
53. J. Rossat-Mignod et al.: Physica C **185-189**, 86 (1991); H. A. Mook et al.: Phys. Rev. Lett. **70**, 3490 (1993); H. F. Fong et al.: *ibid.* **75**, 316 (1995)
54. P. D. Johnson et al.: Phys. Rev. Lett. **87**, 177007 (2001)
55. J. Hwang et al.: cond-mat/0505302
56. M. Eschrig and M. R. Norman: Phys. Rev. Lett. **85**, 3261 (2000), Phys. Rev. Lett. **89**, 277005 (2002), Phys. Rev. B **67**, 144503 (2003)
57. A. Abanov et al.: Phys. Rev. Lett. **89**, 177002 (2002)
58. S. Pailhes et al.: Phys. Rev. Lett. **93**, 167001 (2004)
59. I. Eremin et al.: cond-mat/0409599
60. A. Kaminski et al.: Phys. Rev. Lett. **86**, 1070 (2001)
61. X. J. Zhou et al.: Nature **423**, 398 (2003)
62. A. Koitzsch et al.: Phys. Rev. B **69**, 140507(R) (2004)
63. A. A. Kordyuk et al.: Phys. Rev. Lett **92**, 257006 (2004)
64. T. Valla, A. V. Feferov et al.: Phys. Rev. Lett. **83**, 2085 (1999)
65. Z.-X. Shen and J. R. Schrieffer: Phys. Rev. Lett. **78**, 1771 (1997)
66. S. V. Borisenko et al.: Phys. Rev. Lett. **90**, 207001 (2003)
67. A. A. Kordyuk et al.: Phys. Rev. Lett. **89**, 077003 (2002)
68. T. K. Kim et al.: Phys. Rev. Lett. **91**, 167002 (2003)
69. A. D. Gromko et al.: Phys. Rev. B **68**, 174520 (2003)

70. T. Sato et al.: Phys. Rev. Lett. **91**, 157003 (2003)
71. M. R. Norman, H. Ding: Phys. Rev. B **57**, 11111 (1998)
72. J. Fink et al.: Phys. Rev. B **74**, 165102 (2006)
73. T. K. Kim: The role of inter-plane interactions in the electronic structure of high-T_c cuprates. PhD Thesis, University of Technology, Dresden (2003)
74. S. V. Borisenko et al.: Phys. Rev. Lett. **96**, 067001 (2006)
75. S. V. Borisenko et al.: Phys. Rev. Lett. **96**, 117004 (2006)
76. V. Zabolotnyy et al.: Phys. Rev. Lett. **96**, 037003 (2006)
77. T. P. Devereaux et al.: Phys. Rev. Lett. **93**, 117004 (2004)
78. T. Eckl et al.: Phys. Rev. B **70**, 094522 (2004)
79. M. Kugler et al.: Phys. Rev. Lett. **86**, 4911 (2001)
80. M. Tinkham: *Introduction to Superconductivity* (McGraw-Hill, New York, 1996)

12

High-Resolution Photoemission Spectroscopy of Perovskite-Type Transition-Metal Oxides

H. Wadati, T. Yoshida, and A. Fujimori

Department of Physics and Department of Complexity Science and Engineering, University of Tokyo, Kashiwa, Chiba 277-8561, Japan

Abstract. Transition-metal (TM) oxides have been attracting great interest because of their remarkable physical properties such as metal-insulator transition, colossal magnetoresistance, and high-temperature superconductivity. These properties are considered to originate from strong electron correlation. Nowadays a lot of progress has been made in the photoemission studies of TM oxides with three-dimensional perovskite structures owing to the availability of high-quality single crystals of bulk and epitaxial thin films. Even angle-resolved photoemission has become possible through the progress in synthesizing high-quality samples. Here, we give an overview of high-resolution photoemission studies on perovskite-type Ti, V, Mn, Fe, Ni and Ru oxides, and show how high-resolution photoemission has contributed to the understanding of the electronic structures of these materials.

12.1 Introduction

Most of remarkable electronic properties of solids originate from electron correlation, namely, effects beyond the description within conventional band theory and associated one-electron pictures. Among the so-called strongly correlated electron systems, transition-metal (TM) oxides exhibit richest physical properties such as superconductivity, metal-insulator transition (MIT), colossal magnetoresistance (CMR), and ordering of spin, charge, and orbitals [1]. Although these properties are not unique in TM oxides, they appear at relatively high temperatures in TM oxides, making the materials attractive not only for basic science but also for practical applications. Another attractive aspect of the TM oxides is that material parameters such as band filling and bandwidth can be controlled in a systematic way within a single type of crystal structure such as the perovskite (Fig. 12.1), pyrochlore and spinel types. Among them, the perovskite-type structure including the layered perovskite is the most versatile. In fact, high-temperature superconductivity, MIT, CMR and spin-charge-orbital ordering occur in this class of materials in different regions or even in adjacent regions of materials parameter space and compete with each other.

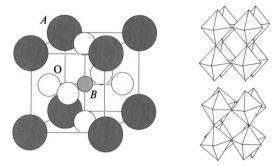

Fig. 12.1. Crystal structure of the perovskite-type oxide ABO_3. *Left panel:* Unit cell of the cubic perovskite. *Right panel:* Arrangement of the octahedral BO_6 building blocks in the cubic and orthorhombically distorted perovskite structures

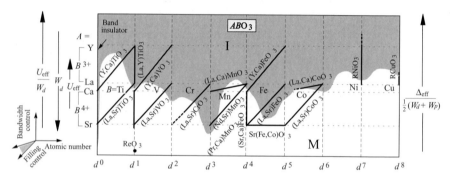

Fig. 12.2. A schematic metal-insulator phase diagram for the filling control and bandwidth control of 3d transition-metal oxides with perovskite structure [1]. The orthorhombic distortion reduces the d-band width W_d and hence increases U_{eff}/W_d or Δ_{eff}/W_d. M: metal, I: insulator

Figure 12.2 shows the electronic phase diagram of the ABO_3 compounds, where B is a TM element with 3d electrons. As the ionic radius of the A-site ion decreases, the orthorhombic GdFeO$_3$-type distortion (Fig. 12.1) increases and the d-band width W_d decreases, thereby increasing the ratio of the effective on-site Coulomb repulsion energy U_{eff} to W_d and driving the system from the metallic towards the insulating side.

12.2 Electronic Structure

Early photoemission studies combined with analyses using the BO_6 cluster model have revealed that the global electronic structure of TM oxides is characterized by three parameters: the on-site Coulomb interaction energy between two TM 3d electrons denoted by U_{eff} (where the effect of the multiplet splitting of local d^n configurations has been included); the charge-transfer energy from the occupied O $2p$ to the empty TM $3d$ orbitals denoted by Δ_{eff}; the p-d hybridization strength denoted by $(pd\sigma)$ [2]. In the case of a Mott insulator, $\Delta_{\text{eff}} > U_{\text{eff}}$, the band gap is of the d-d type, and the compound is called a Mott–Hubbard-type insulator. If $\Delta_{\text{eff}} < U_{\text{eff}}$, the band gap is of the p-d type, and the compound is called a charge-transfer-type insulator. In going from lighter to heavier $3d$ TM elements, the d level is lowered and therefore Δ_{eff} is decreased whereas U_{eff} is increased. Thus Ti and V oxides are usually of the Mott–Hubbard-type while Mn, Fe, Co, Ni, and Cu oxides are of the charge-transfer-type [3,4].

The periodic lattices of the TM oxides lead to the momentum dependence of electronic states, which cannot be treated within the local cluster model. In particular, the momentum dependence of single-particle electronic states, which corresponds to the band structure of the crystal, is important to understand many of the physical properties. It can be studied by angle-resolved photoemission spectroscopy (ARPES). However, it has been well known that, because of the strong Coulomb interaction, conventional band theories such as band-structure calculations using the local-density approximation (LDA) fail in TM oxides, particularly in the Mott insulators. A band-theoretical approach to overcome this problem is the LDA + U method, in which unphysical self-interaction is avoided by introducing the Hartree–Fock-type electron–electron interaction term [6]. In metallic TM oxides, too, the "band structure" is highly non-trivial because of electron correlation: Sharp peaks in ARPES spectra called quasi-particle peaks are generally shifted and broadened compared to the prediction of band-structure calculations, and part of its intensity is distributed as a broad incoherent part. Theoretically, such a behavior has been most clearly demonstrated by dynamical mean-field theory (DMFT) shown in Fig. 12.3 [5], where the density of states (DOS) corresponding to angle-integrated photoemission spectra are displayed.

12.3 Samples

High-quality samples are of course essential part in any photoemission studies of TM oxides. While polycrystalline samples can be used for angle-integrated photoemission spectroscopy, single crystals with atomically flat surfaces are necessary for ARPES. In some cases, single crystals of perovskite-type TM oxides can be cleaved along the [001] surface of the cubic or pseudocubic structure. In particular, when the A-site atom is an alkaline earth element, both the AO and BO_2 planes exposed to the surface by the cleavage are charge neutral, favoring the cleavage along the [001] surface.

Another way of performing ARPES studies of such materials is the use of single-crystal thin films. Recently, high-quality perovskite-type oxide single-crystal thin films grown by the pulsed laser deposition (PLD) method have become available [7,8], and setups have been developed for their in-situ photoemission measurements [9,10]. By using well-defined surfaces of epitaxial

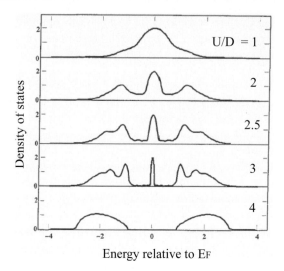

Fig. 12.3. Density of states for the single-band Hubbard model calculated by dynamical mean-field theory [5]. The central peak is due to quasi-particles (coherent part) and the broad peaks at energies $\sim \pm 2$ are due to the incoherent part. The bottom spectrum is that of an insulator. U: on-site Coulomb energy; D: energy unit proportional to the bandwidth

thin films, the band structures of $La_{1-x}Sr_xMnO_3$ and $La_{1-x}Sr_xFeO_3$ have been studied and it has been demonstrated that in situ ARPES measurements on such TM oxide films are one of the best methods to investigate the band structure of TM oxides with three-dimensional crystal structures [11,12].

Horiba et al. [9, 13] performed an in-situ photoemission study on atomically flat high-quality surfaces of $La_{1-x}Sr_xMnO_3$ thin films grown by laser molecular beam epitaxy (laser MBE). Figure 12.4 shows the resonant photoemission spectrum of $La_{0.6}Sr_{0.4}MnO_3$ thin films. The existence of a Fermi edge

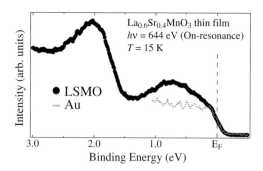

Fig. 12.4. Resonant photoemission spectrum of a $La_{0.6}Sr_{0.4}MnO_3$ thin film [13]; also shown is the spectrum of an Au calibration sample

is clearly seen by comparison with the spectrum of gold, suggesting the importance of in situ photoemission measurements on an atomically flat surface for revealing intrinsic electronic structures.

12.4 Case Studies

12.4.1 $La_{1-x}Sr_xTiO_3$

Light TM oxides such as perovskite-type Ti and V oxides are prototypical Mott–Hubbard-type systems. Those systems have a relatively small number of electrons in the degenerate t_{2g} bands, making the systems normal Fermi liquids on the metallic side and Mott insulators on the insulating side of the MIT. DMFT predicted that the effective mass enhancement occur concomitant with decreasing spectral weight of the coherent part in the vicinity of MIT in Mott–Hubbard systems [5]. Fujimori et al. [14] showed that the d bands of several perovskite-type oxides are split into two characteristic structures, namely, the coherent part around the Fermi level (E_F) corresponding to band-like electronic excitations and the incoherent part 1–2 eV away from E_F corresponding to atomic-like excitations or the remnant of the lower Hubbard band, consistent with the DMFT calculation [5].

The filling-controlled Mott–Hubbard system $La_{1-x}Sr_xTiO_{3+y/2}$ (doped hole concentration: $\delta = x+y$) shows a critical mass enhancement toward MIT (at $\delta \sim 0.06$) of the electronic specific heats γ and the magnetic susceptibility χ_s [15]. Yoshida et al. [16] studied the doping dependence of the electronic structure near E_F using high-resolution photoemission spectroscopy as shown in Fig. 12.5. Within the Pauli-paramagnetic metallic phase ($0.08 \leq \delta \leq 0.65$), both the coherent and incoherent parts became stronger with decreasing δ, corresponding to the increase of the d-band filling [Fig. 12.5(a)]. They demonstrated that the mass-enhancement factor deduced from the bandwidth

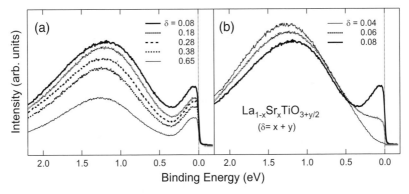

Fig. 12.5. Photoemission spectra of $La_{1-x}Sr_xTiO_{3+y/2}$ in the Ti $3d$ band region in the high doping regime (**a**) and the low doping regime (**b**) [16]

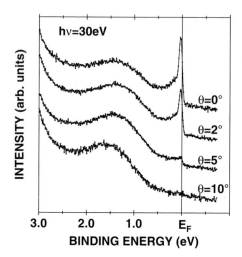

Fig. 12.6. Off-normal-emission spectra of states within the bulk band gap of $SrTiO_3$ as a function of the emission angle from the surface normal (θ) [17]

showed quantitatively the same doping dependence as the specific heat coefficient γ, at least for $0.3 < \delta$. Near MIT ($\delta < 0.2$), γ and χ_s are further enhanced although the band narrowing saturated. This may be explained by an additional band narrowing only in the vicinity of E_F. As the system entered the antiferromagnetic phases ($\delta < 0.08$), the coherent part became weaker and finally vanished in the antiferromagnetic insulating sample ($\delta = 0.04$) [Fig. 12.5(b)].

Lightly electron-doped $SrTiO_3$ was studied in an ARPES measurement [17] of as-grown $SrTiO_3$. Figure 12.6 shows a series of off-normal spectra in the bulk band gap of the as-grown $SrTiO_3$. It is seen that the intensity of the sharp metallic-like state just below E_F is dramatically reduced with increasing emission angle θ from the surface normal. This is consistent with a small electron pocket around the Γ point expected from band theory for the light doping.

12.4.2 $Ca_{1-x}Sr_xVO_3$

In the early photoemission study of a series of bandwidth-controlled Mott–Hubbard systems (ReO_3, $LaTiO_3$, $YTiO_3$, $SrVO_3$, $CaVO_3$, VO_2 [14]), spectral weight was found to be transferred from the upper and lower Hubbard bands to the region near E_F with decreasing U/W (U : d-d Coulomb repulsion energy; W: one-electron d-band width), consistent with the prediction of DMFT (see Fig. 12.3). The interpretation of these spectra, particularly, of the photoemission spectra of $Ca_{1-x}Sr_xVO_3$, however, provoked a lot of discussion and still remains controversial. The early photoemission results with low photon energy showed that, with decreasing x, i.e., with decreasing bandwidth, spectral weight is transferred from the coherent part to the incoherent part [18] in a more dramatic way than the enhancement of γ [19]. Based on a "bulk-sensitive" photoemission study using soft x-rays, Sekiyama et al. [20] claimed that there is no appreciable spectral weight transfer between SrVO$_3$ and CaVO$_3$. The spectra measured at $h\nu = 900$ eV in Fig. 12.7(a)–(c) show small spectral variation with x compared to the earlier results, whereas the coherent spectral weight decreases dramatically with decreasing $h\nu$ for all compounds. This can be interpreted as due to the higher bulk sensitivity of photoemission spectroscopy using higher photon energy in the soft-x-ray region.

According to the universal curve of the photoelectron escape depth [21], photoelectrons excited by low energy photons ($h\nu < 10$ eV) also shows high bulk sensitivity. Interestingly, a recent "bulk-sensitive" photoemission study using a vacuum ultraviolet laser with $h\nu = 7$ eV has shown a suppression of spectral weight near E_F in going from SrVO$_3$ to CaVO$_3$ [22]. Also a systematic

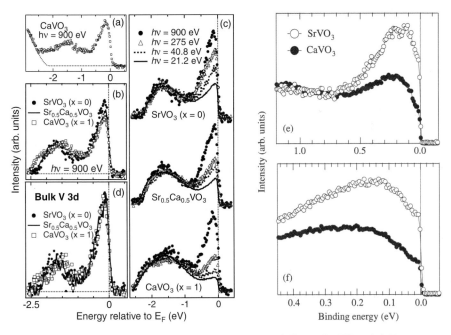

Fig. 12.7. Bulk-sensitive photoemission spectra of Ca$_{1-x}$Sr$_x$VO$_3$. (a) Raw spectrum near E_F of CaVO$_3$ and the fitted tail of the O $2p$ contributions; (b) V $3d$ spectra of Ca$_{1-x}$Sr$_x$VO$_3$ at $h\nu = 900$ eV obtained by subtracting the fitted tails of the O $2p$ contributions from the raw spectra as in (a); (c) $h\nu$ dependence of the V $3d$ spectra normalized to the incoherent spectral weight ranging from -0.8 to -2.6 eV; (d) Bulk V $3d$ spectra of Ca$_{1-x}$Sr$_x$VO$_3$ [20]; (e) photoemission spectra of SrVO$_3$ and CaVO$_3$ measured using a laser source ($h\nu = 6.994$ eV) at 6 K; (f) Expanded view of the coherent part near E_F [22]

suppression of the DOS within ~0.2 eV of E_F was found with decreasing photon energy, i.e., with increasing bulk sensitivity. This suppression of the spectral weight has been interpreted as a superposition of surface and bulk-derived features, suggesting a pseudo-gap feature in the bulk spectra, although such pseudo-gap interpretation seems to be difficult to reconcile with the Fermi-liquid behavior of these compounds.

It should be noted that signature of charge disproportionation into V^{3+} and V^{5+} was also reported in the core-level spectra of $Ca_{1-x}Sr_xVO_3$ [23]. This suggests an electronic phase separation at the surface, leading to a different surface electronic structure from the bulk. Also, in the near-E_F spectra of $Ca_{1-x}Sr_xVO_3$ taken with synchrotron radiation, polarization dependence due to the photoemission matrix elements was found [23]. The discrepancies between the different experiments mentioned above may stem from the difficulty in disentangling the surface *versus* bulk signals and the matrix element effects in the photoemission spectra.

Measurement of band dispersions and Fermi surface by ARPES is a more direct method to reveal the surface and bulk electronic structures separately. Recently, band dispersions and Fermi surfaces of $SrVO_3$ were directly observed by ARPES [24]. The observed spectral-weight distribution near E_F showed cylindrical Fermi surfaces as predicted by LDA band-structure calculations. As shown in Fig. 12.8(b), the enhanced effective electron mass has been obtained from the energy band near E_F. The obtained mass-enhancement factor $m^*/m_b \sim 2$ was consistent with the bulk thermodynamic properties such as γ [19], in accordance with the normal Fermi-liquid behavior of $SrVO_3$. From comparison of the observed spectra with calculated bulk and surface band structures using a Green's function method, it was concluded that the

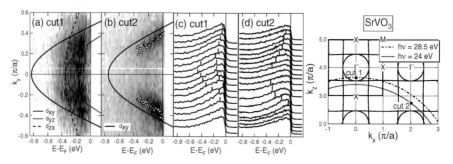

Fig. 12.8. ARPES spectra of $SrVO_3$. Panels (**a**) and (**b**) show intensity plots in E-k space, respectively, for cut 1 and cut 2 illustrated in right panel. LDA energy bands are also superimposed [25]. White dots in (**b**) are peak positions of momentum distribution curves and represents the band dispersion. Panels (**c**) and (**d**) are energy distribution curves corresponding to panels (**a**) and (**b**), respectively. Vertical bars are guides to the eye indicating the positions of the dispersive features [24]

obtained t_{2g} band dispersion reflects the band dispersion of the bulk electronic structure [24].

12.4.3 $La_{1-x}Sr_xMnO_3$ and Related Mn Oxides

Hole-doped perovskite manganese oxides $La_{1-x}Sr_xMnO_3$ have attracted much interest because of their remarkable physical properties, such as CMR and composition- and temperature-dependent MIT. One end member of $La_{1-x}Sr_x$-MnO_3, $LaMnO_3$, is an antiferromagnetic insulator. Hole-doping through the substitution of Sr for La produces a ferromagnetic metallic phase ($0.17 < x < 0.5$) [26].

There have been a lot of reports of photoemission studies on $La_{1-x}Sr_x$-MnO_3. However, the structure near E_F was low in intensity and was not clearly observed in earlier photoemission studies [27,28]. Sarma et al. [29] succeeded in observing the Fermi edge and found that the intensity dramatically decreases at high temperatures as shown in Fig. 12.9. Similar observation was reported for $La_{1-x}Ca_xMnO_3$ [30].

There have been ARPES studies on layered Mn oxides such as La_{2-2x}-$Sr_{1+2x}Mn_2O_7$, and band dispersions and Fermi surfaces have been investigated

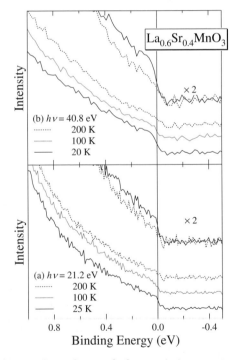

Fig. 12.9. Temperature dependence of photoemission spectra of $La_{0.6}Sr_{0.4}MnO_3$ near E_F [29]

[31–33]. On the contrary, there have been no ARPES studies on $La_{1-x}Sr_xMnO_3$ because bulk samples of $La_{1-x}Sr_xMnO_3$ do not have a cleavage plane. However, by using well-defined surfaces of in situ prepared epitaxial thin films, ARPES study of $La_{1-x}Sr_xMnO_3$ has become possible [10, 11]. Figure 12.10 shows the results of ARPES studies of $La_{1-x}Sr_xMnO_3$ thin films. Figure 12.10(a) shows the ARPES intensity map near E_F of a $La_{0.66}Sr_{0.34}MnO_3$ thin film, where the lower panel is the second-derivative plot [10]. This clearly shows the existence of a band dispersing toward E_F, although it is not clear whether this band crosses E_F or not because the intensity of the band rapidly decreases toward E_F. Figure 12.10(b) shows the band structure near E_F of a $La_{0.6}Sr_{0.4}MnO_3$ thin film [11]. Band A has small but finite spectral weight at E_F, suggesting an E_F crossing of this band. The small spectral weight at E_F and the unusually broad ARPES spectral features are considered to be caused by strong electron–phonon interactions in this system. The lower panel of Fig. 12.10(b) shows comparison between the ARPES results and LDA $+U$ band-structure calculation. The conduction-band minimum in ARPES results is $\sim 0.5\,eV$, while that in calculation is $\sim 1.3\,eV$. From this, one can estimate the effective mass-enhancement factor of $m^*/m_b \sim 2.6$. This value is in good agreement with 2.8 deduced from the electronic specific heat coefficients [34], indicating that the quantitative discrepancy in the conduction-band width be-

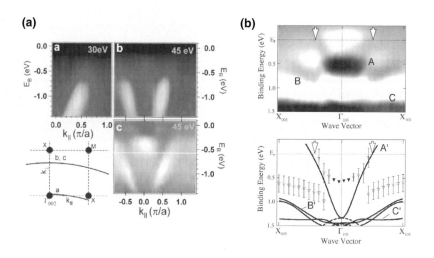

Fig. 12.10. ARPES studies of $La_{1-x}Sr_xMnO_3$ thin films. (a) ARPES intensity map near E_F of a $La_{0.66}Sr_{0.34}MnO_3$ thin film [10]. The lower panel is a second-derivative plot; (b) band structure near E_F of a $La_{0.6}Sr_{0.4}MnO_3$ thin film around the $\Gamma(103)$ point [11]. The lower panel shows comparison between ARPES results and LDA $+U$ band-structure calculation

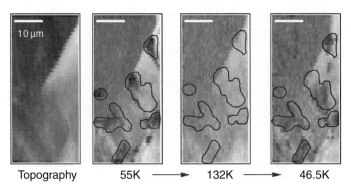

Fig. 12.11. Spectromicroscopic images over $54 \times 25\,\mu m^2$ areas of the surface of $La_{1/4}Pr_{3/8}Ca_{3/8}MnO_3$ for a temperature cycling [35]. *Gray* scale indicates DOS at E_F. *Black* area indicates insulating region while white area indicates metallic region

tween experiment and calculation comes from the renormalization effect due to strong electron correlation and possibly electron–phonon interaction.

It has long been a puzzle whether those manganese oxides show a phase separation and, if so, what the size of the domains is. Sarma et al. [35] performed a spatially-resolved photoemission study on single crystal samples of $La_{1/4}Pr_{3/8}Ca_{3/8}MnO_3$.

Figure 12.11 shows spectromicroscopic images over $54 \times 25\,\mu m^2$ areas of the surface of $La_{1/4}Pr_{3/8}Ca_{3/8}MnO_3$. They confirmed the absence of chemical inhomogeneity by measuring shallow core-level spectra. In Fig. 12.11, one can clearly see the formation of distinct insulating domains embedded in the metallic host at low temperatures. The size of the observed insulating islands was large, on the order of several microns. Surprisingly, these domains exhibited memory effects on temperature cycling. They suggested that such domain formation was initiated by some pinning centers which appeared to be correlated with long-range strains in the specimen.

12.4.4 $La_{1-x}Sr_xFeO_3$

Hole-doped perovskite iron oxides $La_{1-x}Sr_xFeO_3$ have attracted much interest because it undergoes a pronounced charge disproportionation (CD) and an associated MIT around $x \sim 2/3$ [36]. One end member of $La_{1-x}Sr_xFeO_3$, $LaFeO_3$, is an antiferromagnetic insulator, and the insulating phase is unusually wide in the phase diagram especially at low temperatures $0 < x < 0.7$, and even at room temperature $0 < x < 0.5$ [37].

A systematic x-ray photoemission study of bulk polycrystalline $La_{1-x}Sr_x$-FeO_3 samples has been reported by Chainani et al. [38]. Structures in the valence band, however, were not clearly resolved partly because of the limited energy resolution of ~ 0.8 eV. Wadati et al. [39] performed an in-situ photoemission study of $La_{1-x}Sr_xFeO_3$ thin films with well-defined surfaces grown

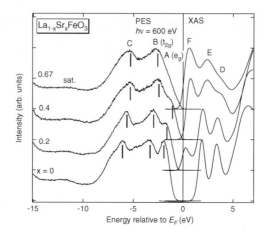

Fig. 12.12. Combined photoemission and O 1s XAS spectra of $La_{1-x}Sr_xFeO_3$ thin films [39]

by laser MBE. Figure 12.12 shows the doping dependence of the combined valence-band photoemission and the O 1s x-ray absorption (XAS) spectra. The structures in the valence band were much more clearly observed than in the scraped bulk samples [38]. From Fig. 12.12, the effect of hole doping could be described in the framework of the rigid-band model as far as the shifts of the spectral features are concerned, whereas structure A (e_g band) showed highly nonrigid-bandlike behavior with transfer of spectral weight from structure A to structure F (hole-induced states) above E_F across the gap at E_F.

It is an interesting question how the electronic structure of $La_{1-x}Sr_xFeO_3$ changes as a function of temperature when CD occurs. Figure 12.13 shows the temperature dependence of the valence-band photoemission spectra of $La_{0.33}Sr_{0.67}FeO_3$. Matsuno et al. [37] observed temperature dependent changes near E_F in the photoemission spectra of $La_{1-x}Sr_xFeO_3$ polycrystals with $x = 0.67$ and found that the intensity at E_F clearly changed across the transition temperature as shown in Fig. 12.13(b). They also reported smaller but finite changes for $x = 0.55$ and 0.80 and suggested that a local CD may occur even away from $x = 0.67$. Wadati et al. [40] performed Fe $2p \rightarrow 3d$ resonant photoemission on in-situ prepared epitaxial thin films, and observed the change of the electronic structure with enhanced Fe 3d contribution. Figure 12.13(b) shows the temperature dependence of the valence-band photoemission spectra of $La_{0.33}Sr_{0.67}FeO_3$ taken at $h\nu = 710$ eV (on Fe $2p \rightarrow 3d$ resonance). The spectra clearly indicate transfer of spectral weight from the vicinity of E_F to structure A, that is, within the e_g band region with decreasing temperature. The energy range in which spectral weight transfer occurs is two orders of magnitude larger than the transition temperature $T_{CD} = 190$ K ($\sim$22 meV). The spectral change was gradual, indicating that CD does not occur abruptly at T_{CD} but occurs gradually from well above T_{CD}, namely,

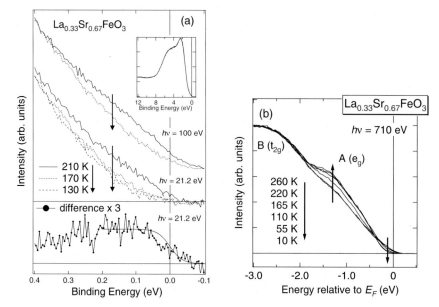

Fig. 12.13. Temperature dependence of the valence-band photoemission spectra of $La_{0.33}Sr_{0.67}FeO_3$. (a) Bulk polycrystals: $h\nu = 100\,eV$ and $21.2\,eV$ [37]; (b) single-crystal thin films: $h\nu = 710\,eV$ [40]

that a local CD occurs well above T_{CD} and continues to develop below T_{CD}, too. The authors also found a similar temperature dependence for $x = 0.4$, and to a lesser extent for $x = 0.2$. The results suggest that the local CD is considered to be extended over a wide composition range as well as a wide temperature range.

An ARPES study of $La_{1-x}Sr_xFeO_3$ has been performed using in situ prepared epitaxial thin films [12].

Figure 12.14 shows the results of ARPES studies of $La_{1-x}Sr_xFeO_3$ thin films. The left panels show the experimental band structure, and the right panels show the result of tight-binding band-structure calculation assuming the G-type AF state corresponding to the spin ordering in $La_{0.6}Sr_{0.4}FeO_3$. The details of the calculation are described in [41]. The appreciable dispersions of the e_g bands, the very weak dispersions of the t_{2g} bands and the width of the O 2p bands have been well reproduced by this calculation. The energy position of the calculated E_F was not in agreement with the experimental E_F, however. This discrepancy corresponds to the fact that this material is insulating up to 70% hole doping while the rigid-band model based on the band-structure calculation gives the metallic state. The authors discussed the origin of the unusually stable insulating state in $La_{1-x}Sr_xFeO_3$ based on the polaronic effect caused by strong electron–phonon interaction.

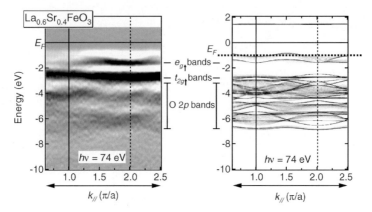

Fig. 12.14. ARPES spectra of $La_{1-x}Sr_xFeO_3$ thin films [12]. The *left panels* show the experimental band structure derived from the second derivatives of the EDC's, and the *right panels* show the result of tight-binding band-structure calculation

12.4.5 $RNiO_3$

Perovskite-type nickel oxides $RNiO_3$ (R = rare earth) form an interesting system in which an MIT occurs in a systematic manner, namely, as a function of the radius of the R ion and hence of the one-electron bandwidth as indicated in Fig. 12.15 [42]. Some $RNiO_3$ with an R ion of intermediate size shows an MIT as a function of temperature.

Vobornik et al. [43] performed a systematic photoemission study of $RNiO_3$ (R = Pr, Nd, Sm, and Eu). The left panel of Fig. 12.16 shows the valence-

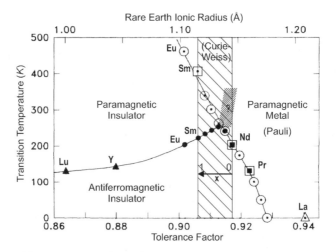

Fig. 12.15. Phase diagram of $RNiO_3$. The *hatched area* indicates the composition range of $Nd_{1-x}Sm_xNiO_3$ [42]

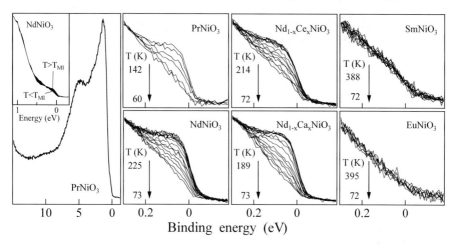

Fig. 12.16. *Left panel*: Valence band of PrNiO$_3$ ($h\nu = 1486.6$ eV); Inset: Enlarged plot of the energy range of ~1 eV of E_F (NdNiO$_3$ with $h\nu = 21.2$ eV); *Remaining panels*: spectra recorded at different temperatures for PrNiO$_3$, NdNiO$_3$, electron doped Nd$_{0.98}$Ce$_{0.02}$NiO$_3$, hole doped Nd$_{0.99}$Ca$_{0.01}$NiO$_3$, SmNiO$_3$, and EuNiO$_3$ ($h\nu = 21.2$ eV) [43]

band photoemission spectra of PrNiO$_3$. The inset shows an enlarged plot of the valence-band spectra at 210 K (metallic state) and at ~70 K (insulating state). The data clearly indicate transfer of the spectral weight from lower to higher binding energies within 0.6 eV below E_F. The energy range in which spectral weight transfer occurs is much larger than the transition temperature $T_{MI} = 130$ K (~11 meV). The remaining panels of Fig. 12.16 show a summary of the temperature-dependent spectra near E_F for the various compounds. In both the Pr and Nd (either pure or doped) compounds, the spectral intensity drops at T_{MI}, and further decreases down to the lowest temperature (72–73 K for NdNiO$_3$ and 60 K for PrNiO$_3$). For SmNiO$_3$ and EuNiO$_3$, they could not explore the MI transition which occurs outside the temperature range of the measurements, and performed all measurements in the insulating state. They could find no changes in the spectra from the highest temperature down to 72 K.

In order to see how the electronic structure evolves from the EuNiO$_3$-type one to the PrNiO$_3$-type one as a function of the size of the R ion, Okazaki et al. [44] measured the photoemission spectra of Nd$_{1-x}$Sm$_x$NiO$_3$, where the MIT and the Néel ordering occur at the same temperature for $x \lesssim 0.4$ (PrNiO$_3$-type) or T_{MI} is higher than the Néel temperature for $x \gtrsim 0.4$ (EuNiO$_3$-type) (see Fig. 12.15). Figure 12.17 shows the photoemission spectra of Nd$_{1-x}$Sm$_x$NiO$_3$ with several compositions at various temperatures. The spectra show finite spectral weight transfer from the region $-(0-0.3)$ to $-(0.3-0.7)$ eV with decreasing temperature with its strength intermediate between $x = 0.0$ and $x = 1.0$. The inset of each panel of Fig. 12.17 shows

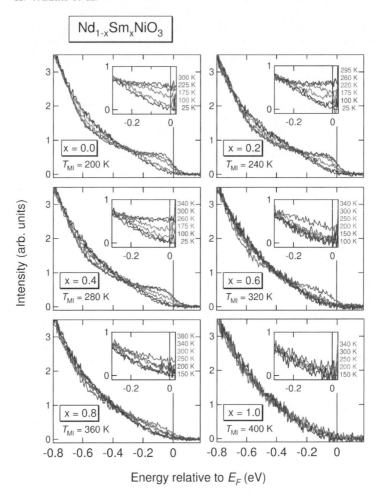

Fig. 12.17. Temperature dependence of the photoemission spectra of $Nd_{1-x}Sm_x NiO_3$ for various x's. The inset for each panel shows spectra divided by the Fermi–Dirac function [44]

the spectra divided by the Fermi–Dirac distribution function to deduce the experimental DOS. For $x \leq 0.4$, the experimental DOS above T_{MI} is almost flat or increases only weakly with energy. Below T_{MI}, the DOS at E_F gradually decreases with decreasing temperature. On the other hand, the spectra for $x > 0.4$ is pseudogaplike already above T_{MI}. From these results, they proposed that a phase boundary or a crossover line exists within the high-temperature metallic phase at $x \sim 0.4$, that is, there is a change in the nature of the electronic correlations in the middle ($x \sim 0.4$) of the metallic phase of the $RNiO_3$ system.

As for Ni perovskites, beyond the range of the phase diagram of Fig. 12.15, because the ionic radius of Bi^{3+} is larger than that of La^{3+}, $BiNiO_3$ had been expected to be metallic. Recently, Ishiwata et al. [45] succeeded in synthesizing $BiNiO_3$ under a high pressure of 6 GPa. Contrary to the expectation, $BiNiO_3$ was found to be an insulating antiferromagnet with localized spins of $S = 1$. The x-ray powder diffraction study revealed that the Bi ions were charge disproportionated into "Bi^{3+}" and "Bi^{5+}." Thus the oxidation state of the Ni ion was concluded to be 2+ rather than 3+. Subsequently, it was reported that the substitution of La for Bi suppressed the charge disproportionation and made the system conducting [46]. Wadati et al. [47] measured the photoemission and XAS spectra of $Bi_{1-x}La_xNiO_3$ to investigate how the electronic structure changes with La doping. Figure 12.18 shows the valence-band spectra of $Bi_{1-x}La_xNiO_3$. One can observe three main structures labeled as A ($\sim -1.7\,eV$), B ($\sim -3.3\,eV$), and C ($\sim -6\,eV$) and the satellite structure at $\sim -11\,eV$. Structure A, which is most pronounced at 600 eV, is therefore attributed to contributions from Ni $3d$. On the other hand, structures B and C are due to the O $2p$ band. At the bottom of Fig. 12.18(a) are shown the calculated spectra for the $[NiO_6]^{10-}$ cluster (Ni^{2+}) and the $[NiO_6]^{9-}$ cluster (Ni^{3+}). Three main structures and the satellite structure are well reproduced in the calculation for both Ni^{2+} and Ni^{3+}. Spectra in the vicinity of E_F are

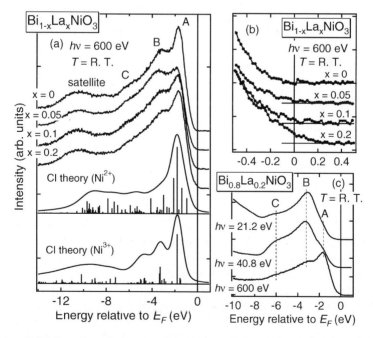

Fig. 12.18. Valence-band spectra of $Bi_{1-x}La_xNiO_3$. (a) $h\nu = 600\,eV$. The calculated spectra assuming Ni^{2+} and Ni^{3+} are presented at the bottom; (b) $h\nu = 600\,eV$ (near E_F); (c) $x = 0.2$ taken at 21.2 eV, 40.8 eV, and 600 eV [47]

shown in Fig. 12.18(b). There is no E_F cutoff for the $x = 0$ sample, consistent with the insulating behavior. Upon La substitution, emission appears at E_F and increases with increasing x, indicating the MIT induced by the La substitution.

12.4.6 $Ca_{1-x}Sr_xRuO_3$

Despite their more extended nature of the $4d$ orbitals of the second-row transition metals than the $3d$ orbitals of the first-row transition metals, ruthenates are found to show various interesting phenomena associated with electron correlation. SrRuO$_3$ is metallic and shows ferromagnetism below $T_C \simeq 160$ K . CaRuO$_3$ remains a paramagnetic metal down to the low temperatures. The effective mass of conduction electrons is strongly enhanced due to electron correlation [48]. Ca$_{1-x}$Sr$_x$RuO$_3$ is metallic in the entire x range [49]. In going from SrRuO$_3$ to CaRuO$_3$, the Ru−O−Ru bond angle is reduced from 165° to 150° and T_C decreases to zero at $x \sim 0.4$.

Photoemission studies of polycrystalline SrRuO$_3$ samples have revealed that the effect of electron correlation is substantial within the Ru $4d$ t_{2g} band [50, 51]. In the measured spectra, the d band structure is much broader and the intensity at E_F is much weaker than the band DOS, indicating that the metallic state in SrRuO$_3$ is highly incoherent.

Park et al. [52] performed photoemission experiment on ex situ prepared Ca$_{1-x}$Sr$_x$RuO$_3$ thin films. Figure 12.19(a) show the photoemission spectra of SrRuO$_3$ taken at various photon energies between 40 eV and 100 eV. Structures A and B around 5–7 eV are assigned to the O $2p$-derived bonding states and structure C around 3 eV to nonbonding state of O $2p$ orbitals. Structure D is the Ru $4d$ t_{2g} state and lies across E_F to form clear Fermi edges. They emphasize that the detailed shapes of the valence band depend on the sample cleaning methods. Takizawa et al. [53] performed an in-situ photoemission study on well-ordered surfaces of Ca$_{1-x}$Sr$_x$RuO$_3$ thin films grown by laser MBE. Figure 12.19(b) shows a combined plot of the valence-band spectra and the O $1s$ XAS spectra. The emission within $\sim$1 eV of E_F with a sharp Fermi edge and the broad band centered at $\sim$1.2 eV peak are assigned to the coherent and incoherent parts of the spectral function, respectively. For O $1s$ XAS, too, two peaks are seen within $\sim$2 eV above the threshold, which they attribute to the coherent and incoherent parts of the unoccupied t_{2g} band. Both in photoemission and XAS, one can see that spectral weight transfer occurs from the coherent part to the incoherent part with Ca doping. They consider that, while the Ru $4d$ one-electron bandwidth does not change with x, the distortion, and hence the splitting of the t_{2g} band, effectively increases electron correlation strength in analogy to Ca$_{1-x}$Sr$_x$VO$_3$.

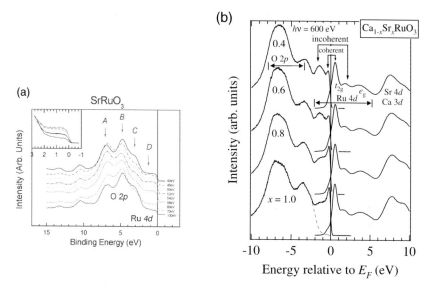

Fig. 12.19. Photoemission studies of $Ca_{1-x}Sr_xRuO_3$ thin films. (a) Valence-band spectra of ex situ prepared $SrRuO_3$ thin films [52]; (b) a combined plot of the valence-band photoemission spectra and the O 1s XAS spectra of $Ca_{1-x}Sr_xRuO_3$ thin films [53]

$SrRuO_3$ has a great potential for future oxide electronic device applications as a material for electrodes. Toyota et al. [54] studied the film-thickness dependence of in-situ photoemission spectra of $SrRuO_3$ layers deposited on $SrTiO_3$ substrates.

Figure 12.20 shows the thickness dependence of the valence-band photoemission spectra of the ultrathin $SrRuO_3$ films. The photoemission spectra for the film thicknesses of 1 − 4 ML clearly exhibit the existence of an energy gap at E_F, while the valence-band maximum of Ru-4d states appears to reach E_F at 5 ML, indicating the occurrence of an MIT at a critical film thickness between 4 and 5 ML. The existence of a Fermi cutoff is clearly seen above 5 ML. With further increase of film thickness from 5 to 15 ML, the DOS at E_F gradually grew and evolved into a sharp peak just at E_F. The evolution of the Ru-4d states seems to saturate at 15 ML. These results show the existence of two critical film thicknesses. One is the film thickness of 4 − 5 ML, where the MIT occurs, consistent with the results of the electrical-resistivity measurements. The other is ~15 ML, where the evolution of Ru 4d-derived states around E_F saturates, corresponding to the saturation of T_C determined from the kink structures in the $\rho - T$ curves.

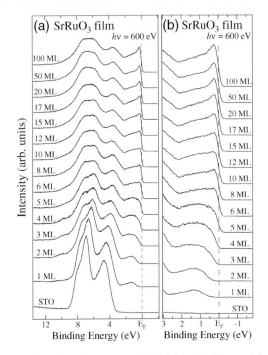

Fig. 12.20. In situ valence-band spectra of SrRuO$_3$ thin films with varying nominal film thickness. (**a**) entire region; (**b**) near E_F [54]

References

1. M. Imada et al: Rev. Mod. Phys. **70**, 1039 (1998)
2. J. Zaanen et al: Phys. Rev. Lett. **55**, 418 (1985)
3. A. E. Bocquet et al: Phys. Rev. B **46**, 3771 (1992)
4. A. E. Bocquet et al: Phys. Rev. B **53**, 1161 (1996)
5. A. Georges et al: Rev. Mod. Phys. **68**, 96 (1995)
6. V. I. Anisimov et al:, Phys. Rev. B **44**, 943 (1991)
7. M. Izumi et al: Appl. Phys. Lett. **73**, 2497 (1998)
8. J. Choi et al: Appl. Phys. Lett. **79**, 1447 (2001)
9. K. Horiba et al: Rev. Sci. Instr. **74**, 3406 (2003)
10. M. Shi et al: Phys. Rev. B **70**, 140407 (2004)
11. A. Chikamatsu et al: Phys. Rev. B **73**, 195105 (2006)
12. H. Wadati et al: Phys. Rev. B **74**, 115114 (2006)
13. K. Horiba et al: Phys. Rev. B **71**, 155420 (2005)
14. A. Fujimori et al: Phys. Rev. Lett. **69**, 1796 (1992)
15. K. Kumagai et al: Phys. Rev. B **48**, 7636 (1993)
16. T. Yoshida et al: Europhys. Lett. **59**, 258 (2002)
17. Y. Aiura et al: Surf. Sci. **515**, 61 (2002)
18. I. H. Inoue et al: Phys. Rev. Lett. **74**, 2539 (1995)
19. I. H. Inoue et al: Phys. Rev. B **58**, 4372 (1998)
20. A. Sekiyama et al: Phys. Rev. Lett. **93**, 156402 (2004)

21. M. Seah and W. A. Dench, Surf. Interf. Anal. **I**(1), 2 (1979)
22. R. Eguchi et al: Phys. Rev. Lett. **96**, 076402 (2006)
23. K. Maiti et al: Phys. Rev. B **73**, 052508 (2006)
24. T. Yoshida et al: Phys. Rev. Lett. **95**, 146404 (2005)
25. K. Takegahara: J. Electron Spectrosc. Relat. Phenom. **66**, 303 (1994)
26. A. Urushibara et al: Phys. Rev. B **51**, 14103 (1995)
27. A. Chainani et al: Phys. Rev. B **47**, 15397 (1993)
28. T. Saitoh et al: Phys. Rev. B **51**, 13942 (1995)
29. D. D. Sarma et al: Phys. Rev. B **53**, 6873 (1996)
30. J.-H. Park et al: Phys. Rev. Lett. **76**, 4215 (1996)
31. D. S. Dessau et al: Phys. Rev. Lett. **81**, 192 (1998)
32. Y.-D. Chuang et al: Science **292**, 1509 (2001)
33. N. Mannella et al: Nature **438**, 474 (2005)
34. T. Okuda et al: Phys. Rev. Lett. **81**, 3203 (1998)
35. D. D. Sarma et al: Phys. Rev. Lett. **93**, 097202 (2004)
36. M. Takano et al: J. Solid State Chem. **39**, 75 (1981)
37. J. Matsuno et al: Phys. Rev. B **60**, 4605 (1999)
38. A. Chainani et al: Phys. Rev. B **48**, 14818 (1993)
39. H. Wadati et al: Phys. Rev. B **71**, 035108 (2005)
40. H. Wadati et al: J. Phys. Soc. Jpn. **75**, 054704 (2006)
41. H. Wadati et al: Phase Transitions **79**, 617 (2006)
42. J. B. Torrance et al: Phys. Rev. B **45**, 8209 (1992)
43. I. Vobornik et al: Phys. Rev. B **60**, R8426 (1999)
44. K. Okazaki et al: Phys. Rev. B **67**, 073101 (2003)
45. S. Ishiwata et al: J. Mater. Chem. **12**, 3733 (2002)
46. S. Ishiwata et al: Phys. Rev. B **72**, 045104 (2005)
47. H. Wadati et al: Phys. Rev. B **72**, 155103 (2005)
48. P. B. Allen et al: Phys. Rev. B **53**, 4393 (1996)
49. F. Fukunaga and N. Tsuda, J. Phys. Soc. Jpn. **63**, 3798 (1994)
50. K. Fujioka et al: Phys. Rev. B **56**, 6380 (1997)
51. J. Okamoto et al: Phys. Rev. B **60**, 2281 (1999)
52. J. Park et al: Phys. Rev. B **69**, 165120 (2004)
53. M. Takizawa et al: Phys. Rev. B **72**, 060404 (2005)
54. D. Toyota et al: Appl. Phys. Lett. **87**, 162508 (2005)

Part VI

High Energy and High Resolution

High-Resolution High-Energy Photoemission Study of Rare-Earth Heavy Fermion Systems

A. Sekiyama, S. Imada, A. Yamasaki, and S. Suga

Department of Material Physics, Graduate School of Engineering Science, Osaka University, Toyonaka, Osaka 560-8531, Japan

Abstract. High-resolution soft x-ray ($h\nu$: 800–1000 eV) photoemission spectra of some rare-earth compounds are presented employing the $3d$–$4f$ enhancement instead of the commonly used $4d$–$4f$ enhancement. The higher photon energy makes the spectra more bulk sensitive. All the data show that the spectra can be well explained by calculations based on the single impurity Anderson model except for the $4f$ spectra of $CeRu_2$.

13.1 Introduction

The effect of the electron correlation on electronic states of rare-earth compounds is one of the most important and essential topics in condensed matter physics because many intriguing phenomena such as superconductivity [1], formation of the Kondo singlet, heavy fermion behavior [2], magnetic and/or quadrupolar ordering [3] have been found. For these phenomena, a finite hybridization between strongly correlated $4f$ states and valence/conduction bands plays an essential role. Since photoemission spectroscopy (PES) is useful to investigate such hybridized $4f$ electronic states, many PES studies of Ce systems have so far been performed. Especially, the technical improvement in photoelectron spectrometer allows one to observe the detailed $4f$ spectral line shapes in the vicinity of the Fermi level (E_F) for decades [4–8]. However, PES using low-energy excitation is surface-sensitive caused by the short photoelectron mean free path λ (≤ 5 Å) [9]. When $\lambda = 5$ Å and a surface thickness is 4 Å, a surface contribution is about 55%. The surface electronic states deviate from the bulk states due to different topological connectivity, which leads to considerably different hybridization strengths. Indeed, it has been recognized that the Ce and Yb $4f$ spectra are remarkably different between the bulk and surface [5, 7, 10]. In order to directly probe the strongly correlated bulk spectral functions, high-resolution high-energy PES is promising because of the longer λ (~15 Å at the photoelectron kinetic energy of ~880 eV). Here we show the recent improvement in the soft x-ray and x-ray PES for probing the bulk states of heavy fermion Ce and Pr compounds [11–18].

13.2 Experimental

In order to realize a breakthrough for studying the bulk electronic states with using PES, we have constructed a high-resolution soft x-ray PES system combined with a varied-line-spacing plane grating monochromator on a twin-helical undulator beam line BL25SU in SPring-8 [19]. At this station, the resolution of ~ 100 meV at an excitation energy near 1000 eV is conventionally realized. The high-resolution soft x-ray PES spectra displayed here were measured at BL25SU in SPring-8 by using a Gammadata-Scienta SES-200 spectrometer. The samples were cooled down to about 20 K by using a closed-cycle He cryostat. The base pressure was about 3×10^{-8} Pa. The sample cleaning (cleavage, fracturing, scraping) can be done at the measuring temperature in situ.

The core-level PES excited at $h\nu = 2450$ and 5450 eV was performed at beamline ID32 in ESRF. The total energy resolution of the core-level PES was set to about 500 meV at $h\nu = 2450$ eV. The samples were cooled and kept at about 180 K by a liquid N_2-cryostat.

13.3 High-Resolution Soft X-ray Photoemission Study of Ce Compounds

13.3.1 Introduction

The photoionization cross-section of the Ce $4f$ orbital is relatively small compared with other orbitals in intermetallic compounds at the low-energy excitations such as $h\nu = 21.2$ and 40.8 eV [20]. Therefore, the Ce $4d - 4f$ resonance photoemission (RPES) has extensively been employed for investigating the Ce $4f$ states owing to strong enhancement of the $4f$ contribution in the spectra. In 1980s, it has been recognized that the Ce $4f$ spectral functions are basically understood by the single-impurity Anderson model (SIAM) [21–23]. However, there has been a long standing controversy in the 1990s with respect to the applicability of SIAM to the $4f$ spectra obtained by the Ce $4d-4f$ RPES at $h\nu \sim 120$ eV [6, 10, 24]. One interpretation infers that SIAM well reproduces the $4f$ spectra in most cases irrespective of the strength of the hybridization or the Kondo temperature (T_K) of the compounds [8] whereas another claims that the experimental $4f$ spectra are essentially inconsistent with SIAM almost without exception [6]. The high-resolution Ce $3d-4f$ RPES ($h\nu \sim 880$ eV) is one of the promising and direct techniques to reveal the bulk $4f$ states due to its longer λ. Here we show the high-resolution bulk $4f$-derived PES spectra of Ce compounds including "low-T_K" $CeRu_2Ge_2$ ($T_K < 1$ K, [25, 26]), CeB_6 ($T_K \sim 3$ K, [27]), $CeRu_2Si_2$ ($T_K \sim 20$ K, [25, 28]), CeNi($T_K \sim 150$ K, [29]) and "very high-T_K" $CeRu_2$ ($T_K > 1000$ K, [13]) obtained by the high-resolution Ce $3d-4f$ RPES, where T_K denotes the Kondo temperature.

13.3.2 Surface-Preparation Dependence of the Spectra

In order to know which surface (fracturing or scraping) is better to obtain the intrinsic photoemission spectra even at the high-energy soft x-ray excitation, we have measured several core-level PES spectra of polycrystalline CePdSn on both fractured and scraped surfaces. CePdSn is known as a reference material to the Kondo semiconductor CeRhAs [30], while $T_{\rm K}$ is as low as <7 K for CePdSn [31, 32] compared with that for CeRhAs. As shown in Fig. 13.1, all core-level peaks in the spectra on the scraped surface displayed here are obviously broader than those on the fractured surface. The gravity of the spectral weight on the fractured surface is shifted to the lower-binding energy

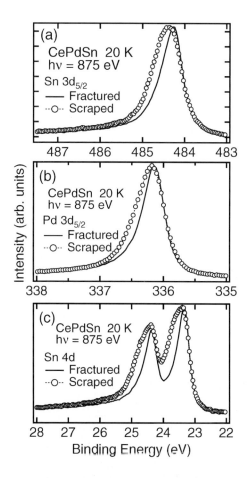

Fig. 13.1. Comparison of the core-level photoemission spectra on the fractured and scraped surfaces for CePdSn; (**a**) Sn $3d_{5/2}$ core level; (**b**) Pd $3d_{5/2}$ core level. (**c**) Sn $4d$ core level. The energy resolution for these spectra were set to 200 meV

side compared with that on the scraped surface. The same tendency has also observed in the Sn $3d_{3/2}$ and Pd $3d_{3/2}$ peaks. The full width of half maximum (FWHM) of the Sn $3d_{5/2}$ and Pd $3d_{5/2}$ peaks in the spectra on the fractured surfaces are estimated as about 630 and 420 meV while those on the scraped surface are about 850 and 620 meV. In the case of an intermetallic compound CePdSn, any complicated structure such as final-state multiplet structure is not expected near the main peaks in the Sn $3d$, $4d$ and Pd $3d$ spectra. Indeed, it seems that there is not any shoulder or pre-peak in the Sn $3d_{5/2}$ as well as Pd $3d_{5/2}$ spectra on the scraped surface while a shoulder seems to be seen at the higher binding energy side (by $\sim$0.3 eV) of the main peak in the Sn $4d$ spectra on the scraped surface. Although the real origin of the broader FWHM of the peak in the spectra on the scraped surface is not clear at present, it is natural to judge that the fractured surface is better than the scraped surface in order to obtain the intrinsic spectra.

We have confirmed that structures near E_F in the valence-band spectra are somewhat smeared for the scraped surface compared with those for the fractured surface. This might be caused by the formation of the damaged scraped surface region, in which the translational symmetry of the crystal is rather broken. The thickness of this region is comparable or larger than the probing depth of the high-energy soft x-ray PES. Therefore, we conclude that the fractured surfaces are better to obtain intrinsic PES spectra if we would like to discuss the electronic states of solids based on the PES spectra within an energy scale of several hundreds meV.

13.3.3 Ce 3d XAS Spectra

Figure 13.2 summarizes the Ce $3d_{5/2}$ core absorption (XAS) spectra of CeRu$_2$, CeRu$_2$Si$_2$, CeB$_6$ and CeRu$_2$Ge$_2$ measured by the total electron yield mode. In order to obtain clean surfaces, the sample surfaces were cleaved or fractured in situ at a measuring temperature of 20 K except for the case of CeB$_6$, which was repeatedly scraped in situ by using a diamond file. The resolution is better than 200 meV. Two peaks at $h\nu$ = 881.6 and 882.6 eV, and some multiplet structures around $h\nu$ = 878–881 eV are observed in the spectra of "low-T_K" compounds CeRu$_2$Si$_2$, CeB$_6$ and CeRu$_2$Ge$_2$. These are ascribed to the $3d^94f^2$ final states representing the $4f^1$ character in the ground state. The same structures are also seen, but quite smeared, in the spectrum of CeRu$_2$. In addition, a broad peak is seen at 887.5 eV, whose intensity is the strongest in CeRu$_2$ among the present Ce compounds. This peak is ascribed to the $3d^94f^1$ final state reflecting the $4f^0$ character or valence-mixing character in the ground state. One notices that the $3d^94f^1$ final-state intensity, which becomes stronger on going from CeRu$_2$Ge$_2$ to CeRu$_2$, corresponds well to the increase of T_K. The smeared $3d^94f^2$ structures in addition to the enhanced $3d^94f^1$ final state seen in CeRu$_2$ are due to the very strong hybridization between the Ce $4f$ and Ru $4d$ states. Such phenomena are also seen in other strongly hybridized system [33]. The $3d^94f^2$ structures are somewhat smeared

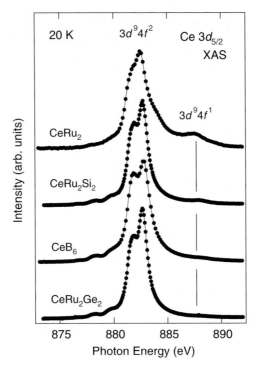

Fig. 13.2. Ce $3d_{5/2}$ core absorption spectra of CeRu$_2$, CeRu$_2$Si$_2$, CeB$_6$ and CeRu$_2$Ge$_2$

in low-T_K CeB$_6$. However, this may be caused by the rough surface of CeB$_6$ obtained by scraping.

13.3.4 Ce 3d-4f Resonance Photoemission of "Low-T_K" Systems

The bulk-sensitive high-resolution Ce $3d-4f$ RPES spectra of CeNi, CeRu$_2$Si$_2$, CeB$_6$ and CeRu$_2$Ge$_2$ are summarized in Fig. 13.3. There are two peaks with comparable intensity in the vicinity of E_F and at 0.2–0.3 eV in the spectra of CeB$_6$ and CeRu$_2$Ge$_2$. The former peak is ascribed to the tail of the Kondo resonance (KR) including its crystalline electric field (CEF) partners ($4f_{5/2}^1$ final states) while the latter corresponds to its spin-orbit partner ($4f_{7/2}^1$ final states). In the case of CeNi and CeRu$_2$Si$_2$ for which T_K is higher than those for CeB$_6$ and CeRu$_2$Ge$_2$, the tail of KR is prominent whereas its spin-orbit partner is markedly suppressed and appears as a shoulder. The relative intensity of the $4f_{5/2}^1$ to $4f_{7/2}^1$ final states becomes stronger on going from low-T_K (CeRu$_2$Ge$_2$) to high-T_K (CeNi) compounds, which is qualitatively consistent with SIAM.

In order to further clarify these bulk Ce $4f$ contributions, we have calculated spectral functions based on the SIAM using the non-crossing

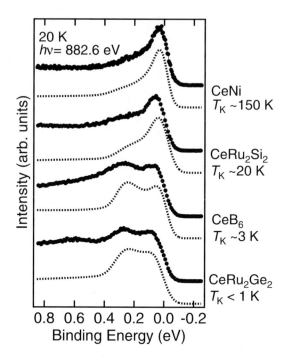

Fig. 13.3. Comparison of the 3d–4f RPES spectra near E_F with the spectra calculated using NCA for CeNi, CeRu$_2$Si$_2$, CeB$_6$ and CeRu$_2$Ge$_2$

approximation (NCA) [34], which has been applied to many Ce and some Yb compounds. The calculated spectra, properly fitted to the experimental 3d–4f on-resonance spectra near E_F, are also shown in Fig. 13.3. Here, the spin-orbit splitting of the 4f levels (the energy difference between the $4f_{7/2}$ level and the center of gravity of the $4f_{5/2}$ level) is adjusted as 310 meV for CeNi and CeRu$_2$X$_2$ (X = Si, Ge), 315 meV for CeB$_6$. Other optimized parameters such as the bare (unhybridized) binding energy of the lowest $4f_{5/2}$ level (ε_{4f}^0), the CEF splitting (Δ_{CEF}), the average hybridization strength defined by $\delta \equiv (\pi/B) \int_0^B \rho v^2(E) dE$ where $\rho v^2(E)$ is the energy dependence of the hybridization strength and B is the bottom binding energy of the hybridized valence band, and the hybridization strength $\rho v^2(E_F)$ at E_F are summarized in Table 13.1. As discussed later, finite values of the CEF splitting are inevitably required for our fitting of the PES spectra. For CeNi and CeRu$_2$X$_2$, the $4f_{5/2}$ level is split into three doublets while it is split into the ground state with four-fold degeneracy (Γ_8) and one doublet (Γ_7) for CeB$_6$. Δ_{CEF}'s of CeNi, CeRu$_2$Ge$_2$ and CeB$_6$ listed in Table 13.1 are set to be consistent with those estimated elsewhere [26, 35, 36]. As for CeRu$_2$Si$_2$, no CEF peaks have been observed within ∼450 K by an experimental neutron-inelastic-scattering study [37]. It is suggested that the CEF splitting should be close to 1000 K to explain the large magnetic anisotropy of CeRu$_2$Si$_2$ [38]. Our optimized $\Delta_{\rm CEF}$'s do not contradict this condition. As shown in Fig. 13.3, the NCA calculations well reproduce the experimental spectra near $E_{\rm F}$. Such a successful calculation indicates the applicability of the SIAM to the "bulk" Ce $4f$ spectra of the weakly and moderately hybridized systems.

Table 13.1. Optimized physical parameters for CeNi, CeRu$_2$Si$_2$, CeRu$_2$Ge$_2$ and CeB$_6$ by the NCA calculation

	ε_{4f}^0 (eV)	$\Delta_{\rm CEF}$ (K)	δ (meV)	$\rho v^2(E_{\rm F})$ (meV)
CeNi	1.58	580, 1200	103	33
CeRu$_2$Si$_2$	1.60	600, 800	72	28
CeB$_6$	1.80	530	73	26
CeRu$_2$Ge$_2$	1.58	500, 800	57	22

In order to evaluate $T_{\rm K}$s, we have simulated magnetic excitation spectra by using NCA with the same parameters as employed for fitting the RPES spectral functions. The obtained $T_{\rm K}$ from the magnetic excitation spectra are 150, ~16, ~8 and <1 K for CeNi, CeRu$_2$Si$_2$, CeB$_6$ and CeRu$_2$Ge$_2$, respectively, which are quite comparable to the real $T_{\rm K}$ of ~150, ~20, ~3 and <1 K. These results suggest that we have properly chosen the parameters for these compounds.

Meanwhile, we have also performed the NCA calculations and estimated $T_{\rm K}$ under an extreme assumption that the CEF splitting is 0 meV, which is equivalent to a sextet $4f_{5/2}$ ground state. We can fit all spectra here even under such a condition with employing different hybridization strength. However, estimated $T_{\rm K}$s with $\Delta_{\rm CEF} = 0$ meV (defined as $T_{\rm K}^h$) are >1000 K for CeNi, >200 K for CeRu$_2$Si$_2$ and of the order of 40 K for CeB$_6$ and CeRu$_2$Ge$_2$. These values deviate too much from actual $T_{\rm K}$s, indicating that the finite CEF effects should be taken into account for the proper analyses of the $4f$ photoemission spectra.

On the other hand, we show that such high $T_{\rm K}^h$s themselves are not fully meaningless. In such a temperature region as the $4f$ state can be regarded as six-fold degenerate, the proper Kondo temperature $T_{\rm K}^h$ is represented as $T_{\rm K}^h = [T_{\rm K}\Delta_{\rm CEF}(1)\Delta_{\rm CEF}(2)]^{1/3}$ for CeNi and CeRu$_2$X$_2$, $T_{\rm K}^h = [T_{\rm K}^2\Delta_{\rm CEF}]^{1/3}$ for CeB$_6$ from a scaling theory [39]. A validity of $T_{\rm K}^h$ is really seen by temperature dependence of the resistivity in Ce$_x$La$_{1-x}$Al$_2$ [40]. Under the present experimental condition, the measuring temperature is far below $\Delta_{\rm CEF}(1)$, $\Delta_{\rm CEF}(2)$ and $\Delta_{\rm CEF}$. However, the finite energy resolution (~100 meV) is beyond these values. Thus $T_{\rm K}^h$ can have a meaning in our experiment. This formula gives $T_{\rm K}^h$ of CeRu$_2$Si$_2$ as ~200 K using $T_{\rm K} = 16$ K, $\Delta_{\rm CEF}(1) = 600$ K, and $\Delta_{\rm CEF}(2) = 800$ K. The $T_{\rm K} \geq 200$ K estimated for CeRu$_2$Si$_2$ for $\Delta_{\rm CEF} = 0$ meV may thus correspond to $T_{\rm K}^h$. As for CeB$_6$, $T_{\rm K}^h$ is estimated as 32 K if we employ our estimated $T_{\rm K} \sim 8$ K. On the other hand, $T_{\rm K}$ for CeRu$_2$Ge$_2$

is derived as $\sim 0.2\,\mathrm{K}$ using $[T_\mathrm{K}^h, \Delta_\mathrm{CEF}(1), \Delta_\mathrm{CEF}(2)] = (40, 500, 800)\,\mathrm{K}$. Thus the above relation of the scaling theory is fairly well satisfied. $T_\mathrm{K}^h \sim 200\,\mathrm{K}$ for CeRu$_2$Si$_2$ may correspond to a broad peak at 220 K in the observed temperature dependence of the thermopower [28]. Therefore our analyses have demonstrated that the NCA fitting to the experimental "bulk and low-T_K" $4f$ spectra can yield the realistic T_K by taking the CEF splitting into account. If T_Ks of these compounds were not known and the CEF were neglected, we might have simply estimated T_K as $> 1000\,\mathrm{K}$ ($= T_\mathrm{K}^h$) for CeNi, $> 200\,\mathrm{K}$ ($= T_\mathrm{K}^h$) for CeRu$_2$Si$_2$, $\sim 40\,\mathrm{K}$ for CeB$_6$ and CeRu$_2$Ge$_2$ by our analyses. Indeed the experimental spectra near E_F of CeB$_6$ and CeRu$_2$Ge$_2$ are mutually very similar. This situation suggests that the Ce $4f$ spectra with the resolution of 100 meV are scaled by T_K^h rather than T_K. By properly considering the crystal field splitting of the $4f_{5/2}$ states, however, one can reproduce the experimental photoemission spectrum measured with the resolution of 100 meV by employing parameters consistent with the low T_K.

13.3.5 Bulk 4f Spectra of a Strongly Valence-Fluctuating CeRu$_2$

Here we show that the bulk $4f$ spectral function of a "very high-T_K" compound CeRu$_2$ is qualitatively different from that for the "low-T_K" ($\leq 150\,\mathrm{K}$) systems, which are well reproduced by SIAM as shown above (spectra of the valence band of CeRu$_2$ are also given in the chapter by Yokoya et al. in this volume). Figure 13.4 demonstrates the high-resolution $3d$–$4f$ RPES spectrum of CeRu$_2$, which is equivalent to the bulk $4f$ spectral function. Neither a peak due to the tail of KR nor its spin-orbit partner is seen in the spectrum while a rather conventional Fermi cut-off, a tiny shoulder in the vicinity of E_F and a broad peak centered at $\sim 0.4\,\mathrm{eV}$ are observed instead.

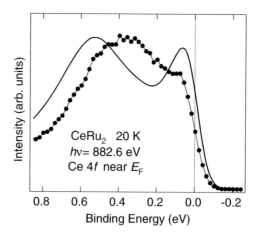

Fig. 13.4. Comparison of the Ce $3d$–$4f$ RPES spectrum near E_F (resolution of $\sim 100\,\mathrm{meV}$) with the calculated Ce $4f$ PDOS

The Ce $4f$ spectrum in a wide energy scale obtained by subtracting the "off-resonance" spectrum from the "on-resonance" spectrum measured at $h\nu = 882.6\,\text{eV}$ is displayed in the lower panel of Fig. 13.5. A strong peak centered at $\sim 0.4\,\text{eV}$ appears while the so-called f^0 final states seen in the surface-sensitive $4d$–$4f$ RPES spectra [11, 41] are strongly suppressed in the bulk-sensitive $4f$ spectrum. It should be noted that the peak near $0.4\,\text{eV}$ is always observed at nearly constant binding energy irrespective of the excitation photon energy and thus cannot be ascribed to an Auger feature. The peak near $0.4\,\text{eV}$ is also seen in the off-resonance spectrum, which mostly reflects the bulk Ru $4d$ states, measured at $h\nu = 875\,\text{eV}$. In addition, a small structure and a prominent peak are seen at 1.5 and $2.5\,\text{eV}$ in the off-resonance spectrum. The peaks at 0.4 and $2.5\,\text{eV}$ are absent in the spectrum taken at $h\nu = 114\,\text{eV}$ on the scraped surface [13] and also in that taken at $h\nu = 115\,\text{eV}$ on the fractured surface by Kang et al. [41]. As discussed later, the peak at $0.4\,\text{eV}$ should be ascribed to the bulk Ru $4d$ contribution strongly hybridized with the bulk Ce $4f$ states. Consequently the absence of the peak in the $4d$–$4f$ off-resonance spectra is due to the surface effect, namely, the Ru $4d$ states in the surface are much less hybridized with the localized surface Ce $4f$ states than in the bulk. It has so far been reported that the off-resonance spectra

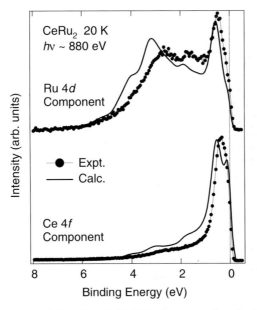

Fig. 13.5. Comparison of the Ce $3d$–$4f$ RPES spectra (resolution of $\sim 200\,\text{meV}$) with the result of the band-structure calculation for CeRu$_2$. The *upper panel* compares the "off-resonance" spectrum with the Ru $4d$ PDOS. The *lower panel* comparers the $4f$ spectrum obtained by subtracting the "off-resonance" spectrum from the "on-resonance" spectrum with the Ce $4f$ PDOS

dominated by the non Ce $4f$ components show negligible surface effects for weakly hybridized Ce compounds [10]. When the hybridization in the bulk is weak enough, the hybridization is also weak in the surface and it is natural that the valence-bands line shape does not change much between the bulk and surface. However, our results have demonstrated that even the $4d$–$4f$ off-resonance spectrum does not faithfully represent the "bulk" valence-band structures in the very strongly hybridized case.

In order to further clarify the electronic states of CeRu$_2$, we compare the experimentally obtained $3d$–$4f$ RPES spectra with the results of the band-structure calculation [42], which well explains the Fermi surfaces of CeRu$_2$ obtained from an experimental de Haas–van Alphen measurement [43]. Here, the calculated partial density of states (PDOS) has been convoluted with use of both Lorentzian broadening with the maximum full width at half maximum (FWHM) of 0.4 eV and Gaussian broadening with a fixed FWHM corresponding to the instrumental resolution. The Lorentzian width originates from a lifetime of the photoemission final states, and thus the Lorentzian contribution is considered at each binding energy (E_B), whose FWHM is assumed to increase with E_B reflecting the shorter lifetime. The band-structure calculation fairly well reproduces the Ce $4f$ spectrum with respect to the observed $4f$ band width. The upper panel of Fig. 13.5 shows the $3d$–$4f$ off-resonance spectrum in comparison with a broadened Ru $4d$ PDOS obtained by the band-structure calculation. The peak positions and band width of the experimental off-resonance spectra agree qualitatively with those of the theoretical Ru $4d$ PDOS. Especially, the peak at 0.4 eV is well reproduced by the calculation.

A detailed comparison of the $3d$–$4f$ on-resonance spectrum with the Ce $4f$ PDOS in the near-E_F region is shown in Fig. 13.4, where E_F denotes the Fermi level. Although there are some discrepancies between the spectra, the calculation fairly well explains the essential features of the experimental spectrum. In particular, the broad peak near 0.4 eV in the Ce $4f$ spectrum, which is not simply predicted from SIAM, corresponds well to the band-structure calculation. A peak in the vicinity of E_F in the $4f$ PDOS may correspond to the tiny shoulder near E_F in the experimental spectrum. The fact that the peak appears at $\sim$0.4 eV in both bulk Ru $4d$ and Ce $4f$ PDOS's proves that the appearance of the 0.4 eV peak in both experimental spectra is not fortuitous but due to the very strong Ru $4d$ – Ce $4f$ hybridization. To date, there are many reports that bulk Ce $4f$ spectra cannot be explained by band-structure calculations and should be reproduced by calculations based on SIAM in relatively weakly hybridized cases. Contrary to these, our present results clarify that the band-structure calculation is appropriate for explaining the Ce $4f$ spectral function rather than SIAM in the case of very strongly hybridized CeRu$_2$. At least, band-structure calculation is a good starting point to analyze bulk $4f$ PES spectra of very strongly hybridized systems. It should be noted that Weschke et al. have claimed that a $4f$ inverse-photoemission spectrum of CeRh$_3$, which is also a very strongly hybridized system, is fairly reproduced by a band-structure calculation [44].

Although the calculated $4f$ PDOS based on the one-electron band theory reproduces the the PES spectrum fairly well, some quantitative disagreements are seen. This band calculation [42] does not give an appropriate electronic specific-heat coefficient $\gamma \approx 30\,\mathrm{mJ/(mol \cdot K^2)}$ [45], suggesting that there is still the electron correlation effect not taken into account in the band calculation. Indeed, a reasonable γ can be deduced by a calculation using a method of perturbation expansion with respect to electron–electron interactions, where a mass-enhancement factor of the f electrons has been calculated as 5.77 [42]. Therefore we can conclude that the discrepancies between the experimental and calculated spectra seen in Fig. 13.4 originate from the Ce $4f$ electron correlation effect. A general $4f$ spectral calculation based on the Kondo lattice model, in which the correlation effect would be fully considered, is surely desirable in order to solve the discrepancy.

13.4 High-Energy Photoemission Study of Pr Compounds

13.4.1 Introduction

Heavy-fermion properties observed in many Ce, Yb and U compounds emerge when the hybridization between the conduction band in the vicinity of the Fermi level and the f state (c–f hybridization) is moderate. The $4f$ electrons in Pr are more localized and less hybridized with conduction electrons than in Ce. No heavy-fermion Pr compound was known until the discovery of $PrInAg_2$ with a large specific-heat coefficient reaching $\sim 6.5\,\mathrm{J/mol\,K^2}$ [46]. Recently, the heavy electron mass has been found in $PrFe_4P_{12}$ under high magnetic field [47]. In both $PrInAg_2$ and $PrFe_4P_{12}$, the crystal-electric field ground state is suggested to be a non-Kramers doublet [46,48,49], which is nonmagnetic but has an electric quadrupolar degree of freedom. Therefore, the heavy-fermion behaviors in these Pr compounds may result from the quadrupolar Kondo effect [50,51], which was first applied to U compounds and is in contrast to the usual spin Kondo effect applied to Ce and Yb compounds.

$PrFe_4P_{12}$ is one of the Pr-based filled skutterudites PrT_4X_{12} (T = Fe, Ru, Os; X = P, Sb). In PrT_4X_{12}, $PrRu_4P_{12}$ shows the metal-insulator transition at $\sim 64\,\mathrm{K}$ [52] whereas $PrRu_4Sb_{12}$ and $PrOs_4Sb_{12}$ are known as a conventional [53] and a heavy-fermion [54] superconductor, respectively. $PrFe_4P_{12}$ is particularly interesting due to the phase transition at around 6.5 K [55] and the Kondo-like behavior. A large electronic specific-heat coefficient of $C_{\mathrm{el}}/T \sim 1.2\,\mathrm{J/K^2 mol}$ is found under 6 T [47], which suggests the Kondo temperature T_{K} of the order of 10 K. These facts suggest the following scenario; the quadrupolar degree of freedom of the Pr $4f$ state due to the non-Kramers twofold degeneracy leads to the quadrupolar Kondo effect, and the phase transition at 6.5 K resulting in the antiquadrupolar ordering is driven by the lifting of the quadrupolar degeneracy. In order for the quadrupolar Kondo effect to take place, c–f hybridization must be appreciably strong. On the other hand, $PrSn_3$ is also known as a heavy fermion compound with the $AuCu_3$-type crystal structure [56]. T_K of $PrSn_3$ seems to be several ten K judging from the temperature dependence of the resistivity.

Here we report the results of (1) bulk-sensitive Pr $3d$–$4f$ RPES for $PrFe_4P_{12}$, $PrRu_4P_{12}$ and $PrRu_4Sb_{12}$, by which the Pr 4f contributions in the bulk spectra are clearly shown, and (2) Pr $3d$ core-level spectra of $PrFe_4P_{12}$ and $PrSn_3$ by using hard x-ray ($h\nu$: 2450–5450 eV).

13.4.2 High-resolution $3d$–$4f$ Resonance Photoemission of Heavy Fermion Pr-based Skutterudites

For the measurement of Pr $3d$–$4f$ XAS and RPES, single crystals of $PrFe_4P_{12}$ and $PrRu_4Sb_{12}$, and polycrystals of $PrRu_4P_{12}$ were employed, which were fractured in situ at the measuring temperature of 20 K except for the temperature dependence measurement. The total energy resolution of the PES measurement was set to $\sim$80 meV in the high-resolution mode and $\sim$130 meV, otherwise. The Pr $3d$–$4f$ XAS spectrum for $PrFe_4P_{12}$ is shown in the inset of Fig. 13.6(a). This spectrum reflects the predominant Pr^{3+} ($4f^2$) character in the initial state [57]. Spectra of $PrRu_4P_{12}$ and $PrRu_4Sb_{12}$ were also quite similar to this spectrum. Valence-band PES spectra were measured at three photon energies. On-resonance spectra were taken at 929.4 eV, around the x-ray absorption maximum. Off-resonance spectra were taken at 921 and 825 eV, which were quite similar in shape. The on- and off- (921.0 eV) resonance spectra are compared in the main panel of Fig. 13.6(a). We consider that the Pr $4f$ contribution is mainly enhanced in the on-resonance spectra, and therefore that the difference between the on- and off-resonance spectra mainly reflects the Pr $4f$ spectrum.

The off-resonance spectra taken at 825 eV with better statistics are shown in Fig. 13.6(b) in a magnified intensity scale. The valence band between E_F and binding energy (E_B) of $\sim$7 eV is expected to be composed of Pr $5d$ and $4f$, T d, and X p orbitals. Among these, main contribution to the off-resonance spectrum (more than 60%) comes from the T d states according to the photoionization cross-section [20]. The off-resonance spectral features are reproduced in the theoretical off-resonance spectra based on the FLAPW (full-potential linearized augmented plane wave) and LDA (local density approximation)+U band structure calculation (see Fig. 13.7(b)) [58], where the parameter for the on-site Coulomb interaction U of Pr $4f$ electron is set as 0.4 Ry (5.4 eV).

The on-resonance spectra shown in Fig. 13.6(a) are characterized by two features. First, the on-resonance spectra have various multiple peak structures, where peaks (or structures) are indicated by arrows, in contrast to the calculated Pr f PDOS (see Fig. 13.7(a)) that has a strong peak and small structures near E_F for all the three compounds. This feature will be interpreted in the next paragraph taking into account the hybridization between

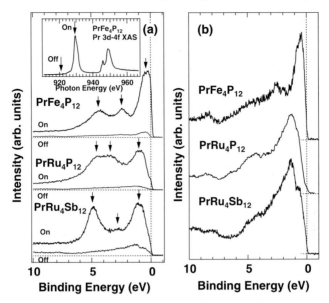

Fig. 13.6. (a) On- and off-resonance spectra normalized by the photon flux. Inset: Pr 3d–4f XAS spectrum for $PrFe_4P_{12}$. *Arrows* show the energies at which the spectra in the main panel were taken; (b) off-resonance spectra taken at 825 eV in an enlarged intensity scale

the valence band and the Pr $4f$ states (v–f hybridization) in the *final* states of PES. Second, the intensity near E_F, i.e., between E_F and $E_B \sim 0.3$ eV, is much stronger in $PrFe_4P_{12}$ than in other two systems. Such strong intensity at E_F is not found in other reported Pr $4f$ spectra. This result reflects the strong c–f hybridization in the *initial* state of $PrFe_4P_{12}$ as discussed later.

Multiple peak structures observed for various Pr compounds have been interpreted in terms of the v–f hybridization [59–61]. We adopt the cluster model [62], i.e., the simplified version of SIAM [21]. The part of the valence band that hybridizes strongly with the $4f$ state is expected to be similar between PrT_4X_{12} and LaT_4X_{12}. La f PDOS of LaT_4X_{12} at a certain energy correspond roughly to the v–f hybridization strength at that energy since La f states below E_F comes only from the hybridization with the valence band. As the first approximation, we replace the La f PDOS with two levels, v_1 and v_2, the energies of which, $E_B(v_k)$, are shown by the arrows in Fig. 13.7(a). We now assume that the initial Pr $4f$ state is $|f^2\rangle$. Although it turns out later that deviation from this state is appreciable in $PrFe_4P_{12}$, this is a good approximation when discussing the overall spectral features. Then the final states of Pr $4f$ PES are linear combinations of $|f^1\rangle$, $|(f^2)^*\underline{v_1}\rangle$, and $|(f^2)^*\underline{v_2}\rangle$, where $\underline{v_k}$ denotes a hole at v_k. Since the resulting f^2 state includes all the excited states, it is denoted as $(f^2)^*$ so as to distinguish it from the initial ground state f^2. The average excitation energy $E((f^2)^*) - E(f^2)$ is ~ 1.4 eV

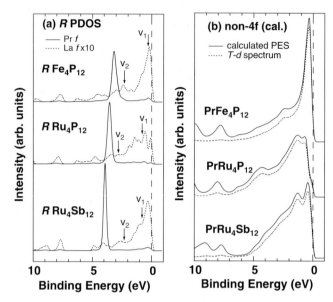

Fig. 13.7. Calculated photoemission spectra based on band structure calculation. Density of states is multiplied by the Fermi–Dirac function for 20 K and is broadened by the Gaussian with the full width at half maximum of 80 meV. (**a**) Calculated Pr (*solid line*) and La (*dashed line*: magnified ten times) f spectra of PrT_4X_{12} and LaT_4X_{12}; (**b**) calculated off-resonance spectra (*solid lines*), where partial density of states except for Pr f are multiplied by the cross sections [20] and summed up. *Dashed lines* show the contribution of the T d state

according to the atomic multiplet calculation [57]. The main origin of this excitation energy is found to be the exchange interaction. The energies of the bare $|(f^2)^*v_k\rangle$ with respect to the initial state $|f^2\rangle$ are hence $E_B(v_k) + [E((f^2)^*) - E(\overline{f^2})]$ and are shown by the thin open and filled bars in the upper panels of Figs. 13.8 (a)–(c). We take the remaining three parameters, E_B of the bare $|f^1\rangle$ (E_0), the hybridization between $|f^1\rangle$ and $|(f^2)^*v_k\rangle$ (V_k), to be free parameters, and numerically solve the 3×3 Hamiltonian matrix. When the parameters are set as in the upper panels of Figs. 13.8(a)–(c), the three final states are obtained as shown in the lower panels. At each of the tree eigen-energies for the final states is placed a vertical bar with a length proportional to the Pr 4f excitation intensity, which is proportional to the weight of the $|f^1>$ state. The line spectra reproduce qualitatively well the experimentally observed system dependence in the energy positions and intensity ratios of the three-peak structures of the on-resonance spectra (see Fig. 13.6(a)). The present analysis revealed the character of each final state. For example, the final state with the smallest E_B is the bonding state between $|f^1\rangle$ and $|(f^2)^*v_1\rangle$. The trend in the E_B of bare $|f^1\rangle$ corresponds to some extent with the trend in the peak position of Pr f PDOS in Fig. 13.7(a). The origin of these trends could be understood as E_B of the Pr 4f electron

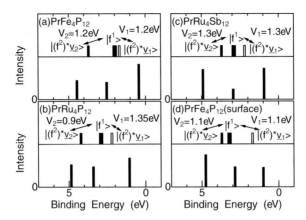

Fig. 13.8. Pr $4f$ spectrum reproduced by the cluster model for (**a**) PrFe$_4$P$_{12}$; (**b**) PrRu$_4$P$_{12}$; (**c**) PrRu$_4$Sb$_{12}$, and (**d**) surface of PrFe$_4$P$_{12}$. *Upper panels:* Binding energies of bare $|f^1\rangle$ and $|(f^2)^*v_k\rangle$ final states are shown by the bars and the effective hybridization between $|f^1\rangle$ and $|(f^2)^*v_k\rangle$ are given by V$_k$ ($k=1,2$). *Lower panels:* The resultant final states are shown. The *bars* show the $4f$ spectral weights in the final states. For details, see text

becomes smaller when the negative X ion comes spatially closer to the Pr atom through PrRu$_4$Sb$_{12}$, PrRu$_4$P$_{12}$ and PrFe$_4$P$_{12}$.

The present Pr $4f$ spectrum of PrFe$_4$P$_{12}$ obtained from the bulk-sensitive $3d$–$4f$ RPES is qualitatively different from that obtained from the surface-sensitive $4d$–$4f$ RPES [63]. The surface-sensitive spectrum also has a three peak structure but the peak at $E_{\rm B} \sim 4.5$ eV is the strongest and the intensity at $E_{\rm F}$ is negligible. The origin of the difference is the increase of the localization of $4f$ electrons at the surface, in other words, the increase of the $4f$ binding energy and the decrease of the hybridization. By making such changes in E_0 and V_k, the surface-sensitive spectrum is reproduced (see Fig. 13.8(d)).

The high-resolution Pr $3d$–$4f$ RPES spectra near $E_{\rm F}$ are shown in Fig. 13.9(a). The most prominent feature is the strong peak of PrFe$_4$P$_{12}$ at $E_{\rm B} \simeq 100$ meV. The Pr $4f$ spectra of PrRu$_4$P$_{12}$ and PrRu$_4$Sb$_{12}$, on the other hand, decrease continuously with some humps upon approaching $E_{\rm F}$. Spectral features similar to PrRu$_4$P$_{12}$ and PrRu$_4$Sb$_{12}$ have been found for very localized Ce systems such as CePdAs, in which Ce $4f$ takes nearly pure $4f^1$ state [64]. This indicates that pure $4f^2$ state is realized in PrRu$_4$P$_{12}$ and PrRu$_4$Sb$_{12}$. On the other hand, spectral similarity between PrFe$_4$P$_{12}$ and Kondo Ce compounds (see Subsect. 13.3.4 and [11,12]) suggests that the Pr $4f^2$-dominant Kondo state, with the finite contribution of $4f^1$ or $4f^3$ state, is formed in PrFe$_4$P$_{12}$. The present energy resolution of $\sim$80 meV exceeds the characteristic energy $k_{\rm B}T_{\rm K} \sim 1$ meV for PrFe$_4$P$_{12}$. KR has been observed even in such cases, for example, for CeRu$_2$Si$_2$ ($T_{\rm K} \sim 20$ K) (see Subsect. 13.3.4

and [12]) and YbInCu$_4$ ($T_{\rm K} \sim 25$ K for $T > 42$ K) [65] with energy resolution of $\sim$100 meV.

In the Kondo Ce (Yb) system, KR is accompanied by the spin-orbit partner, the $E_{\rm B}$ of which corresponds to the spin-orbit excitation, $J = 5/2 \to 7/2$ ($J = 7/2 \to 5/2$), of the $4f^1$ ($4f^{13}$)-dominant state [21]. KR in Pr would then be accompanied by satellites corresponding to the excitation from the ground state ($^3H_4$) to excited states ($^3H_5$, $^3H_6$, $^3F_2$, and so on) of the $4f^2$ states. Fig. 13.9(a) shows that the on-resonance spectrum of PrFe$_4$P$_{12}$ have structures at $\sim$0.3 and $\sim$0.6 eV which correspond to the lowest few excitation energies.

KR is expected to depend upon temperature reflecting the temperature dependence of the $4f$ occupation number. In fact, a temperature dependence was found as the temperature approaches the suggested $T_{\rm K} \sim 10$ K as shown in Fig. 13.9(b). The temperature dependence was reproducible in both heat-up and cool-down processes. The temperature dependence is characterized not only by the narrowing of the $\sim$0.1 eV structure but also by the increase of the weight of all the structures at $\sim$ 0.1, $\sim$0.3, and $\sim$0.6 eV. Although the former can be attributed at least partly to the thermal broadening, the latter should

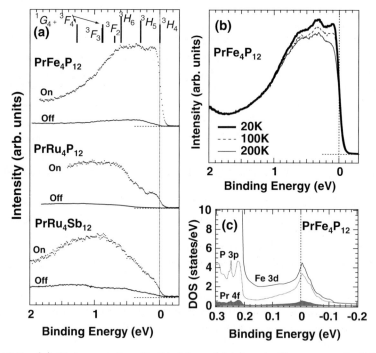

Fig. 13.9. (a) High-resolution Pr $3d$–$4f$ on- (*dots*) and off-resonance (*solid lines*) spectra near $E_{\rm F}$ at 20 K. The *vertical lines* show the energy positions of the atomic $4f^2$ multiplets with the ground state set at $E_{\rm F}$; (b) temperature dependence of the on-resonance spectrum of PrFe$_4$P$_{12}$; (c) calculated partial density of states

be attributed to intrinsic temperature dependence of the excitation spectrum. Therefore it is quite possible that the ∼0.1 eV structure is the KR and the ∼0.3 and ∼0.6 structures are its satellite structures.

The temperature dependence can be a vital clue to check whether the observed structure is the Kondo peak itself or the tail of the Kondo peak centered above E_F. These cases correspond respectively to the $c_2|f^2\rangle + c_3|f^3\rangle$ or $d_1|f^1\rangle + d_2|f^2\rangle$ initial states, where the hole or electron in the valence or conduction band is not denoted explicitly. The NCA calculation based on the SIAM for the Ce system [66] shows that, as temperature is lowered, the Kondo tail is sharpened [67] but the *weights* of both the Kondo tail and its spin-orbit partner *decrease* when the spectra are normalized in a similar way as in Fig. 13.9(b). This contradicts the present temperature dependence for $PrFe_4P_{12}$. On the other hand, for Yb systems, it is well known that the intensities of both the Kondo peak itself and its spin-orbit partner increase with decreasing temperature [7]. Since this is consistent with the $PrFe_4P_{12}$'s temperature dependence, we believe that the observed structure is the Kondo peak itself, and therefore that the initial state is dominated by $c_2|f^2\rangle + c_3|f^3\rangle$. We consider that the Kondo peak at around $k_B T_K \sim 1$ meV is broadened due to the energy resolution of ~ 80 meV resulting in the observed structure at ∼100 meV.

The microscopic origin of the c–f hybridization is considered to be the P $3p$ – Pr $4f$ mixing since the nearest neighbors of the Pr atom are the twelve P atoms. The large coordination number definitely enhances the effective p–f mixing. It has been pointed out that the calculated P p PDOS of RFe_4P_{12} shows a sharp peak in the vicinity of E_F [58, 68]. This is also the case for $PrFe_4P_{12}$ as shown in Fig. 13.9(c). Therefore, the large P $3p$ PDOS at E_F together with the large effective P $3p$ – Pr $4f$ mixing is interpreted to be the origin of the Kondo state in $PrFe_4P_{12}$.

13.4.3 Hard X-Ray Core-level Photoemission of Pr Compounds

Pr $3d$ core-level spectra of $PrFe_4P_{12}$ and $PrSn_3$ measured at $h\nu = 2450$ eV are shown in Fig. 13.10(a), in which the spectrum of Pr metal is also displayed as a reference. In order to obtain these spectra, a single crystal of $PrFe_4P_{12}$ was fractured in situ at the measuring temperature of 180 K whereas $PrSn_3$ and Pr metal were scraped in situ. For all materials we observed double peak structures originating from the $3d_{5/2}$ and $3d_{3/2}$ components split by the spin-orbit interaction of the core hole [63,69,70]. At the first glance, the branching ratios between these components are different for these materials, which is discussed later. The multiplet structure of $PrSn_3$ and especially Pr metal is more remarkable than that of $PrFe_4P_{12}$, reflecting the localization of the $4f$ electrons. In the $3d_{5/2}$ component region one can see the intensity of the $|3d^9 4f^3\rangle$ final state on the lower binding energy side of the $|3d^9 4f^2\rangle$ final state for all materials. In particular, the spectrum of $PrFe_4P_{12}$ has the significant intensity of the $|3d^9 4f^3\rangle$ final state, suggesting the strong hybridization between

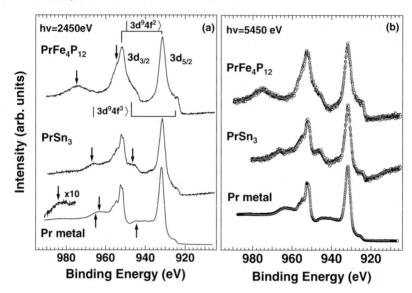

Fig. 13.10. (a) Pr $3d$ core-level spectra of the *fractured* PrFe$_4$P$_{12}$, *scraped* PrSn$_3$, and *scraped* Pr metal measured at $h\nu = 2450$ eV. The spectral intensities are normalized to the intensity of the Pr $3d_{5/2}$ peak. *Arrows* indicate the positions of expected energy loss structures. In the Pr metal, the structures shown by *up-arrows* (*down-arrows*) are attributed to the energy loss components of the $3d_{5/2}$ ($3d_{3/2}$) peak; (b) Pr $3d$ core-level spectra of PrFe$_4$P$_{12}$, PrSn$_3$ and Pr metal taken at $h\nu = 5450$ eV. *Dots* denote the experimentally obtained the Pr $3d$ core-level spectra. *Thick solid lines* show the calculated spectra

the Pr $4f$ and conduction electron states in comparison with PrSn$_3$ and Pr metal. In the $3d_{3/2}$ region various spectral shapes and different intensities are observed in three materials. In the Pr metal, even a dip structure is seen at 948 eV. Figure 13.10(b) shows the Pr $3d$ core-level spectra measured at $h\nu = 5450$ eV with higher bulk sensitivity due to the longer escape depth of the photoelectron in solids. The Sn $3s$ peak, which is located at about 885 eV in the spectrum of PrSn$_3$, shows strong satellites at about 899 and 913 eV due to the increase of the photoionization cross section of the Sn $3s$ state relative to that of Pr $3d$ states [20]. The intensity of the $|3d^94f^3\rangle$ final state relative to the $|3d^94f^2\rangle$ final state does not change drastically between these two $h\nu$s.

In Fig. 13.10 some weak satellites can be seen on the higher binding energy sides of the main peaks. Corresponding satellites are also clearly seen in other core-level spectra such as the P $3s$ of PrFe$_4$P$_{12}$, Sn $3d$ of PrSn$_3$, and Pr $3p$ spectra of Pr metal, in which the satellites are located at 22.8, 14.3, and 10.5 eV above the main peaks for these three materials, respectively. These energies are in agreement with the calculated bulk plasmon energies, for which the free electron gas model is employed. In Pr metal, another satellite is seen at 31.5 eV above the main peak. Arrows in Fig. 13.10(a) indicate the energies of expected energy loss structures which are identified from the spectra of other core levels and positioned relative to the $3d_{5/2}$ and $3d_{3/2}$ main peaks. The energy positions in the Pr $3d$ core-level spectra of all materials well correspond to those in other core levels. Therefore, we conclude that the satellites except for that at 31.5 eV in Pr metal originate from the energy loss process related to the bulk plasmon excitation. We note that any spectral intensities of surface plasmons, which can occur near $1/\sqrt{2}$ of the bulk plasmon excitation energy, are not observed. The higher-energy loss peak in Pr metal may originate from different excitations.

Figure 13.11 shows the calculated Pr $3d$ core-level spectrum based on the SIAM in comparison with the measured $PrFe_4P_{12}$ spectrum for $h\nu = 2450$ eV. The Pr $3d$ core-level spectrum contains large energy loss contributions. Therefore, the calculated spectrum is composed of the calculated Pr $3d$ core-level spectrum and its energy loss spectrum as shown in the bottom of Fig. 13.11. Before summing both spectra and adding the Shirley background [71], the calculated Pr $3d$ core-level spectrum is broadened by a Lorentzian and a Gaussian (with FWHM of $\gamma_{main} = 2.0$ eV and $\Gamma = 0.5$ eV, respectively) accounting for the lifetime of the $3d$ core hole and the total energy resolution of the experimental setup, respectively. The energy loss spectrum is generated through an adequate broadening of the Pr $3d$ core-level spectrum to reproduce the intensity of the energy loss structure (Lorentzian width $\gamma_{pl} = 10.4$ eV) with shifting the energy by the bulk plasmon excitation energy to the higher binding energy side of the Pr $3d$ core-level spectrum. The calculated spectrum reproduces well the experimental Pr $3d$ core-level spectrum including the relative intensity of the multiplet structures. The intraatomic Coulomb repulsive energy between the $4f$ electrons U_{ff} in calculations is fixed to 7.70 eV for consistency with the energy separation between the $|4f^1\rangle$ and $|4f^3\rangle$ peaks in the Pr $3d$–$4f$ RPES spectra and bremsstrahlung isochromat [72] spectra, respectively.

We discuss the parameters obtained from the comparison between the experimental and calculated results. The effective hybridization strengths V_{eff} are very different among these materials, as summarized in Table 13.2. The V_{eff} in $PrFe_4P_{12}$ is larger than that in the localized Pr metal. That is consistent with other experimental and theoretical results, suggesting the strong hybridization between the Pr $4f$ and the P $3p$ states [47]. Even in comparison with $PrSn_3$, $PrFe_4P_{12}$ seems to have a stronger c–f hybridization strength. This could be due to the unique crystal structure of filled skutterudite, where Pr atom is located at the center of the icosahedron of the P atoms. The relative ratio of the configuration $|f^n\rangle$ ($n = 1$–3) and the number of the $4f$ electrons n_f are also listed in Table 13.2. The ratio of $|4f^3\rangle$ to $|4f^2\rangle$ configuration tends to increase with $h\nu$ from $h\nu = 1253.6$ eV ($E_k \simeq 300$ eV) to 2450 eV ($E_k \simeq 1500$ eV), suggesting the number of the $4f$ electron in the bulk is larger than that in the topmost surface due to stronger hybridization in the bulk. n_f in $PrFe_4P_{12}$ deviates considerably from the nominal trivalent electron number ($n_f = 2$), which is in strong contrast to the case of the localized Pr system. In Ce compounds, the $|4f^{n-1}\rangle$ configuration contributes from 0.3% to

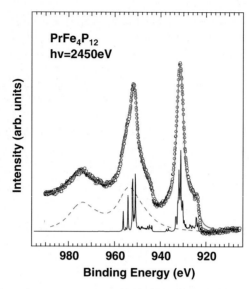

Fig. 13.11. Calculated Pr 3d core-level spectra of PrFe$_4$P$_{12}$. *Thick solid line* denotes the sum of the calculated spectra, being the Pr 3d core-level spectrum and plasmon contribution in addition to a Shirley background. *Bristle-shaped solid line* shows the calculated Pr 3d core-level spectrum before being broadened by the Lorentzian and Gaussian. *Broken line* shows the broadened Pr 3d core-level spectrum with the energy shift corresponding to the energy loss contribution. *Open circles* show the Pr 3d core-level spectrum experimentally observed at $h\nu = 2450$ eV

40% in mixing with the $|4f^n\rangle$ configuration due to the c–f hybridization [64]. For the present Pr Kondo systems our theoretical calculations suggest that a large fraction of the $|4f^{n-1}\rangle$ configuration mixed into $|4f^n\rangle$ would lead to a T_K below 1 K [64], which is inconsistent with other experimental results [47]. On the other hand, our measurements have shown that the mixture of the $|4f^{n+1}\rangle$ to the $|4f^n\rangle$ configuration is dominant in the Pr Kondo systems, as shown in Table 13.2. The deviation from the nominal $4f$ electron number in PrFe$_4$P$_{12}$ ($n_f = 2.07$) is comparable to that in the Ce(Pd$_{1-x}$Cu$_x$)$_3$ systems, where n_f is 0.93 for $x = 0.033$ estimated by bulk-sensitive valence band PES and SIAM calculations with NCA [66]. Both of them are known to have similar T_K (several ten K). This obviously indicates that the mixture between the $|4f^n\rangle$ and $|4f^{n+1}\rangle$ should play an important role in revealing the Kondo resonance in Pr Kondo systems.

Acknowledgments

This work was performed in collaboration with M. Hedo, Y. Haga, Y. Ōnuki, R. Settai, S. Araki, T. Nanba, S. Kunii, H. Sugawara, H. Sato, C. Sekine, I. Shirotani, Y. Saitoh, S. Ueda, S. Tanaka, H. Harima, M. Tsunekawa, C. Dallera,

Table 13.2. Fitted effective hybridization strength ($V_{eff} = \sqrt{12}V$), the charge transfer energy (Δ), calculated Pr $4f$ configurations and the $4f$ electron numbers in the initial state. Errors are listed in ()

Sample	$h\nu$ (eV)	V_{eff}	Δ	f^1 (%)	f^2 (%)	f^3 (%)	n_f	Ref.
PrFe$_4$P$_{12}$	1253.6	–	–	–	95	5	2.05	[63]
	2450	1.2(0.1)	−4.0	0.8	91.8(0.3)	7.4(0.3)	2.07	
	5450	1.2(0.1)	−4.0	0.8	91.5(0.4)	7.7(0.3)	2.07	
PrSn$_3$	1253.6	–	–	–	98	2	2.02	[63]
	2450	0.7	−4.0	0.3	96.6(0.3)	3.1(0.3)	2.03	
	5450	0.8(0.1)	−4.0	0.4(0.1)	96.0(0.3)	3.6(0.3)	2.03	
Pr metal	1253.6	–	–	–	99	1	2.01	[63]
	2450	0.6	−4.5	0.3	98.0(0.3)	1.7(0.2)	2.02	
	5450	0.6	−4.5	0.3	98.0(0.3)	1.7(0.2)	2.02	

L. Braicovich and T. -L. Lee. We are grateful to the staff of SPring-8, especially to T. Muro and T. Matsushita for supporting the experiments. This work was supported by a Grant-in-Aid for COE Research (10CE2004), 21st Century COE Research (G18), and Creative Scientific Research (15GS0213) from the Ministry of Education, Culture, Sports, Science and Technology (MEXT), Japan. The XAS and PES measurements were performed with the approval of the Japan Synchrotron Radiation Research Institute.

References

1. R. R. Joseph et al.: Phys. Rev. B **6**, 3286 (1972)
2. Y. Ōnuki and T. Komatsubara: J. Magn. Magn. Mater. **63–64**, 281 (1987)
3. T. Fujita et al.: Solid. State. Commun. **35**, 569 (1980)
4. F. Patthey et al.: Phys. Rev. B **42** 8864 (1990)
5. E. Weschke et al.: Phys. Rev. B **44**, 8304 (1991)
6. J. J. Joyce et al.: Phys. Rev. Lett. **68**, 236 (1992)
7. L. H. Tjeng et al.: Phys. Rev. Lett. **71**, 1419 (1993)
8. M. Garnier et al.: Phys. Rev. Lett. **78**, 4127 (1997)
9. S. Tanuma et al.: Surf. Sci. **192** L849 (1987)
10. L. Duò et al.: Phys. Rev. B **54**, R17363 (1996)
11. A. Sekiyama et al.: Nature **403**, 396 (2000)
12. A. Sekiyama et al.: J. Phys. Soc. Jpn. **69**, 2771 (2000)
13. A. Sekiyama et al.: Solid State Commun. **121**, 561 (2002)
14. A. Sekiyama and S. Suga: Physica B **312**, 634 (2002)
15. A. Sekiyama et al.: Acta Physica Polonica B **34**, 1105 (2003)
16. A. Yamasaki et al.: Phys. Rev. B **70**, 113103 (2004)
17. A. Sekiyama et al.: J. Electron Spectr. Relat. Phenom. **144–147**, 655 (2005)
18. A. Yamasaki et al.: J. Phys. Soc. Jpn. **74**, 2538 (2005)
19. Y. Saitoh et al.: Rev. Sci. Instrum. **71**, 3254 (2000)

20. J. J. Yeh and I. Lindau: At. Data Nucl. Data Tables **32**, 1 (1985)
21. O. Gunnarsson and K. Schönhammer, Phys. Rev. B **28**, 4315 (1983)
22. N. E. Bickers et al.: Phys. Rev. Lett. **54**, 230 (1985)
23. J. W. Allen et al.: Adv. Phys. **35**, 275 (1986)
24. A. B. Andrews et al.: Phys. Rev. B **53**, 3317 (1996)
25. A. Loidl et al.: Physica B **156–157**, 794 (1989)
26. A. Loidl et al.: Phys. Rev. B **46**, 9341 (1992)
27. N. Sato et al.: J. Phys. Soc. Jpn. **54**, 1923 (1985)
28. A. Amato et al.: J. Magn. Magn. Mat. **76–77**, 263 (1988)
29. S. Araki et al.: J. Phys. Soc. Jpn. **68**, 3334 (1999)
30. T. Takabatake et al.: J. Magn. Magn. Mater. **177–181**, 277 (1998)
31. D. T. Adroja et al.: Solid State Commun. **66**, 1201 (1988)
32. T. Takabatake et al.: Phys. Rev. B **41**, 9607 (1990)
33. J. C. Fuggle et al.: Phys. Rev. B **27** 4637 (1983)
34. Y. Kuramoto: Z. Phys. B **53**, 37 (1983)
35. E. Zirngiebl et al.: Phys. Rev. B **30**, 4052 (1984)
36. R. Felten et al.: J. Magn. Magn. Mat. **63–64**, 383 (1987)
37. A. Severing et al.: Phys. Rev. B **39**, 2557 (1989)
38. P. Haen et al.: J. Low Temp. Phys. **67**, 391 (1987)
39. K. Hanzawa et al.: J. Magn. Magn. Mat. **47-48**, 357 (1985)
40. Y. Ōnuki et al.: J. Phys. Soc. Jpn. **53**, 2734 (1984)
41. J.-S. Kang et al.: Phys. Rev. B **60**, 5348 (1999)
42. S. Tanaka et al.: J. Phys. Soc. Jpn. **67**, 1342 (1998)
43. M. Hedo et al.: J. Phys. Soc. Jpn. **64**, 4535 (1995)
44. E. Weschke et al.: Phys. Rev. Lett. **69**, 1792 (1992)
45. A. D. Huxley et al.: J. Phys.: Condens. Matter **5**, 7709 (1993)
46. A. Yatskar et al.: Phys. Rev. Lett. **77**, 3637 (1996)
47. H. Sugawara et al.: Phys. Rev. B **66**, 134411 (2002)
48. Y. Nakanishi et al.: Phys. Rev. B **63**, 184429 (2001)
49. Y. Aoki et al.: Phys. Rev. B **65**, 064446 (2002)
50. D. L. Cox: Phys. Rev. Lett. **59**, 1240 (1987)
51. T. M. Kelley et al.: Phys. Rev. B **61**, 1831 (2000)
52. C. Sekine et al.: Phys. Rev. Lett. **79**, 3218 (1997)
53. N. Takeda and M. Ishikawa: J. Phys. Soc. Jpn. **69**, 868 (2000)
54. E. D. Bauer et al.: Phys. Rev. B **65**, 100506R (2002)
55. M. S. Torikachvili et al.: Phys. Rev. B **36**, 8660 (1987)
56. R. Settai et al.: J. Phys. Soc. Jpn. **69**, 3983 (2000)
57. B. T. Thole et al.: Phys. Rev. B **32**, 5107 (1985)
58. H. Harima and K. Takegahara: Physica B **312-313**, 843 (2002)
59. R. D. Parks et al.: Phys. Rev. Lett. **52**, 2176 (1984)
60. S. Suga et al.: Phys. Rev. B **52**, 1584 (1995)
61. Yu. Kucherenko et al.: Phys. Rev. B **65**, 165119 (2002)
62. A. Fujimori: Phys. Rev. B **27**, 3992 (1983)
63. H. Ishii et al.: J. Phys. Soc. Jpn. **71**, 156 (2002)
64. T. Iwasaki et al.: Phys. Rev. B **65**, 195109 (2002)
65. H. Sato et al.: J. Synchrotron Rad. **9**, 229 (2002)
66. S. Kasai et al.: unpublished
67. F. Reinert et al.: Phys. Rev. Lett. **87**, 106401 (2001)
68. H. Sugawara et al.: J. Phys. Soc. Jpn. **69**, 2938 (2000)
69. F. U. Hillebrecht and J. C. Fuggle: Phys. Rev. B **25**, 3550 (1982)
70. S. Suga et al.: Phys. Rev. B **52**, 1584 (1995)
71. D. A. Shirley: Phys. Rev. B **5**, 4709 (1972)
72. A. Yamasaki: Ph.D. thesis, Osaka University (2002)

14

Hard X-Ray Photoemission Spectroscopy

Y. Takata

RIKEN SPring-8 Center, Sayo-gun, Hyogo 679-5148, Japan
takatay@spring8.or.jp

Abstract. An instrument for high energy (6 keV), high resolution (75 meV) photoemission experiments is described. The specifications allow data accumulation times similar to those in the low photon energy regime. The power of the instrument is demonstrated with data of core level and valence band spectra taken for Au and Si in a few minutes. The large probing depth at these high energies makes the spectra very surface insensitive. Data are presented for a semiconductor interface, magnetic semiconductors, an f-electron system with a valence transition and $3d$-electron systems.

14.1 Introduction

Photoemission spectroscopy (PES) has been used extensively to experimentally determine electronic structure of core levels and valence bands (VBs) [1]. However, conventional PES is surface sensitive because of short inelastic mean free paths (IMFPs) [2–4]. In order to attain larger probing depths of VBs than that in vacuum ultraviolet (VUV) spectroscopy, soft x-ray (SX) VB-PES using synchrotron radiation (SR) has recently become attractive [5]. However, it is obvious that SX-PES is still surface sensitive, because, for example, the IMFPs of a valence electron are only 1.3 and 2 nm for Au and Si at a kinetic energy (KE) of 1 keV, respectively [4]. In the case of core levels, smaller KEs than those of VBs enhance the surface sensitivity, making it rather difficult to probe bulk character [6].

In contrast to the above-mentioned surface sensitive PES techniques, the IMFP values of a valence electron for Au and Si increase to 5.5 and 9.2 nm, respectively at 6 keV [4], which lies in the range of hard x-rays. The straightforward way to realize an intrinsic bulk probe is to increase KE of photoelectrons by use of hard x-rays. The first feasibility test of hard x-ray (HX)-PES was done by Lindau et al. in 1974 using a 1st generation SR source. However, the feeble signal intensity even of the Au $4f$ core level excluded the possibility of studies of VBs [7]. What has prevented HX VB-PES is the rapid decrease in subshell photo-ionization cross section (σ). The σ values for Au $5d$ (1×10^{-5} Mb) and Si $3p$ (3×10^{-5} Mb) at 6 keV are only 1–2% of those at 1 keV [8].

In order to realize HX-PES with high-energy resolution and high throughput, both high-brilliance SR and a high performance electron energy analyzer are required. After 2nd generation SR became available, there have been a few reports on core level photoemission and resonant Auger spectroscopy using several keV x-rays [9, 10]. In the last few years, unprecedented high-flux and high-brilliance SR at third generation facilities such as ESRF, APS and SPring-8 has stimulated us to develop HX-PES with high-energy resolution and high throughput. The results of the feasibility test at the excitation energy of 6 keV done at SPring-8 demonstrated the capability to probe intrinsic bulk electronic structure of both core levels and VBs [11, 12]. HX-PES has also been developed at ESRF [13–15]. All these experimental achievements indicate that HX-PES will contribute significantly to the study of electronic structure of solids [9–15]. In 2003, the first workshop on HX-PES was held at ESRF, and the potential of HX-PES for study of the depth-resolved electronic structure, buried layers, interfaces, ultrashallow junctions and the bulk electronic structure of strongly correlated electron systems has been discussed and recognized [16]. Here, we describe the HX-PES experimental work carried out at SPring-8 and the achieved state of the art, as well as typical applications to a variety of materials.

14.2 Experimental Aspects

The essential problem to overcome and realize HX-PES is weak signal intensity due to small σ values as pointed out above. Of course, intensity of x-rays and detection efficiency of an electron energy analyzer are critical factors. In addition to these, configuration of the experimental setup also influences the signal intensity. Figure 14.1 shows IMFPs up to KE of 10 keV for several materials [4]. IMFPs at the KE of 6 keV range from 4 to 15 nm and are almost 5 times larger than those at 1 keV. However, these values are much shorter than the x-ray attenuation length (30 μm for Si and 1 μm for Au at 6 keV). In order to avoid wasting x-rays in the region deeper than the electron escape depth, grazing incidence of x-rays relative to the sample surface is desirable.

The detection angle of a photoelectron relative to the polarization vector of x-rays also plays a role in gaining photoelectron intensity. When we use linearly polarized light as an excitation source, photoelectrons from free atoms show an angular distribution depending on the asymmetry parameter β (see Eq. (5) in [8]) as shown in Fig. 14.2. For HX-PES, the typical photon energy is above 6 keV, and almost all subshells have positive β values [8]. In this case, the photoelectron intensity has a maximum along the direction parallel to the polarization vector. This behavior is considered applicable even to solids.

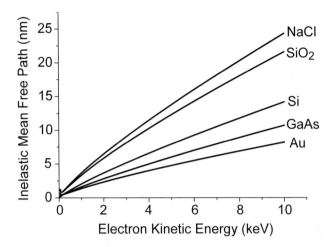

Fig. 14.1. Inelastic mean free paths for electron kinetic energies up to 10 keV, for Au, GaAs, Si, SiO$_2$, and NaCl [4]

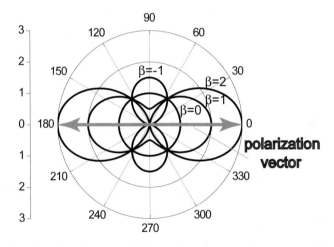

Fig. 14.2. Angular distribution of photoelectrons from free atoms. For positive asymmetry parameter β, the intensities have a maximum along the direction of the polarization vector

On the other hand, to achieve large probing depth, photoelectrons should be detected along the direction close to the normal of the sample surface.

Following these considerations, an HX-PES apparatus with the configuration shown in Fig. 14.3 [12, 17] has been constructed at an x-ray undulator beamline BL29XU [18] in SPring-8. The lens axis of the analyzer is placed perpendicular to the x-ray beam and the incidence angle relative to the sample surface is typically set to about 1° for samples with a flat surface. The first-version electron energy analyzer (a modified SES-2002) has recently been

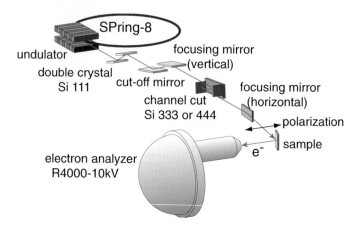

Fig. 14.3. Schematic of experimental setup including x-ray optics at the beamline BL29XU in SPring-8 [17]

replaced by a newly developed one, R4000-10kV (Gammadata-Scienta Co.), and the measurable kinetic energy has been extended from 6 to 10 keV. In addition to the analyzer, a sample manipulator with a motorized XYZΘ stage, a flow-type He cryostat for sample cooling, two turbo-molecular pumps, and CCD cameras are equipped on the measurement chamber. The whole system including load-lock and preparation chambers is mounted on a position adjustable stage. The design is made as compact as possible so as to carry the apparatus into the experimental hutch. The vacuum of the measurement chamber is 10^{-8} Pa, and the lowest sample temperature achieved is 20 K.

In order to realize high energy resolution and high throughput, x-ray optics dedicated for HX-PES is essential. Figure 14.3 shows the schematic of the optics at BL29XU. X-rays from an undulator are premonochromatized with a Si 111 double-crystal monochromator. A channel-cut Si monochromator is placed downstream to realize high energy resolution. The Bragg angle is fixed at 85°, and Si 333 and 444 reflections are used for 5.95 and 7.35 keV x-rays, respectively. The incident energy bandwidth is less than 60 meV. Horizontal and vertical focusing mirrors are installed and the spot size at the sample position is 60 (vertical) × 70 (horizontal) μm^2 with the x-ray intensity of 10^{11} photons/sec. Details of x-ray optics for HX-PES are described by Ishikawa et al. [19]. The fine focus considerably increases the photoelectron intensity because the lens system of the analyzer magnifies the spot size on the sample surface by 5 times at the entrance slit of the hemispherical analyzer.

14.3 Performance and Characteristics

In this section, typical spectra are shown to demonstrate the high energy resolution and high throughput of HX-PES and also to characterize this method.

14.3.1 High Throughput and High Energy Resolution

Figure 14.4 shows HX-PES spectra of a Au plate measured at 35 or 20 K with the excitation energy of 5.95 keV. In the Au $4f$ core level spectrum measured with the analyzer pass energy (E_p) of 200 eV (thick solid curve in (a)), the signal-to-noise ratio is very good even with short accumulation times of 30 sec [17]. The total instrumental energy resolution including the x-ray band width was 280 meV in this conventional setting. The dotted spectrum measured with $E_p = 50$ eV can be fitted with a pure Lorentzian function (thin solid curve) with FWHM of 335 meV, indicating the experimental resolution is much less than the lifetime broadening. The top spectrum in Fig. 14.4(b) is the VB spectrum measured at 5.95 keV with $E_p = 200$ eV in the early stage of the instrumental development [12]. After the improvements of the apparatus as described in Sect. 14.2, the same kind of spectra can be measured within several minutes [17], indicating the throughput is quite high even for VB

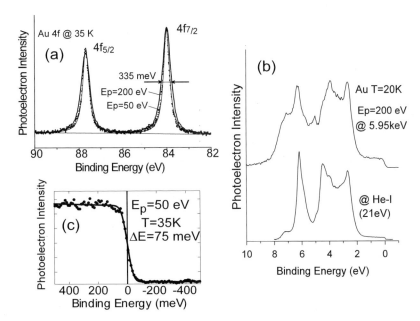

Fig. 14.4. (a) Au $4f$, (b) valence band and (c) Fermi-edge spectra of a Au plate measured at 35 K or 20 K with 5.95 keV excitation. Thick *solid curve* in (a) and *top curve* in (b) were measured spectra with E_p of 200 eV. *Dotted* Au $4f$ spectrum in (a) and Fermi-edge profile in (c) were measured with $E_p = 50$ eV. The total energy resolution of 75 meV ($E/\Delta E = 79000$) was evaluated for 5.95 keV photoelectrons by fitting the Fermi-edge profile

measurements. A comparison with the VUV spectrum (bottom) measured at 21 eV with the energy resolution of 30 meV shows that all the features are also observed in the HX-PES spectrum. The difference in relative intensities between these spectra is simply due to change in the ratio of $\sigma(\text{Au }5d)$ and $\sigma(\text{Au }6s)$. A similar study was reported recently for the VB spectra of Ag [20]. Figure 14.4(c) shows Fermi-edge profile at 35 K with $E_p = 50$ eV. By fitting this profile, the total instrumental energy resolution at 5.95 keV is determined to be 75 ± 2 meV ($E/\Delta E = 79000$). At 7.35 and 9.92 keV, the highest resolution of 90 and 93 meV was realized, respectively [21]. At ESRF, Torelli et al. also achieved almost the same energy resolution of 71 ± 7 meV at 5.93 keV [15].

14.3.2 Surface Insensitivity

The capability of HX-PES to probe bulk states of reactive surfaces has been demonstrated by the spectra of a Si(100) surface with a thin-SiO$_2$ layer [12]. Figure 14.5(a) shows the Si $2p$ (binding enegy, BE ~ 100 eV) spectrum of 0.8 nm-SiO$_2$/Si(100) measured at 7.94 keV. The peak intensity of the surface SiO$_2$ layer is negligibly small ($\sim 2\%$) in comparison with that of substrate Si. Negligible surface contribution is also confirmed for the VB spectrum in Fig. 14.5(b). Comparing the HX spectrum with the SX (0.85 keV) spectrum, the structures marked by arrows in the SX spectrum are due to the 0.58 nm-surface SiO$_2$ layer. These features almost vanish in the HX spectrum.

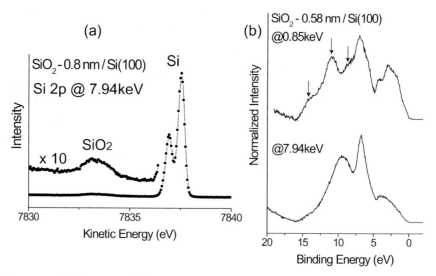

Fig. 14.5. (a) Si $2p$ and (b) valence band spectra of a Si(100) surface with a thin-SiO$_2$ layer measured at 7.94 keV. SX valence band spectrum measured at 0.85 keV is also shown as a reference. *Arrows* indicate the structure originating from the surface SiO$_2$ layer

The "surface insensitivity" of HX-PES enables us to investigate the intrinsic bulk state of thin films which are beyond the reach of "surface sensitive" PES. This is because surface-sensitive PES necessarily requires surface cleaning and preparation procedure. It should be noted that these Si $2p$ and VB spectra can be obtained within short acquisition times of 30 and 300 sec, respectively.

14.3.3 Large Probing Depth

In addition to "surface insensitivity", the large probing depth of HX-PES extends the applicability to embedded layers and interfaces in nano-scale buried layer systems. Figure 14.6 shows the Sr $2p_{3/2}$ spectra of a bare SrTiO$_3$ (STO) substrate and the substrate covered with a thin layer (20 nm) of La$_{0.85}$Ba$_{0.15}$MnO$_3$ (LBMO) measured at 5.95 keV [17]. The Sr $2p_{3/2}$ (BE $\sim$ 1940 eV) photoelectrons from the substrate with the KE of 4010 eV are still observable through the 20 nm thick overlayer. The small KE difference between these two samples is attributed to band bending. From the intensity variation, the IMFP value of electrons with KE = 4010 eV in LBMO is estimated as 4 nm. Concerning the probing depth of HX-PES, Sacchi et al. have recently determined the effective attenuation length over the KE range from 4 to 6 keV in Co, Cu, Ge and Gd$_2$O$_3$, and showed the use of HX-PES for studying buried layers and interfaces [22]. Dallera et al. have recently reported a study on AlAs layer buried under different thickness of GaAs, emphasizing the role of HX-PES in non-destructive analysis of buried layers [14].

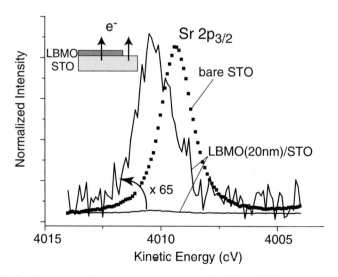

Fig. 14.6. Sr $2p_{3/2}$ core level spectra of bare SrTiO$_3$ (STO) substrate and the substrate covered with a thin layer (20 nm) of La$_{0.85}$Ba$_{0.15}$MnO$_3$ (LBMO) measured at 5.95 keV [17]. The Sr $2p_{3/2}$ (BE = 1940 eV) photoelectrons from the substrate with KE of 4010 eV are still observable through the 20 nm thick overlayer

14.4 Applications

The performance and the characteristics of HX-PES described in Sect. 14.3 has expanded the range of applicable targets. It has enabled us to investigate depth-resolved electronic structure, buried layers, interfaces, ultrashallow junctions and the bulk electronic structure of strongly correlated electron systems. This method has also given us new knowledge of core level PES. In this section, typical studies with HX-PES are presented.

14.4.1 Interfacial Chemical Structure in Si High-k Insulator System

Non-destructive chemical state analysis of layered materials is becoming more important for the current and future technologies of nano-science and engineering. HX-PES with larger probing depth than that for the PES using Al Kα (1486.6 eV) or Mg Kα (1253.6 eV) source greatly widens the applicability to various materials with nano-layered structures and nano-particles. As an example, we show a HX-PES study on high-k CMOS gate dielectrics [11], which urgently need investigation for future Si-ULSI devices.

Figure 14.7 shows Si 1s spectra of $HfO_2/SiO_2/Si(100)$ structure before and after rapid thermal annealing (RTA) for 5 sec at 1000 °C measured at 5.95 keV with the take-off-angle TOA (relative to the sample surface) of 30°. The sample was prepared by atomic layer deposition (ALD) of a 4nm thick HfO_2 layer on a 0.8 nm thick chemical oxide (SiO_2) layer. The spectrum of a sample with 1.32 nm SiO_2 on Si(100) is also shown as a reference. After deposition of the HfO_2 film, the Si 1s peak for the intermediate layer appears at about 0.6 eV lower BE and is broader than for the SiO_2 peak. This peak is attributed to Hf silicate, indicating that formation of Hf silicate has already taken place during the HfO_2 deposition. RTA enhances the intensity of the Hf silicate peak. This result suggests that silicate formation is related to diffusion of Si atoms into the deposited layers from the Si substrate. A detectable increase in the spectral intensity appearing on the low BE side of the substrate peak (as shown by the dotted curve in Fig. 14.7) indicates the formation of Hf-Si bonds by annealing. Figure 14.8(a) shows TOA dependence of the Si 1s spectra of the annealed $HfO_2/SiO_2/Si(100)$. As shown in the inset, the broad Hf silicate peak consists of two components, and the TOA dependence of the integrated intensity of these components is plotted in Fig. 14.8(b). The lower BE component (I) dominates at smaller TOA, while the higher BE component (II) dominates at larger TOA. It is deduced from this result that annealing leads to the formation of a two-layer structure. The component (I) comes from the layer near the surface while the component (II) comes from the layer near the substrate.

This kind of analysis of Si-based materials has usually been done by measuring Si 2p spectra because Si 1s electrons cannot be ionized in the conven-

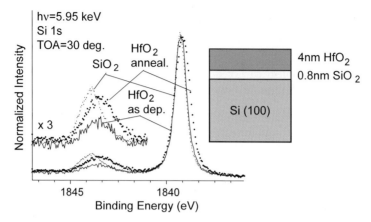

Fig. 14.7. Si 1s spectra of as-deposited (*solid curve*) and annealed (*dotted curve*) $HfO_2/SiO_2/Si(100)$ measured at 5.95 keV. As a reference, the spectrum of 1.32 nm $SiO_2/Si(100)$ are shown [11]

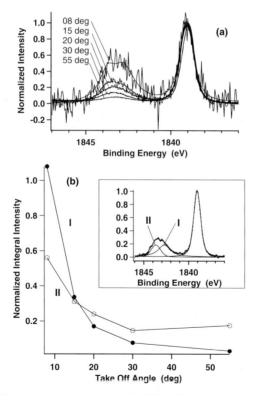

Fig. 14.8. (a) Si 1s spectra of annealed $HfO_2/SiO_2/Si(100)$ for various take-off angles. The smooth curves indicate curve fits using Voigt functions; (b) the chemical shifted Si 1s peaks were decomposed into two peaks (I and II) as shown in the inset. The integral intensity of each component is plotted as a function of take-off angle [11]

tional SX-PES. The advantage of HX-PES of Si 1s is that the deconvolution procedure of superposing spin-orbit components is not required.

14.4.2 Hybridization of Cr 3d-N 2p-Ga 4s in the Diluted Magnetic Semiconductor $Ga_{1-x}Cr_xN$

Hole-mediated ferromagnetism based on the Zener model has produced reliable estimates of the Curie temperature (T_C) for diluted magnetic semiconductors (DMSs) like $Ga_{1-x}Mn_xAs$ [23]. This theory basically assumes that the ferromagnetism is induced by interactions between the local moments of the transition-metal atoms mediated by itinerant holes in the material. Recently, very stable room-temperature ferromagnetism of Cr-doped GaN was confirmed experimentally. Since doped transition metals introduce deep levels in wide-band-gap semiconductors, the applicability of the hole-mediated ferromagnetism is questioned. To elucidate this point, Kim et al. investigated the electronic structure of $Ga_{1-x}Cr_xN$ using HX-PES [24].

VB spectra of $Ga_{0.899}Cr_{0.101}N$ (open circles) and undoped GaN (filled circles) measured at 5.95 keV are shown in Fig. 14.9. The solid line shows the difference spectrum. Cr doping clearly introduces new electronic states in the

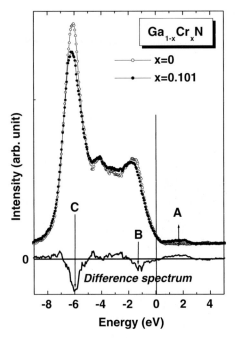

Fig. 14.9. Valence band PES spectra of undoped GaN (*open circles*) and $Ga_{0.899}Cr_{0.101}N$ (*filled circles*) measured at 5.95 keV. These spectra are obtained by subtracting the trivial background from the experimental spectra. The *solid line* at the bottom panel shows the difference spectrum [24]

band gap (A), and causes small changes in the VB structure (B and C). The intensity of the in-gap states (A) proportionally increased with increase of the Cr concentration in the range from $x = 0$ to 0.101 as shown in Fig. 14.10(b), indicating that the in-gap states are closely related to the Cr $3d$ orbitals.

Figure 14.10(a) shows the total and Cr $3d$ partial density of states (DOS) of Cr-doped GaN obtained by the density-functional-theory (DFT) [24]. There are two sharp up-spin bands in the band gap. By considering the localized model, the origin of these bands can be explained as follows: since the Cr atoms in Ga sites are tetrahedrally bonded with four N atoms, $3d\varepsilon(d_{xy}, d_{yz},$ and d_{zx} orbital) and $3d\gamma(d_{x2-y2}$ and d_{z2} orbital) states are separated into nonbonding (**e**), bonding (**t^b**), and antibonding (**t^a**) states in the tetrahedral crystal field. The two sharp up-spin bands correspond to the **e** and **t^a**. The **t^b** is merged with the valence band and the down-spin band overlaps the conduction band. Because of Fermi energy (E_F) positioned at **t^a**, the spin-up and -down states are separated. Thus the material is spin polarized.

Based on the first-principles calculation, the Cr $3d$ mainly contributes to the in-gap states as shown in Fig. 14.10(a). However, the σ value of Cr $3d$ (4×10^{-7} Mb) is very small compared to that of Ga $4s$ (5.4×10^{-5} Mb) for $Ga_{0.899}Cr_{0.101}N$ at 5.95 keV. The Ga $4s$ and Cr $3d$ contributions to the in-gap states is compared in Fig. 14.10(c). These are obtained from the partial DOSs multiplied by respective σ values. The ratio of Ga $4s$ to Cr $3d$ contributions to the area of the in-gap states is estimated to be 3.7. These results indicate that the new electronic states in the band gap is dominantly of Ga $4s$ nature and is spin-polarized the same as the Cr $3d$.

The influence of Cr doping was also investigated by core-level PES at 5.95 keV. Open circles, solid gray circles, and solid black circles in Fig. 14.11(a) show N $1s$ spectra of undoped GaN and $Ga_{1-x}Cr_xN$ ($x = 0.063, 0.101$). The main peak decreases and the tail in the low BE region increases with Cr doping. The rate of the main peak decreases and the increase in the tail feature is proportional to the Cr concentration. Namely, Cr doping causes a decrease of the N $1s$ in the matrix and introduces a new chemically shifted component at the low BE region. Because the electronegativities (EN) of the Ga, Cr, and N atoms are 1.81, 1.56, and 3.07, respectively, the EN difference of the Cr-N bond is larger than that of Ga-N bond. Therefore, the N atoms bonded with Cr are more strongly shielded by electrons compared to the N atom bonded with Ga atoms. Accordingly, Cr doping causes a chemical shift of the N $1s$ state to low binding-energy. Figure 14.11(b) shows Ga $2p_{3/2}$ core-level spectra of these samples. Intensity of the peak decreases with Cr doping. Decrease of the intensity is almost linearly proportional to the increase of Cr content. The lineshape in difference spectra is evidently asymmetric, suggesting the existence of an unresolved chemical shifted component at the low BE side. This is reasonably expected from the EN differences between Ga-Cr and N-Cr. the linear increase of the full width at half maximum (FWHM) with increasing Cr content (inset in Fig. 14.11(b)) is a further corroboration of the existence of increasing chemical shift component.

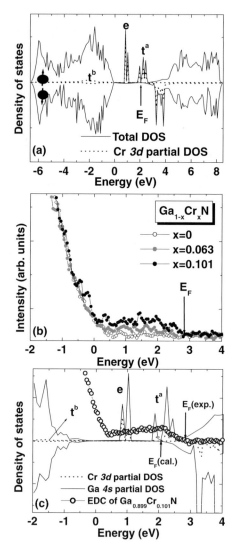

Fig. 14.10. (a) Calculated total DOS and Cr 3d partial DOS based on the DFT; (b) PES spectra near the valence band maximum of the undoped GaN (*open circles*) and the $Ga_{1-x}Cr_xN$ ($x = 0.063$ (*solid gray circles*), 0.101 (*solid black circles*). The intensity of the new energy state in the band gap depends on the Cr concentration; (c) the comparison between the valence spectra (open circles) of the $Ga_{0.899}Cr_{0.101}N$ and the quantitatively scaled partial DOS's both of Cr 3d (*dotted line*) and Ga 4s (*solid line*) [24]

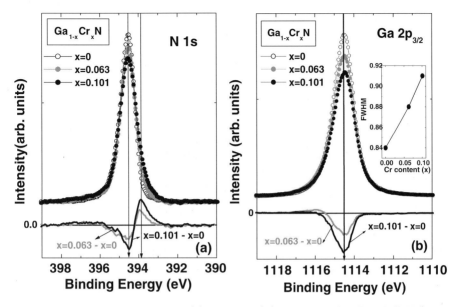

Fig. 14.11. Core-level spectra of (**a**) N 1s and (**b**) Ga $2p_{3/2}$ of undoped GaN (*open circles*) and $Ga_{1-x}Cr_xN$ ($x = 0.063$ (*solid gray circles*), 0.101 (*solid black circles*)) measured at 5.95 keV. The *right* inset in (**b**) shows the FWHM variation of the Ga $2p_{3/2}$ spectra with Cr concentration [24]

These VB and core-level spectra and theoretical results lead us a hypothesis that the ferromagnetic interaction between distinct Cr atom is mediated by the Cr 3d-N 2p-Ga 4s hybridization. Similar results were also obtained for $Ga_{1-x}Mn_xN$ [12].

As demonstrated here, HX-PES can probe intrinsic electronic properties of MBE-grown thin films, whose ideal clean surface usually cannot be prepared after exposure to atmosphere. HX-PES studies have also been reported for other technologically important semiconductor thin films [14, 25].

14.4.3 Temperature Induced Valence Transition of YbInCu$_4$

PES has widely been used for valence determination, however, small probing depth of VUV- and SX-PES sometimes obstructs us to probe the intrinsic bulk property even for the sample with a clean surface. The study of valence transition of YbInCu$_4$ is a typical case (see also the contribution by Shimada, this volume). Sato et al. applied HX-PES for this study, and successfully determined the Yb valence transition of the bulk state [26].

YbInCu$_4$ has attracted great interest because of the first-order valence transition at $T_V = 42$ K [27]. In accordance with the valence transition, abrupt changes in the lattice volume, electrical resistivity, magnetic susceptibility, and other physical properties are observed with no change of the crystal structure.

In the high-temperature phase, the Yb valence (z) is generally believed to be $z \sim 3$ from the Curie-Weiss susceptibility, and the sharp valence change at T_V is estimated to be $\Delta z \sim 0.1$, that is, $z \sim 2.9$ in the low-temperature phase from the lattice expansion.

Reinert et al. performed detailed temperature-dependent VUV-PES measurements at $h\nu = 43$ eV [28]. The Yb valence, estimated from the intensity ratio of the Yb^{2+} and Yb^{3+} $4f$-derived structures in the valence bands, changes from $z \sim 2.85$ at 220 K to $z \sim 2.56$ at 20 K rather continuously even on crossing through the valence transition. Based on the small probing depth, they proposed the existence of surface region with different physical properties from the bulk. In SX (800 eV)-PES results by Sato et al., the valence transition became clearer than in the VUV-PES because of the larger probing depth, but the Yb valence was still smaller than that expected from thermodynamic data [29].

Figure 14.12(a) shows the temperature dependence of the Yb $3d$ HX-PES spectra of YbInCu$_4$ measured at 30 and 50 K with the excitation energy of 5.95 keV. The sample was prepared by fracturing the single crystal under UHV condition. The Yb $3d$ spectra are split into the $3d_{5/2}$ and $3d_{3/2}$ regions due to the spin-orbit interaction, and each region consist of a Yb^{2+} single peak and Yb^{3+} multiplet structures. The Yb^{2+}- and Yb^{3+}-derived features are well separated, and the hybridization effect is negligible in contrast to the VB spectra. It should be noticed that the drastic change in intensity is clearly observed from 50 to 30 K, across the valence transition.

The Yb valence was estimated from the intensity ratio of the Yb^{2+}- and Yb^{3+}-derived Yb $3d$ structures, and the results are shown by circles in Fig. 14.12(b). For comparison, the results of VB spectra with VUV (diamonds) [28] and SX (squares) [29] PES are also shown. Probing depth for VUV-, SX- and HX-PES are expected to be ~ 0.5, 1.5 and 7.5 nm. The Yb valence at 220 K is $z \sim 2.90$, almost constant down to 50 K, and then sharply drops to $z \sim 2.74$ from 50 to 30 K through the valence transition. It should be noticed that the results for the Yb valence are the closest to the thermodynamic results, and the change through the valence transition is the sharpest among the three PES measurements. These results indicate that the Yb $3d$ HX-PES spectra with large probing depth are almost free from contribution of the surface.

The Yb valence of YbInCu$_4$ was also studied by the Yb L_{III} x-ray absorption spectroscopy (XAS) [27]. However, Yb^{2+}- and Yb^{3+}-derived spectra overlap with each other, and the accuracy of the deconvolution analysis is rather limited. On the other hand, Dallera et al. reported that bulk-sensitive Yb $L\alpha$ resonant inelastic x-ray scattering (RIXS) experiments can detect the sharp valence transition of YbInCu$_4$ [30]. Because of the resonance enhancement, however, the relative intensity between the Yb^{2+} and Yb^{3+} components, which also overlap in the RIXS spectrum, does not provide the Yb valence directly and one has to rely on the XAS spectrum as a reference of the Yb valence. In contrast to XAS and RIXS, one can regard Yb $3d$ HX-PES as the

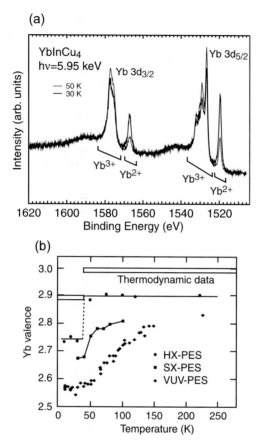

Fig. 14.12. (a) Temperature-dependent Yb $3d$ core-level spectra of YbInCu$_4$ measured at 30 and 50 K with the excitation energy of 5.95 keV. Spectra are classified into the $3d_{5/2}$ and $3d_{3/2}$ regions and both parts clearly separate further into the Yb^{2+} and Yb^{3+} components. Between 50 and 30 K, a remarkable change is observed, reflecting the valence transition at $T_V = 42$ K [26]; (b) temperature-dependent Yb valence derived form the Yb $3d$ core level spectra in (a) [26], in comparison with the VUV-PES [28] and SX-PES [29] results. Thermodynamic results are also shown

straightforward method for the quantitative estimation of the Yb valence with high accuracy because the Yb^{2+}- and Yb^{3+}-derived structures are completely separated. Using the characteristics of core-level HX-PES, valence transition of EuNi$_2$(Si$_{0.20}$Ge$_{0.80}$)$_2$ [31] and charge order of Na$_{0.35}$CoO$_2 \cdot 1.3$H$_2$O [32] were investigated. Core-level studies of f-electron systems based on Pr and Yb have also been carried out at ESRF [33, 34].

Figure 14.13 shows the temperature dependent VB spectra of YbInCu$_4$ measured at 5.95 keV. Two peaks derived from the Yb^{2+} $4f_{7/2}$ states at 0.1 eV and Yb^{2+} $4f_{5/2}$ states at 1.45 eV are significantly enhanced in the spectrum

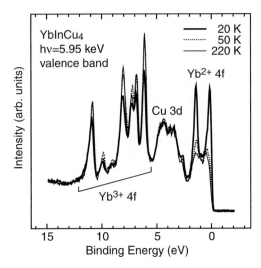

Fig. 14.13. Temperature-dependent valence-band spectra of YbInCu$_4$ measured at 220, 50, and 20 K with the excitation energy of 5.95 keV. From 50 to 20 K, the feature of the spectra exhibits a remarkable change, reflecting the valence transition at $T_V = 42$ K

at 20 K compared with those at 50 and 220 K, while the Yb^{3+} 4f multiplet structures at 5.5–12 eV reduce in intensity. The amount of the enhancement for the Yb^{2+} 4f peaks is quite remarkable in comparison with the VUV-PES [28] and SX-PES [29] experiments, again indicating that the present spectra are almost free from contribution from surface region.

14.4.4 Bulk Screening in Core Level Spectra of Strongly Correlated Electron Systems

HX-PES has also been applied to 3d transition-metal (TM) compounds with strong electron correlations. HX-PES spectra of TM 2p core levels show new characteristic low BE features in their metallic phase, which are absent in the SX-PES results. Theoretical calculations revealed that the feature is well-screened state in the bulk and reflects metallic DOS at E_F as described below.

The first example is Mn 2p core-level spectra of La$_{1-x}$Sr$_x$MnO$_3$ (LSMO) thin films. LSMO exhibits a rich phase diagram and unusual physical properties with hole doping [35]. Using PES, temperature-dependent half-metallic ferromagnetism, charge and orbital ordering, and its connection with the electronic structure and colossal magnetoresistance of the manganites have been clarified [36, 37]. Nevertheless, the change in the Mn 2p spectra of LSMO was still not conclusive [38]. Horiba et al. measured Mn 2p HX-PES spectra of LSMO thin films and compared them to those obtained with SX excitation [39].

Figure 14.14 shows Mn 2p core-level spectra of a non-doped LaMnO$_3$ thin film at 300 K measured with different probing depths by changing the photon energy and emission angle θ [39]. The sample was grown on a SrTiO$_3$ (STO) substrate by laser molecular beam epitaxy (laser MBE). The estimated probing depths at (i) $h\nu = 800$ eV, $\theta = 0°$ (SX spectra), (ii) $h\nu = 5.95$ keV, $\theta = 75°$, and $h\nu = 5.95$ keV, $\theta = 10°$ (HX spectra) are 6, 20 and 70 Å, respectively. The HX-PES spectra were measured for the sample transferred into the PES chamber from air without surface cleaning procedure. On the other hand, the SX-PES spectrum was measured using an in-situ technique [38]. Nevertheless, there are obvious differences between the SX- and HX-PES spectra. In the SX-PES spectrum, a shoulder structure at the BE of about 642 eV is clearly observed. The intensity of this shoulder-like structure at the high BE side of

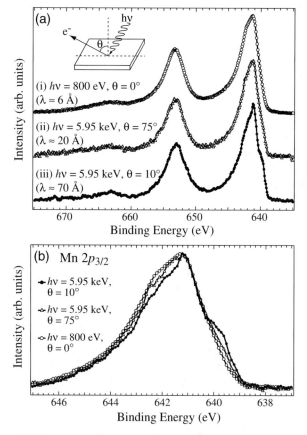

Fig. 14.14. (a) Mn 2p core-level spectra of LaMnO$_3$ measured with different probing depth by changing photon energy (800 eV and 5.95 keV) and emission angle ($\theta = 0°$, 10° and 75°) [39]. The inset shows the experimental configuration; (b) the Mn 2$p_{3/2}$ region on an expanded scale

the Mn $2p_{3/2}$ systematically decreases with increasing probing depth, indicating that this feature has a surface electronic structure component, which is minimized in the HX-PES spectrum.

A more remarkable difference between HX- and SX-PES spectra is a new shoulder structure at the low BE side of Mn $2p_{3/2}$ main peak in the HX-PES spectra. The intensity of this feature systematically increases with increasing probing depth, in contrast to the high BE side shoulder of main peak. The data indicate that this feature is derived from the bulk electronic structure. This bulk-derived feature has not been observed in SX-PES measurements [36,38].

Figure 14.15 shows the temperature dependence of Mn $2p_{3/2}$ spectra with various hole concentrations measured at $h\nu = 5.95\,\mathrm{keV}$ with $\theta = 10°$. In the $x = 0.2$ and $x = 0.4$ spectra, the bulk-derived low BE feature exhibits a drastic increase of intensity and becomes a sharp peak structure at 40 K. On the other hand, the low BE feature of $x = 0$ and $x = 0.55$ shows little change in the intensity and the spectral shape. Concerning the temperature-dependent physical properties of LSMO thin films, the $x = 0.2$ compound shows an insulator-to-metal transition between 300 and 40 K [38]. The $x = 0.4$ compound shows metallic behavior at all temperatures below 300 K, but the metallicity increases on stabilizing the ferromagnetic (FM) state at low temperature, and is attributed to its half-metallic nature. On the other hand, while the $x = 0$ stoichiometric compound is insulating, it also is effectively hole doped due to excess oxygen. As is well known, excess oxygen is easily introduced during the growth of thin films, and the electronic structure is significantly changed by the existence of excess oxygen [40]. For $x = 0.55$, the material is in the antiferromagnetic (AFM) metal phase. Taking these physical properties into account, the bulk-derived low BE features are "well-screened"

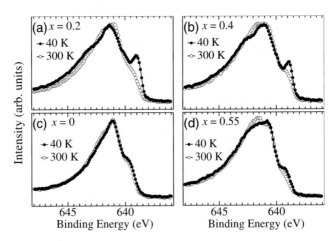

Fig. 14.15. Temperature dependence of Mn $2p_{3/2}$ spectra of $\mathrm{La}_{1-x}\mathrm{Sr}_x\mathrm{MnO}_3$ thin films with various hole concentrations: (a) $x = 0.2$; (b) $x = 0.4$; (c) $x = 0$; (d) $x = 0.55$ [39]

states and assumed to be related to the doping-induced DOS responsible for the ferromagnetism and metallicity.

The assumption was confirmed by model cluster calculations using a MnO_6 ($3d^4$) cluster with D_{4h} symmetry [39]. In addition to the usual model which includes the Mn $3d$ and ligand O $2p$ states, new states at E_F (labeled C in the inset of Fig. 14.16) were introduced. These new states represent the doping-induced states which develop into a metallic band at E_F, but are approximated as a level for simplicity. Very recently, a similar model using dynamic mean field theory has been successfully applied to calculate core-level spectra in a series of ruthenates across the metal-insulator transition, but in the absence of ligand states [41]. In the present study, four configurations, namely $3d^4$, $3d^5\underline{L}$, where $\underline{L}$ is a hole in the ligand O $2p$ states, and $3d^3\text{C}$ and $3d^5\underline{\text{C}}$ which represent charge transfer (CT) between DOS at E_F and the Mn $3d$ state are considered. The cluster calculation was carried out for a high-spin configuration, consistent with the magnetic moment estimated from susceptibility measurements [42]. The experimental spectra were fitted

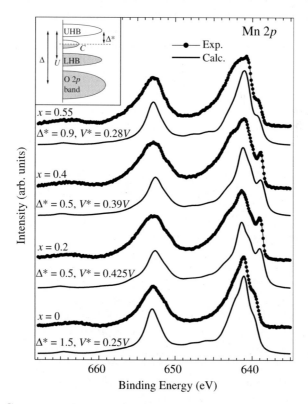

Fig. 14.16. Comparison between the cluster calculation and HX-PES spectra of the Mn $2p$ core level [39]. The inset shows a schematic diagram of energy levels on the valence band

by changing two parameters: the CT energy (Δ^*) between Mn $3d$ and the new C states and the hybridization (V^*) between Mn $3d$ and the new C states. Except for these two parameters, all other parameter values are fixed and determined from previous work [43]: the $d-d$ Coulomb interaction of Mn $3d$ states $U = 5.1\,\text{eV}$, the CT energy between Mn $3d$ and ligand O $2p$ states $\Delta = 4.5\,\text{eV}$, the hybridization between Mn $3d$ and ligand O $2p$ states $V = 2.94\,\text{eV}$, the crystal field splitting 10Dq $= 1.5\,\text{eV}$, and the Coulomb interaction between Mn $3d$ and Mn $2p$ core hole states $U_{dc} = 5.4\,\text{eV}$.

Figure 14.16 shows the comparison between the HX spectra and the optimized calculations. For all x values, the calculation reproduces well the intensity and position of the well-screened feature of HX-PES spectra. The well-screened feature in the calculation is analyzed to originate from the $2p^5 3d^5 \underline{C}$ configuration of the final state, and increases in intensity with increasing V^*. The cluster calculations indicate a larger hybridization strength V^* with the coherent states, or increase in delocalization, for the FM compositions ($x = 0.2$ and 0.4) compared to the AFM compositions ($x = 0$ and 0.55). This is consistent with the known half-metallic ferromagnetism which is stabilized with an increase in hybridization, for the manganites upon doping [44]. This also suggests an analogy with the Kondo coupling between f states and conduction band states with $V^*(E_\text{F}) \propto \sqrt{D(E_\text{F})}$ [45], where $D(E_\text{F})$ is DOS at E_F. It is also noted that the high BE side of the Mn $2p_{3/2}$ main peak, particularly for high doping, does not match with the calculations. This disagreement is due to the Mn^{4+}-derived state appearing at the high BE side of main peak with hole doping, and is not included in the calculations.

The next example is Cu $2p$ core-level spectra of high temperature superconductors, La$_{2-x}$Sr$_x$CuO$_4$ (LSCO) and Nd$_{2-x}$Ce$_x$CuO$_4$ (NCCO). While many Cu $2p$ core level SX-PES of LSCO and NCCO have been performed, the spectra show very little change upon doping [46–49]. This leads to another significant issue: the presence of the predicted Zhang–Rice singlet (ZRS) in Cu $2p$ PES of the insulating cuprates [50], which is considered very important for superconductivity but had not been observed by core level PES. Recently, Taguchi et al. measured Cu $2p$ HX-PES spectra of LSCO and NCCO, and revealed the intrinsic electronic character [51].

Figure 14.17(a) shows Cu $2p_{3/2}$ spectra measured at $h\nu = 5.95$ (HX) and 1.5 (SX) keV of NCCO, La$_2$CuO$_4$ (LCO), and LSCO. Figure 14.17(b) shows the Cu $2p_{3/2}$ HX-PES spectra of undoped Nd$_2$CuO$_4$ (NCO), LCO, and hole-doped LSCO. NCCO and LSCO show a superconducting T_c of 22 and 36 K, respectively. All the samples were fractured in UHV. The spectra of NCCO and LSCO ware measured at 35 K while LCO and NCO were measure at room temperature. The NCO Cu $2p_{3/2}$ spectrum consists of a main peak at 933.5 eV ($2p^5 3d^{10}\underline{\text{L}}$) and a broad satellite centered at 943 eV ($2p^5 3d^9$ state), and is very similar to earlier SX-PES [48,49]. The HX-PES spectra of NCCO, LCO, and LSCO (Figs. 14.17 (a) and (b)) are clearly different and provide new results:

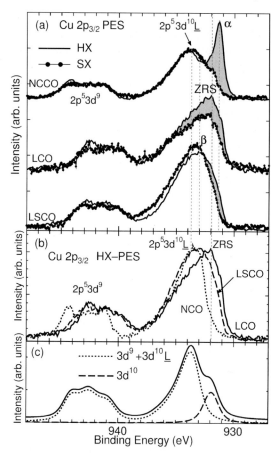

Fig. 14.17. (a) Comparison between experimental Cu $2p$ HX-PES (*solid line*) and SX-PES (*line with symbols*) for electron-doped NCCO, undoped LCO, and hole-doped LSCO. *Shaded regions* highlight the differences in HX-PES; (b) experimental Cu $2p$ HX-PES comparison of NCO, LCO, and LSCO; (c) calculated spectra with $3d^9 + 3d^{10}\underline{L}$ (*dotted line*, represents the NCO spectrum) and with $3d^{10}$ (*dashed line*) [51]

(i) The LCO HX-PES spectrum (Fig. 14.17(b)) shows a main peak at $\sim$932 eV and a shoulder at $\sim$933.5 eV. Using a multisite cluster model (MSCM) [50], it was shown that even for the insulating parent, the Cu $2p_{3/2}$ spectrum has a low BE ZRS feature due to nonlocal screening, while the $2p^5 3d^{10}\underline{L}$ state occurs at higher BE. But earlier SX-PES showed only a single peak at 933.5 eV due to the $2p^5 3d^{10}\underline{L}$ state. HX-PES clearly shows that, in LCO, the peak at 933.5 eV is the $2p^5 3d^{10}\underline{L}$ and the new feature at $\sim$932 eV is the ZRS peak. This is the first observation of the ZRS feature in Cu $2p$ PES of LCO. The ZRS feature is missing in the NCO HX-PES (Fig. 14.17(b)).

(ii) The LSCO spectrum (Fig. 14.17(b)) shows clear changes compared to LCO. The ZRS feature is retained on hole-doping, but is weakened compared to LCO, and additional spectral weight is seen at higher BE (feature β). Since the main peak width (nearly 4 eV) is very large in LSCO, it consists of more than a single state ($2p^53d^{10}\underline{L}$ and β features).

(iii) The HX-PES for NCCO (Fig. 14.17(a)) shows a sharp low BE feature α which is not observed in NCCO SX-PES as well as NCO HX-PES. Its energy position is different from the ZRS feature in LCO. More importantly, since the ZRS feature is missing in undoped NCO, its origin is different and discussed in the framework of the impurity Anderson model (IAM) later. In terms of the Ce content ($x = 0.15$), a maximum of 15% of the spectral intensity can arise due to the formally Cu^{1+} $3d^{10}$ state, in contrast to the observed intensity ($\sim$30%). We calculated spectra for NCO (using $3d^9 + 3d^{10}\underline{L}$ states) in the IAM [49], and for NCCO (using a linear combination of $3d^9 + 3d^{10}\underline{L}$ and $3d^{10}$ with a relative weight of 85% and 15%), respectively (Fig. 14.17(c)). While the calculation for NCO matches the data, the calculated intensity of feature α does not match the experimental HXPES data for NCCO. This indicates that a simple $3d^{10}$ state due to electron-doping cannot explain the observed high intensity of feature α.

The ZRS state observed in LCO, the hole-doping induced changes in LSCO, and the feature α in electron-doped NCCO can be explained by MSCM with a nonlocal screening effect [50]. However, NCO (the undoped parent for the electron-doped system) does not show the ZRS feature (Fig. 14.17(b)). Since the IAM calculation works for NCO (Fig. 14.17(c)), and the calculations are simpler than MSCM calculations, IAM calculations was performed in the D_{4h} local symmetry including intra-atomic multiplets, although it is clear that the IAM calculation cannot reproduce the ZRS feature in LCO. The details are almost the same as in the LSMO case. The essential feature is the charge transfer from doping-induced states at E_F (labeled C) to the upper Hubbard band (UHB). Δ^* is the CT energy between $3d^9$ and $3d^{10}\underline{C}$.

The calculated and experimental results are shown in Fig. 14.18 for LSCO and NCCO. The calculations reproduce well the main peaks and satellite structure. The sharp peak at low BE in NCCO originates from core hole screening by doping-induced states at E_F, the $2p^53d^{10}\underline{C}$ state. The most important parameter is Δ^*, which represents the energy difference between the UHB and doping-induced states. The small value of $\Delta^* = 0.25$ eV obtained for NCCO indicates that the doping-induced states lie just below the UHB, whereas a large value of $\Delta^* = 1.35$ eV for LSCO describes the situation for doping-induced states lying near the top of the VB.

As shown by the TM $2p$ core-level spectra of LSMO thin films and the cuprates, HX-PES with large probing depth enables us to probe intrinsic bulk properties. The well screened bulk feature was also observed in the HX-PES

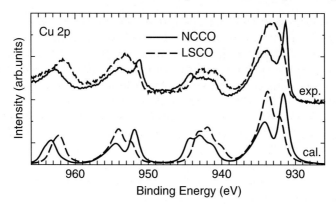

Fig. 14.18. IAM calculations for the Cu $2p$ core level PES of NCCO and LSCO (*lower panel*) compared with experiments (*upper panel*). The broad satellite around 944 eV is due to multiplets of $2p^5 3d^9$. The O $2p$ band width broadens the $2p^5 3d^{10}\underline{L}$ peak around 933.5 eV [51]

spectra of V $1s$ and $2p$ for V_2O_3 [52], V $2p$ for $V_{1.98}Cr_{0.02}O_3$ [53], Cu $2p$ for $Bi_2Sr_2CaCu_2O_{8+\delta}$ [53], and Mn $2p$ for $La_{1-x}Ba_xMnO_3$ [54].

14.5 Summary

HX-PES with high throughput and high energy resolution has been realized for core-level and VB studies using high-energy and high-brilliance SR at the beamline BL29XU in SPring-8. In addition to x-ray optics and an electron analyzer dedicated for HX-PES, optimized experimental configuration, such as photoelectron detection along the direction parallel to x-ray polarization and grazing incidence of the well-focused x-ray beam, strongly improved photoelectron intensity. When we set the total instrumental energy resolution to be about 250 meV, not only core-level but also VB spectra of Au and Si can be measured within several tens of seconds to several minutes, demonstrating the high throughput of the experimental system. The most important characteristics of HX-PES, i.e., surface insensitivity and large probing depth, were confirmed. The highest total energy resolution achieved is 75 meV at 5.95 keV, and ~90 meV at 7.35 and 9.92 keV.

HX-PES has been applied to studies of Si high-k dielectrics, diluted magnetic semiconductors, f-electron systems with valence transition, and $3d$ transition-metal compounds with strong electron correlation. All the results confirm the capability of HX-PES to probe depth-resolved electronic structure, buried layers, interfaces, ultrashallow junctions and the bulk electronic structure of strongly correlated electron systems. We believe that HX-PES will become a standard method to study electronic properties of various materials in the field of basic science and technologies.

Concerning future perspectives, improvement of the energy resolution down to ~40 meV can be expected to be done soon. Further improvement down to 10 meV is a challenge because the development of a new power supply system based on alternative technology will be necessary. It should be noted that x-ray band width of 120 µeV has been achieved at 14.41 keV [19,55]. Angle resolved PES of VB is also very attractive even with the present resolution of ~70 meV. For this purpose, high angular resolution less than 0.1° is required because first Brillouin zone shrinks at several keV. It is also interesting to develop spin polarized HX-PES.

Acknowledgements

The development of HX-PES in SPring-8 was initiated by T. Ishikawa after successfully achieving an ultra-high-resolution x-ray monochromator with ~100 µeV resolution at 14.41 keV. T. Ishikawa, and his group members, M. Yabashi, K. Tamasaku, Y. Nishino and D. Miwa have designed and prepared the x-ray optics dedicated for the method. The author expresses thanks to them immensely. The author appreciates collaborations with M. Arita, M. Awaji, A. Chainani, R. Eguchi, T. Hattori, K. Horiba, E. Ikenaga, N. Kamakura, T. Kawai, J.J. Kim, K. Kobayashi, H. Makino, M. Matsunami, H. Namatame, H. Nohira, H. Sato, K. Shimada, S. Shin, M. Taguchi, Y. Takeda, A. Takeuchi, T. Takeuchi, H. Tanaka, M. Taniguchi, T. Tokushima, S. Tsuda, K. Yamamoto T. Yao and T. Yokoya. The author thanks Gammadata-Scienta Co., in particular, S. Södergren, B. Wannberg, M. Wang and B. Åhman, for their very valuable help in developing analyzers for HX-PES.

References

1. S. Hüfner: *Photoelectron Spectroscopy*, 3rd edn (Springer-Verlag, Berlin-Heidelberg 2003)
2. M. P. Seah and W. A. Dench: Surf. Interf. Anal. **1**, 2 (1979)
3. S. Tanuma et al.: Surf. Interf. Anal. **21**, 165 (1993)
4. The electron inelastic mean free paths were estimated using NIST Standard Reference Database 71, "NIST Electron Inelastic-Mean-Free-Path Database: Ver. 1.1". It is distributed via the web site http://www.nist.gov/srd/nist71.htm, and references therein
5. A. Sekiyama et al.: Nature **403**, 396 (2000)
6. L. Braicovich et al.: Phys. Rev. B **56**, 15047 (1997)
7. I. Lindau et al.: Nature **250**, 214 (1974)
8. J. J. Yeh and I. Lindau: Atom. Data Nucl. Data Tables **32**, 1 (1985)
9. W. Drube et al.: AIP conference proceedings **705**, 1130 (2002)
10. J. Danger et al.: Phys. Rev. Lett. **88**, 243001 (2002)
11. K. Kobayashi et al.: Appl. Phys. Lett. **83**, 1005 (2003)
12. Y. Takata et al.: Appl. Phys. Lett. **84**, 4310 (2004)

13. S. Thiess et al.: Solid State Commun. **132**, 589 (2004)
14. C. Dallera et al.: Appl. Phys. Lett. **85**, 4532 (2004)
15. P. Torelli et al.: Rev. Sci. Instrum. **76**, 23909 (2005)
16. The proceedings of the workshop on Hard X-ray Photoelectron Spectroscopy has been edited by J. Zegenhagen and C. Kunz: Nucl. Instrum. Methods. A **547** (2005)
17. Y. Takata et al.: Nucl. Instrum. Methods. A **547**, 50 (2005)
18. K. Tamasaku et al.: Nucl. Instrum. Methods. A **467/468**, 686 (2001)
19. T. Ishikawa et al.: Nucl. Instrum. Methods. A **547**, 42 (2005)
20. G. Panaccione et al.: J. Phys.: Condens. Matter **17**, 2671 (2005)
21. K. Kobayashi: Nucl. Instrum. Methods. A **547**, 98 (2005)
22. M. Sacchi et al.: Phys. Rev. B **71**, 155117 (2005)
23. T. Dietl et al.: Phys. Rev B **63**, 195205 (2001)
24. J. J. Kim et al.: Phys. Rev B **70**, 161315 (2004)
25. K. Kobayashi et al.: Jpn. J. Appl. Phys. **43**, L1029 (2004)
26. H. Sato et al.: Phys. Rev. Lett. **93**, 246404 (2004)
27. I. Felner et al.: Phys. Rev. B **35**, 6956 (1987)
28. F. Reinert et al.: Phys. Rev. B **58**, 12808 (1998)
29. H. Sato et al.: Phys. Rev. B **69**, 165101 (2004)
30. C. Dallera et al.: Phys. Rev. Lett. **88**, 196403 (2002)
31. K. Yamamoto et al.: J. Phys. Soc. Jpn. **73**, 2616 (2004)
32. A. Chainani et al.: Phys. Rev. B **69**, 180508 (2004)
33. A. Yamasaki et al.: J. Phys. Soc. Jpn. **74**, 2045 (2005)
34. S. Suga et al.: J. Phys. Soc. Jpn. **74**, 2880 (2005)
35. *Colossal Magnetoresistive Oxides*, Advances in Condensed Matter Scienece Vol. 2, edited By Y. Tokura (Gordon and Breach, Amsterdam 2000)
36. T. Saitoh et al.: Phys. Rev. B **51**, 13942 (1995); T. Saitoh et al.: Phys. Rev. B **56**, 8836 (1997)
37. D. D. Sarma et al.: Phys. Rev. B **53**, 6873 (1996)
38. K. Horiba et al.: Phys. Rev. B **71**, 155420 (2005); and references therein
39. K. Horiba et al.: Phys. Rev. Lett. **93**, 236401 (2004)
40. J.-H. Park et al.: Phys. Rev. Lett. **76**, 4215 (1996)
41. H.-D. Kim et al.: Phys. Rev. Lett. **93**, 126404 (2004)
42. A. Urushibara et al.: Phys. Rev. B **51**, 14103 (1995)
43. M. Taguchiand and M. Altarelli: Surf. Rev. Lett. **9**, 1167 (2002)
44. W. E. Pickett and D. J. Singh: Phys. Rev. B **53**, 1146 (1996); T. Mizokawa and A. Fujimori: Phys. Rev. B **54**, 5368 (1996)
45. O. Gunnarsson and K. Schönhammer: in *Handbook on the Physics and Chemistry of Rare Earths*, edited by K. A. Gschneidner, Jr. L. Eyring, S. Hüfner (Elsevier Science Publishers B. V., New York, 1987), Vol. 10
46. A. Fujimori et al.: Phys. Rev. B **35**, 8814 (1987)
47. Z-X. Shen et al.: Phys. Rev. B **36**, 8414 (1987)
48. T. R. Cummins and R. G. Egdell: Phys. Rev. B **48**, 6556 (1993)
49. A. Koitzsch et al.: Phys. Rev. B **66**, 24519 (2002)
50. M. A. van Veenendaal et al.: Phys. Rev. B **47**, 11462 (1993); M. A. van Veenendaal and G. A. Sawatzky: Phys. Rev. B. **49**, 3473 (1994)
51. M. Taguchi et al.: Phys. Rev. Lett. **95**, 177002 (2005)
52. N. Kamakura et al.: Europhys. Lett. **68**, 557 (2004)
53. M. Taguchi et al.: Phys. Rev. B **71**, 155102 (2005)
54. H. Tanaka et al.: cond-mat/0410223; to appear in Phys. Rev. B
55. M. Yabashi et al.: Rev. Sci. Instr. **72**, 4080 (2001)

Index

Abrikosov-Suhl resonance *see* Kondo resonance
ADC *see* angular distribution curve
Ag   152, 153, 164, 168
  stripes   159
Ag(111)   34, 154
  surface state   5, 8
Ag/Cu(111)   44, 152
  band gap   44
Ag/Pt(997)   159
AlAs   379
alkali metals   166, 167
alkaline earths   166, 167
alkylbenzene   235
analyzer   15
  electrostatic   15
  field-ionization   226
  hemispherical electron-energy   86
  time-of-flight   219, 222
  zero kinetic energy (ZEKE)   223, 224, 228
angular distribution curve   39
aniline   223, 231
  -Ar   231, 234, 235
  -Ar_2 234
  -$Ar_{1,2}$ 225
(aniline)$^+$ 234
anisole   231
  -Ar   231
anthracene   227, 231
  -Ar   231
  -Ar_2 231
  -Ar_3 231

  -Ar_4 232
  -Ar_5 232
antiferromagnetic correlation   286
Ar   221
  -NO   221
argon   7, 231
ARUPS *see* ultraviolet photoelectron spectroscopy
atomic chains   147, 151, 152, 163, 179, 181
Au   4, 87, 164–166, 168, 172, 173, 175–177, 191, 377, 395
  chains   165, 166, 171–173
Au(11 9 9)   162
Au(111)   34, 42, 157, 176, 177
  herringbone reconstruction   43
  spin-orbit splitting   42
  vicinal surface   154
Au(23 23 21)   154–156, 160, 161
Au(887)   156, 157
Auger
  electron   302
  process   305
  spectroscopy   374
azulene   232
  -Ar   232
  -$Ar_{1,2}$ 234

Ba_8Si_{46} 201
$Ba_{1-x}K_xBiO_3$ 197
backfolding   43, 117, 125, 128
background intensity   37
band
  filling   171, 327

fractional filling 171, 173
free-electron-like 150, 153, 171, 177
gap 72, 383
spin-plit 177
band-structure calculation 360, 362
  LDA+U 336
bandwidth 327
Bardeen, Cooper, and Schrieffer 38, 187
BaVS$_3$ 122
BCS see Bardeen, Cooper, and Schrieffer, 39
  character 64
  density of states 3
  mechanism 77
  superconductor 246, 265
  theory 187, 189, 272, 275–277
Be(0001)
  surface state 60
Bechgaard salts 132
benzene 221, 223
  -Ar 235, 236
  -N$_2$ 232
benzonitrile 232
  -Ar 232, 235
  -Ar$_2$ 232
  -Ar$_{1,2}$ 234
Bethe ansatz 134
Bi$_2$Sr$_2$Ca$_2$Cu$_3$O$_{10+\delta}$ 273
Bi$_2$Sr$_2$Ca$_2$Cu$_3$O$_{10}$ 272
Bi$_2$Sr$_2$Ca$_{n-1}$Cu$_n$O$_{2n+4}$ 272, 273
  "kink" in energy dispersion 272
  optimally doped 273
  overdoped 273
  underdoped 273
Bi$_2$Sr$_2$CaCu$_2$O$_8$ 298
Bi$_2$Sr$_2$CaCu$_2$O$_{8+\delta}$ 75, 77, 78, 80, 244, 247–249, 251, 252, 261, 274
Bi$_2$Sr$_2$CaCu$_2$O$_{8+\delta}$ 395
Bi$_{1-x}$La$_x$NiO$_3$ 343
(Bi,Pb)$_2$Sr$_2$CaCu$_2$O$_{8+\delta}$ 307–309, 314, 317, 318, 320
  optimally doped 312
  overdoped 310
  underdoped 310
bilayer splitting 309–311, 315, 316
BiNiO$_3$ 343
Born approximation 61
Bose–Einstein distribution 26, 59, 91

bosonic excitations 303
BQP see quasi-particle
Brillouin zone 58, 60, 64, 65, 68, 69, 76, 79, 149, 158, 169, 171, 176, 180
buried layers 380

C$_{10}$H$_8$ 232
C$_{14}$H$_{10}$ 231
C$_4$H$_4$N$_2$ 232
C$_{60}$ 199
C$_6$H$_5$–C≡CH 232
C$_6$H$_5$–C≡N 232
C$_6$H$_5$–CH=CH$_2$ 232
C$_6$H$_5$–F 232
C$_6$H$_5$–NH$_2$ 231
C$_6$H$_5$–OCH$_3$ 231
C$_6$H$_5$–SCH$_3$ 233
C$_9$H$_7$N 232
Ca$_2$CuO$_2$Cl$_2$ 251–253, 255, 260
Ca$_{1-x}$Sr$_x$RuO$_3$ 344, 345
  thin films 344, 345
Ca$_{1-x}$Sr$_x$VO$_3$ 332–334, 344
Ca$_{1.9}$Na$_{0.10}$CuO$_2$Cl$_2$ 265
Ca$_{2-x}$Na$_x$CuO$_2$Cl$_2$ 248, 250–252, 256, 258–264, 266, 268
Ca(Al,Si)$_2$ 206
CaRuO$_3$ 344
catalysis 151
cation spectroscopy 215, 221
CaVO$_3$ 332, 333
CDW see charge density wave
Ce 94, 98
  α-phase 97
  α–γ phase transition 94, 105
  α-phase 94–96, 98, 99
  γ-phase 97
  γ-phase 95, 96, 98, 99
  compounds 44
  metal 85, 88, 94, 95, 97, 105
CeCu$_2$Si$_2$ 45–48
CeCu$_6$ 46–48
CEF see crystal electric field
CeNi$_2$Ge$_2$ 48
CePtSn 99–101
CeRhAs 99, 100, 102–105
CeRhSb 99–102, 104, 105
CeRu$_2$ 209, 210
CeRu$_2$Si$_2$ 48
CeSi$_2$ 46–48

Index   401

charge
  density wave   56, 63, 64, 116, 148, 149, 163, 177, 178, 197
    gap   68, 70, 71
    state   64, 66, 68–70, 265
    transition   68
    ordering   268
charge-transfer
  -type insulator   245, 247, 296, 329
  energy   328
  gap   252, 257, 299
chemical potential   248, 249, 267
CMR   see colossal magnetoresistance
Co   151, 152, 179, 379
  atomic chains   151, 152
coherence length   126
coherent peak   79
collective excitations   81
colossal magnetoresistance   327
complex oxides   55
conductivity   180
Cooper
  minimum   180
  pairs   187
core-level photoemission   362, 367
correlation
  antiferromagnetic   286
  effects   132, 141
  satellite   2
Coulomb
  interaction   56, 119, 132, 134, 299, 392
  on-site interaction energy   328
  on-site repulsion   299
  on-site repulsion energy   328
Cr   383
crystal electric field   355, 357
  splitting   46, 48, 392
Cu   153, 379
Cu(10 10 11)   157–159
  superlattice   157
Cu(111)   6, 32–35, 42–44, 156–159
  -Ag   153
  surface state   4, 6, 60
  vicinal   153
Cu(119)   159
Cu(335)   152, 160, 161
Cu(443)   156
Cu(445)   156

Cu(775)   160, 161
CuO_2 plane   76, 244, 297–299
CuO_6 octahedron   244
cuprates   76, 257, 261
  high-T_c 243
  hole-doped   261
  lightly doped   262
  prototypical   248
Curie temperature   71

d-band   328
d-d Coulomb interaction   392
d-wave
  superconductivity   265
  symmetry   76
DDMRG   see density-matrix renormalization group 134
de Haas–van Alphen
  effect   26
  measurement   89, 93, 94, 360
Debye model   6, 27–28, 32, 33, 35, 37
  energy   33, 59, 60, 62, 75, 78, 79
  self-energy   28
  spectral function   29, 30
  temperature   187
defect scattering   37
Δ_{eff} 329
density of states   73, 74
  majority-spin   74
  minority-spin   74
  phonon   25
  spin-projected   73
  total   74
density-matrix renormalization group   134
deoxyribonucleic acid   175
depth-resolved electronic structure   380
dichalcogenides   63, 64, 66, 70
DMFT   see dynamical mean-field theory
DNA   see deoxyribonucleic acid
dynamic photoelectron spectroscopy   220
dynamical mean-field theory   99, 329
Dynes function   191, 194, 198, 201, 204

EDC   see energy distribution curve

Index

effective electron mass 75, 77, 170, 177, 334
Einstein mode 255, 256, 311
Einstein, Albert 13, 295
electron
  correlation 22, 93, 327, 329, 344, 351
  dispersion curve 3
  doping 249, 252, 394
  escape depth 374
electron–boson
  coupling 243
electron–electron
  correlations 257
  coupling 25, 36, 74
    contribution to the linewidth 33
    parameter β 37
    self-energy 25
  interaction 2, 7, 61, 62, 90, 94, 115, 119, 257, 296, 315
  repulsion 247
electron–hole
  excitation 314
  pair 302
electron–impurity
  interaction 61, 90
  scattering 90
electron–ion interaction 2
electron–lattice interaction 257, 260
electron–magnon
  coupling 74, 259
  interaction 71, 74
  scattering 74, 75
electron–phonon
  coupling 25, 26, 37, 59, 60, 63, 64, 66, 70, 73–75, 118, 127, 135, 169, 187, 188, 193, 256, 257, 259, 262, 320
  constant 59, 60, 62, 73, 93, 193
  contribution 37, 38, 61, 62
  Debye model 27
  Eliashberg function 25
  in Shockley states 32
  observation in three-dimensional solids 35
  parameter 169
  renormalization 73
  renormalized dispersion 30
  self-energy 26, 39
  interaction 2, 7, 56, 61, 71, 75, 78, 90, 92, 94, 296, 336

electron–spin interaction 296
electronic
  correlation 22, 259
  structure, bulk 380
  structure, depth-resolved 380
  susceptibility 117
Eliashberg
  analysis 198
  calculation 198
  coupling constant 59
  equation 59, 61, 79
  function 25–27, 37, 39, 57, 78, 81, 192
  phonon interaction 192
  theory 39, 41
energy
  distribution curve 29, 30, 302
  gap 117, 123, 285, 286
escape depth 16
ESRF 352
ethylbenzene 232
  -Ar 232, 235
EuNi$_2$(Si$_{0.20}$Ge$_{0.80}$)$_2$ 387
EuNiO$_3$ 341
  -type 341
exchange
  processes 74
  splitting 4, 74, 75, 88

f-spins 74
FDD see Fermi–Dirac distribution
Fe 71, 75
Fe(001) 75
Fe(CO)$_5$ 221
Fermi
  edge see Fermi–Dirac distribution
  level 66, 149, 163, 181, 298
  liquid 33, 36, 37, 55, 61, 62, 115, 253, 256, 258, 262, 302, 315, 331
    marginal 301, 303, 312, 314, 315
  momentum 71
  surface 64, 66, 70, 71, 73, 80, 85, 89, 149, 150, 153, 168, 170, 174, 176–178, 189, 203, 205, 262–264, 272, 273, 278, 283, 284, 287–291, 298, 299, 308–310, 315, 316
    hole-like 273, 289
    instability 64
    map 16, 43

nesting   64, 70, 117, 124
   nesting vector   64
   renormalisation   26, 48
   shadow-   310
   velocity   26, 71, 77, 170
non-Fermi liquid   76
Fermi's golden rule   22
Fermi–Dirac distribution   4, 17, 23, 26, 46, 58, 59, 91, 193, 291, 300, 302, 342
ferromagnetic order   152
fluorescence spectrum   221
fluorobenzene   232
   -Ar   232, 235
   -Ar$_2$ 232
   -Ar$_{1,2}$ 234
   -N$_2$ 232
fractional band filling   171, 173
Franck–Condon   256, 268
   broadening   248, 251–253, 256, 257, 262, 268
   principle   254
free molecules   8
fullerides   199

Ga   383
Ga$_{0.899}$Cr$_{0.101}$N   382, 383
Ga$_{1-x}$Cr$_x$N   382–385
Ga$_{1-x}$Mn$_x$As   382
Ga$_{1-x}$Mn$_x$N   385
GaAs   379
gadolinium   71
GaN   382, 383
   band gap   383
gas discharge lamp   15
Gd   71, 73
Gd$_2$O$_3$ 379
Gd(0001)   71–73
GdFeO$_3$ 328
Ge   379
gold   see Au
golden rule   see Fermi's golden rule
Green's function   23, 89, 300, 301

Hall effect   298
   quantum Hall effect   171
hard x-ray photoemission spectroscopy   104, 373

HAXPES   see hard x-ray photoelectron spectroscopy
heavy fermion   351, 361
   compound   46
HfO$_2$ film   380
HfO$_2$/SiO$_2$/Si(100)   380, 381
high-temperature superconductor   9, 30, 36, 55, 56, 75, 77, 79, 172–174, 188, 243, 244, 247, 248
   doped   303
   electron-doped   283, 289
   hole-doped   297
   optimally doped   297
   over-doped   289, 297, 309
   phase diagram   297
   structure   297
   under-doped   297, 309, 315
   undoped   298, 299, 303
hole
   binding energy   74
   doping   246, 249, 262, 394
holon   120, 137, 142, 148
HTSC   see high-temperature superconductor
Hubbard
   band   249, 331, 332
      lower   249, 252, 253, 255, 257, 259, 260, 262
   gap   169
   model   119, 132, 134, 330
hybridization   392
hydrogen   63
   adsorption   63
   impurity centers   62
(2-hydroxy-pyridine)$^+$ 231
(9-hydroxyphenalenone)$^+$ 231

IMFP   see mean free path
impurity scattering   61, 63, 74
in-gap states   383
incommensurate potentials   123
indole   232
   -Ar   232
insulator
   charge-transfer   245, 247, 296
   Mott   329, 331
   Mott–Hubbard-type   296, 314, 329
inverse phtotoemission   181
iron   71, 75

K_2NiF_4 253
K_3C_{60} 199, 200
$K_{0.3}MoO_3$ 122
kinematic compression 25
kink 66, 68, 69, 77, 78, 89, 92, 261, 262, 277–279, 281–283, 304, 311, 313, 316, 317
Kohn anomaly 118
Kondo
  effect 44
  minimum 45
  peak 169
  resonance 44, 46, 96, 355, 366, 367
  semiconductor 99, 106
  temperature 45, 46, 48, 96, 99, 352, 357, 362, 370
Koopmans' theorem 2
KR see Kondo resonance
Kramers–Kronig
  relation 92
  transformation 57, 62, 303, 310
Kronig–Penney model 156, 157

$\bar{L}$-gap 32
La_2CuO_4 244, 245, 252, 253, 255, 257, 260, 272, 392
$La_{0.33}Sr_{0.67}FeO_3$ 338, 339
  polycrystals 339
  thin films 339
$La_{0.66}Sr_{0.34}MnO_3$ 336
$La_{0.6}Sr_{0.4}FeO_3$ 339
$La_{0.6}Sr_{0.4}MnO_3$ 330, 336
  thin films 336
$La_{0.85}Ba_{0.15}MnO_3$ 379
$La_{1-x}Ca_xMnO_3$ 335
$La_{1-x}Sr_xFeO_3$ 330, 337–339
  polycrystals 337, 338
  thin films 337–340
$La_{1-x}Sr_xMnO_3$ 16, 330, 335, 336, 388, 390
  thin films 336, 390
$La_{1-x}Sr_xTiO_3$ 331
$La_{1-x}Sr_xTiO_{3+y/2}$ 331
$La_{1/4}Pr_{3/8}Ca_{3/8}MnO_3$ 337
$La_{2-2x}Sr_{1+2x}Mn_2O_7$ 335
$La_{2-x}Ba_xCuO_4$ 244
$La_{2-x}Sr_xCuO_4$ 77, 78, 245, 246, 248–253, 256, 258–264, 266–268, 392

$LaFeO_3$ 337
$LaMnO_3$ 389
  thin films 389
laser 4, 7, 8, 16
  Nd:YAG 229
$LaTiO_3$ 332
LDA see local-density approximation
LDA+U method see local-density approximation
lead see Pb
LED see light emitting diode
LEED see low energy electron diffraction
Li purple bronze 130
$Li_{0.9}Mo_6O_{17}$ 122, 130
light emitting diode 151
local-density approximation 329
  +U method (LDA+U) 329
  band-structure calculation 316
  band-structure calculation (LDA+U) 283, 336
low energy electron diffraction 43
Luttinger
  theorem 26, 171, 310
  volume 266

m-chlorophenol
  cis 232
  $trans$ 232
magnetic
  excitation 56, 313, 314
  moment, orbital 152
  resonance mode 80
magnetism 86, 149, 163, 174
many-body effect 16, 32, 33, 44, 56, 115, 308
mass
  enhancement 57, 77, 79, 94
  enhancement factor 26, 30, 93
  renormalization 57, 60, 75, 77, 80, 127
mean free path 85
  inelastic (IMFP) 86, 373, 375
mean-field theory 119
metal-insulator
  phase diagram 328
  transition 106, 116, 118, 178, 327, 390
metallic chain structures 148

MgB$_2$ 203, 206, 207, 209
mid-gap state   250, 251
miscut   151
MIT   *see* metal-insulator transition
Mo   75
Mo(110)   60, 62
  surface   320
  surface state   61, 315
molecular hydrogen   3
molybdenum   *see* Mo, 75
momentum distribution
  curve   29, 30, 302
  function   31
Mott
  -insulating state   306
  insulator   76, 138, 245, 248, 252, 253,
    256–258, 263, 268, 329, 331
Mott–Hubbard-type
  insulator   296, 314, 329
  systems   331, 332
MPI   *see* multiphoton ionization
multiphoton ionization   216–218

N   383
n-propylbenzene   231, 232, 235, 236
nanostripes   154
nanowires   122
naphthalene   221, 232
  -Ar   232
(naphthalene)$^+$ 231
Nb   189, 193, 194
Nb$_3$Al   196, 198
Nb$_3$Ge   195
NbS$_2$ 64, 70
NbSe$_2$ 64, 65, 69–71, 205, 206
NbSe$_3$ 122, 124
NCA   *see* non-crossing approximation
Nd$_2$CuO$_4$ 252, 253, 255, 272, 392
Nd$_{0.98}$Ce$_{0.02}$NiO$_3$ 341
Nd$_{0.99}$Ca$_{0.01}$NiO$_3$ 341
Nd$_{1-x}$Sm$_x$NiO$_3$ 340–342
Nd$_{2-x}$Ce$_x$CuO$_4$ 245, 246, 249, 252,
    272, 392
NdNiO$_3$ 341
nesting   70, 156
neutron scattering   48, 77, 280, 297,
    313, 321
NH$_3$ 221
Ni   88, 89, 91–93

6 eV satellite   88
borocarbides   201
metal   2, 5, 85, 88
Ni(110)   88, 90, 91
NO   219–221, 225, 228
NO$^+$ 225, 228, 229, 231
(NO)$_2$ 221
NO-Ar   225, 231
noble gases   42
nodal direction   76, 77, 81
anti-nodal direction   76, 79
non-crossing approximation   45, 47, 48,
    356, 367, 370

O^{18}-O^{16} substitution   78
O$_2$ 221
one-color
  REMPI-based technique   236
  scheme   221
optical
  conductivity   288
  modes   70
orbital magnetic moment   152

p-d hybridization strength   329
p-dimethoxybenzene
  -Ar
    cis   232
    trans   232
  -Ar$_2$
    cis   232
    trans   232
  cis   232
  trans   232
p-phenylenediamine   232
PAM   *see* periodic Anderson model
Pauli
  principle   302
  repulsion   42
Pb   38–40, 87, 189, 191, 193–196
Pb(110)   37, 39
Peierls
  distortion   177
  gap   174, 177
  instability   116
  scenario   178
  transition   64, 148, 178
periodic Anderson model   102
perovskite-type   326

nickel oxides 340
structure 327
PFI *see* pulsed field ionization
PFI-ZEKE *see* pulsed field ionisation
phenylacetylene 232
 -Ar 225, 232
phonon
 acoustical branches 70
 excitation 56, 303, 315
 mode 77
 optical branches 70
 spectrum 70, 80
photoelectron spectroscopy
 hard x-ray 104
 molecular 215, 216
 REMPI-based 216
 two-color ZEKE 221
photoemission
 angle-resolved 149
 Hamiltonian 22
 inverse 181
 principle 14
 process 56
 spin-resolved 71, 73, 74
 surface contribution 23
 theory of the spectrum 22
 two-photon 181
photohole 24, 55, 56, 66, 71, 74
plasmon 368, 369
 excitation 303
polaron 254, 255
polaronic broadening 129
$Pr_{1-x}La Ce_x CuO_4$ 272, 292
$PrNiO_3$ 341
pseudogap 75, 76, 121, 122, 125, 135, 243, 247, 265, 287, 288, 297, 309, 311, 317, 321, 322
Pt 164
Pt(997) 151, 152
pulsed field ionization 221, 223, 224, 226, 227
 zero kinetic energy (ZEKE) 21
pulsed-electric-field ionization *see* pulsed field ionization
pyrimidine 232
 -Ar 232
 -Ar_2 232

quantum

Hall effect 171
critical point 297
phase transition 267
well 154
quasi-particle 30, 55, 69, 90, 119, 127, 195, 272, 287, 301, 302, 315, 329, 330
 Bogoliubov 272, 273, 275–278
 dispersion 286
 heavy-mass 287
 lifetime 60
 mass enhancement 301
 peak 31, 61, 63, 195, 287, 288, 329
 renormalization 30, 301
 renormalized band 287
 spectral weight 32, 315
 weight 301
quasiparticle 253, 255, 256, 258, 261
 Fermi-liquid-like lineshape 253
 pole 253, 255
 velocity 258

Raman scattering 48
rare gas 4
 adsorbate 42
rare-earth
 compound 85, 351
 nickel oxides 340
Rashba hamiltonian 177
Rb_2CsC_{60} 199
Rb_3C_{60} 199
RE *see* rare-earth
REMPI *see* resonantly enhanced multiphoton ionization
ReO_3 332
resolution
 sub-wavenumber 221, 222
resonance photoemission (RPES) 352, 355, 362
resonantly enhanced multiphoton ionization 215, 216, 218, 221, 226, 227, 231, 236, 237
 two-color experiment 219
resonating
 valence bond 259
 valence bond state 137
rigid-band
 approximation 309
 model 338

picture 249, 251
RNiO$_3$ 340, 342
Rydberg states 221, 226, 228

s-f Hamiltonian 74
s-wave gap 246
satellites 22, 30, 88
SBZ *see* surface Brillouin zone
scanning tunneling
  microscope 32
  spectroscopy 154
scattering
  electron–electron 61, 62, 74, 79
  electron–impurity 25, 33, 34, 49, 61, 63, 74, 75
  electron–magnon 74, 75
  electron–phonon 61, 71, 75, 78
  inelastic 74
  rate 56
  spin flip 75
Schrödinger equation 156
SDW *see* spin density wave
self-energy 2, 23, 25, 55, 57–59, 61, 62, 66, 68, 79, 88–90, 92, 300, 301, 303, 305, 310, 311, 313
  complex 56
  corrections 55, 66
  effect 56
  effect, spin-resolved 71
semiconductor surfaces 163
shadow
  band 124, 142
  Fermi surface *see* Fermi surface
Shockley state 4, 7, 32–35, 41–44, 62, 71, 72, 74, 75, 159, 163
  Fermi surface 43
  free-electron-like 153
  photoemission linewidth 33
Si 149, 172–175, 177, 378, 395
  adatoms 167, 170
  chains 174
Si(100) 166, 378
Si(111) 164–168, 180
  5×2-Au 175
  7×7 165
  $\sqrt{21} \times \sqrt{21}$-(Ag+Au) 150, 170
  $\sqrt{3} \times \sqrt{3}$-Ag 150, 168, 170
  $\sqrt{3} \times \sqrt{3}$-Au 168
  -Au 180

4×1-In 171
5×2-Au 165, 166, 171–174
5×4-In 179
7×7 149, 165, 169
Si(335)
  -Au 167
Si(553)
  -Au 150, 167, 170–172, 174, 175, 178, 180
Si(553)-Au 177
Si(557) 164
  -Au 164–167, 175–178, 181
Si(775)
  -Au 167
Si-Au orbitals 177
SIAM *see* single-impurity Anderson model
silicon 149
  clathrate 200
single
  impurity Anderson model 44, 95, 352, 355, 356, 363, 367, 369, 370, 394
  particle
    excitation 148
  particle picture 22
single-crystal thin film 329
single-electron dispersion 33
SiO$_2$ 378
  peak 380
skutterudite 361, 362
SmNiO$_3$ 341
soft x-ray photoemission 351
solar cell 151
solid
  quasi one-dimensional 8
  quasi-low dimensional 36
  three-dimensional 35, 37
specific heat 73
spectral function 1, 23, 57, 119, 121, 301, 304, 305, 310, 311, 321
spin
  chains 148, 179
  density wave 118, 126, 148
  excitation 56, 71, 77, 79–81, 303
  flip scattering 75
  fluctuation scenario 321
  resolved self-energy effect 71
  susceptibility 80

spin-charge separation   119, 131, 148, 175, 177, 178, 180
spin-orbit
  interaction   95, 176, 177
  parameter   74
  satellite   45
  splitting   42, 46, 175, 177
spin-Peierls phase   138
spin-split
  bands   177
spinon   120, 131, 137, 142, 148
SPring-8   352
$Sr_2CuO_2Cl_2$ 252, 253, 255, 257, 260
Sr_2CuO_3 136
Sr_2RuO_4 253, 255
$SrCuO_2$ 136
$SrRuO_3$ 344–346
  on $SrTiO_3$ 345
  thin films   345, 346
$SrTiO_3$ 332, 345, 379, 389
  substrate   345, 389
$SrVO_3$ 332–334
stepped surface   148
stilbene
  trans   231
STM   *see* scanning tunneling microscope
stripes   262
strong-coupling   195
  analysis   194
  theory   188
strongly correlated electron systems   247, 327
styrene   232
  -Ar   225, 232
sudden approximation   23, 24
superconducting
  gap   64, 76, 80, 273, 274, 276, 278, 287, 290, 296, 306, 313
    d-wave anisotropy   243
    anisotropy   76
    energy   283, 305, 319
    symmetry   272, 288
  state   76, 79–81, 248, 283, 304, 305, 313, 315, 316, 318–320
superconductivity   25, 38, 55, 64, 70, 75, 76, 78, 86, 148, 149, 163, 174, 180, 187, 189, 244, 271, 272, 282, 287, 296, 297, 327

d-wave   265
  conventional   306
  in the dichalcogenides   64
  order parameter   76, 315
  transition temperature   60, 63, 76, 79, 246
superconductor   2, 7, 246
  conventional   87, 319
  normal-state gap   247
  normal-state properties   246
  optimally doped   76, 78, 246
  order parameter   247
  overdoped   246
  underdoped   77, 80, 81, 246
superlattice   65, 153, 157
surface
  band   155
  band structure   148
  Brillouin zone   43
  noble metal   151, 179
  phonons   6
  photoemission   23
  quality   81
  resonance   60, 160
  sensitivity   16
  states   *see* Shockley state
  stepped   148
synchrotron radiation   16, 85, 316

t-J model   253, 256, 257, 260
TaS_2 64
$(TaSe_4)_2I$ 122
$TaSe_2$ 64, 65, 68–71
thin film
  single-crystal   329
thioanisole   233
  -Ar   233
  -Ar_2 233
three-step model   24
tight-binding
  band-structure   298, 299, 311, 339
  band-structure calculation   316, 339, 340
  parameters   310
time-of-flight   20, 216, 219, 221
  electron analyzer   216, 219
  spectrometer   4, 20
TiOCl   137
$TiTe_2$ 36, 37

$Tl_2Ba_2CuO_{6+\delta}$ 266
$(tolane)^+$ 231
toluene   233
  -Ar   233, 235, 236
$(toluene-Ar)^+$ 235
Tomonaga–Luttinger liquid   116, 121, 127, 135
  line shapes   130
transition metal
  oxide   326, 327
transport
  1D   163
  2D   163
  studies   56
$(tropolone)^+$ 231
TTF-TCNQ   132
tunneling
  spectra   194
  spectroscopy   102, 187
two-color
  process   219
  REMPI experiment   222
  REMPI-based technique   236
two-dimensional system   25, 55
two-photon photoemission   181

U_{eff} 328, 329
ultrashallow junctions   380
umklapp   122

V_3Si 2, 3, 196
van Hove singularity   64
very low energy electron diffraction   38
vicinal surfaces   43, 151, 153, 156, 158–161, 165
VO_2 332

work function   13, 300

Xe   221
Xe/Cu(111)   43

$Y(Ni_{0.8}Pt_{0.2})_2B_2C$ 201, 202
Yb valence   385, 386
$YBa_2Cu_3O_7$ 321
$YBa_2Cu_3O_{6+\delta}$ 76, 81, 280, 281
$YBa_2Cu_3O_{7-\delta}$ 244, 248
YbB_{12} 106, 108
$YbInCu_4$ 16, 104, 106, 107, 385–388
  valence transition   106, 385
YNi_2B_2C 201, 202
$YTiO_3$ 332

zero kinetic energy   215–217, 219, 222–229, 231, 235, 236
zero-phonon line   254
Zhang–Rice singlet   251, 257, 392
$ZrTe_3$ 208, 209

Lecture Notes in Physics

For information about earlier volumes please contact your bookseller or Springer LNP Online archive: springerlink.com

Vol.669: G. Amelino-Camelia, J. Kowalski-Glikman (Eds.), Planck Scale Effects in Astrophysics and Cosmology

Vol.670: A. Dinklage, G. Marx, T. Klinger, L. Schweikhard (Eds.), Plasma Physics

Vol.671: J.-R. Chazottes, B. Fernandez (Eds.), Dynamics of Coupled Map Lattices and of Related Spatially Extended Systems

Vol.672: R. Kh. Zeytounian, Topics in Hyposonic Flow Theory

Vol.673: C. Bona, C. Palenzula-Luque, Elements of Numerical Relativity

Vol.674: A. G. Hunt, Percolation Theory for Flow in Porous Media

Vol.675: M. Kröger, Models for Polymeric and Anisotropic Liquids

Vol.676: I. Galanakis, P. H. Dederichs (Eds.), Half-metallic Alloys

Vol.677: A. Loiseau, P. Launois, P. Petit, S. Roche, J.-P. Salvetat (Eds.), Understanding Carbon Nanotubes

Vol.678: M. Donath, W. Nolting (Eds.), Local-Moment Ferromagnets

Vol.679: A. Das, B. K. Chakrabarti (Eds.), Quantum Annealing and Related Optimization Methods

Vol.680: G. Cuniberti, G. Fagas, K. Richter (Eds.), Introducing Molecular Electronics

Vol.681: A. Llor, Statistical Hydrodynamic Models for Developed Mixing Instability Flows

Vol.682: J. Souchay (Ed.), Dynamics of Extended Celestial Bodies and Rings

Vol.683: R. Dvorak, F. Freistetter, J. Kurths (Eds.), Chaos and Stability in Planetary Systems

Vol.684: J. Dolinšek, M. Vilfan, S. Žumer (Eds.), Novel NMR and EPR Techniques

Vol.685: C. Klein, O. Richter, Ernst Equation and Riemann Surfaces

Vol.686: A. D. Yaghjian, Relativistic Dynamics of a Charged Sphere

Vol.687: J. W. LaBelle, R. A. Treumann (Eds.), Geospace Electromagnetic Waves and Radiation

Vol.688: M. C. Miguel, J. M. Rubi (Eds.), Jamming, Yielding, and Irreversible Deformation in Condensed Matter

Vol.689: W. Pötz, J. Fabian, U. Hohenester (Eds.), Quantum Coherence

Vol.690: J. Asch, A. Joye (Eds.), Mathematical Physics of Quantum Mechanics

Vol.691: S. S. Abdullaev, Construction of Mappings for Hamiltonian Systems and Their Applications

Vol.692: J. Frauendiener, D. J. W. Giulini, V. Perlick (Eds.), Analytical and Numerical Approaches to Mathematical Relativity

Vol.693: D. Alloin, R. Johnson, P. Lira (Eds.), Physics of Active Galactic Nuclei at all Scales

Vol.694: H. Schwoerer, J. Magill, B. Beleites (Eds.), Lasers and Nuclei

Vol.695: J. Dereziński, H. Siedentop (Eds.), Large Coulomb Systems

Vol.696: K.-S. Choi, J. E. Kim, Quarks and Leptons From Orbifolded Superstring

Vol.697: E. Beaurepaire, H. Bulou, F. Scheurer, J.-P. Kappler (Eds.), Magnetism: A Synchrotron Radiation Approach

Vol.698: S. Bellucci (Ed.), Supersymmetric Mechanics – Vol. 1

Vol.699: J.-P. Rozelot (Ed.), Solar and Heliospheric Origins of Space Weather Phenomena

Vol.700: J. Al-Khalili, E. Roeckl (Eds.), The Euroschool Lectures on Physics with Exotic Beams, Vol. II

Vol.701: S. Bellucci, S. Ferrara, A. Marrani, Supersymmetric Mechanics – Vol. 2

Vol.702: J. Ehlers, C. Lämmerzahl, Special Relativity

Vol.703: M. Ferrario, G. Ciccotti, K. Binder (Eds.), Computer Simulations in Condensed Matter Systems: From Materials to Chemical Biology Volume 1

Vol.704: M. Ferrario, G. Ciccotti, K. Binder (Eds.), Computer Simulations in Condensed Matter Systems: From Materials to Chemical Biology Volume 2

Vol.705: P. Bhattacharyya, B.K. Chakrabarti (Eds.), Modelling Critical and Catastrophic Phenomena in Geoscience

Vol.706: M.A.L. Marques, C.A. Ullrich, F. Nogueira, A. Rubio, K. Burke, E.K.U. Gross (Eds.), Time-Dependent Density Functional Theory

Vol.707: A.V. Shchepetilov, Calculus and Mechanics on Two-Point Homogenous Riemannian Spaces

Vol.708: F. Iachello, Lie Algebras and Applications

Vol.709: H.-J. Borchers and R.N. Sen, Mathematical Implications of Einstein-Weyl Causality

Vol.710: K. Hutter, A.A.F. van de Ven, A. Ursescu, Electromagnetic Field Matter Interactions in Thermoelastic Solids and Viscous Fluids

Vol.711: H. Linke, A. Månsson (Eds.), Controlled Nanoscale Motion

Vol.712: W. Pötz, J. Fabian, U. Hohenester (Eds.), Modern Aspects of Spin Physics

Vol.713: L. Diósi, A Short Course in Quantum Information Theory

Vol.714: Günter Reiter and Gert R. Strobl (Eds.), Progress in Understanding of Polymer Crystallization

Vol.715: Stefan Hüfner (Ed.), Very High Resolution Photoelectron Spectroscopy

Printing: Krips bv, Meppel
Binding: Stürtz, Würzburg